RSRM '87

ADVANCES IN REMOTE SENSING
RETRIEVAL METHODS

Studies in
Geophysical Optics and Remote Sensing
Series Editor: *Adarsh Deepak*
Published Volumes and Volumes in Preparation

Proceedings of the

Workshop on Advances in Remote Sensing Retrieval Methods, held in Williamsburg, Virginia, 15–18 December 1987

Sponsored by the

NASA Office of Space Sciences Application and the Air Force Geophysics Laboratory

In cooperation with

NASA Langley Research Center

Organized by the

Science and Technology Corporation (STC) Meetings Division under partial funding from contract NAS1–18397

Any opinions, findings, conclusions and/or recommendations are those of the authors and do not necessarily reflect the views of the sponsors.

RSRM '87

ADVANCES IN REMOTE SENSING RETRIEVAL METHODS

Edited by

Adarsh Deepak
Science and Technology Corporation
Hampton, Virginia

Henry E. Fleming
National Oceanic and Atmospheric Administration
National Environmental Satellite, Data, and Information Service
Washington, D.C.

and

John S. Theon
NASA Headquarters
Washington, D.C.

A. DEEPAK Publishing   1989
A Division of Science and Technology Corporation
Hampton, Virginia USA

A. DEEPAK Publishing
A Division of Science and Technology Corporation
101 Research Drive
Hampton, Virginia 23666–1340   USA

Library of Congress Cataloging-in-Publication Data

RSRM '87: advances in remote sensing retrieval methods/edited by
Adarsh Deepak, Henry E. Fleming, and John S. Theon.
p. cm.
Papers presented at a workshop held December 15–18, 1987 in
Williamsburg, Virginia.
Includes bibliographical references.
ISBN 0–937194–13–1
1. Remote sensing—Congresses.   I. Deepak, Adarsh.
II. Fleming, Henry E.   III. Theon, John S.   IV. Title: Advances in
remote sensing retrieval methods.
G70.4.R77   1989
621.36'78—dc20   89–36947
                            CIP

Printed in the United States of America

TABLE OF CONTENTS

SESSION C.1. MATHEMATICAL INVERSION TECHNIQUES
Chair: *Jean I.F. King*, Air Force Geophysics Laboratory

SESSION C.2. MAXIMUM ENTROPY TECHNIQUES
Chair: *Grace Wahba*, University of Wisconsin-Madison

SESSION C.3. SIMULTANEOUS RETRIEVAL OF SEVERAL PARAMETERS
Chair: *Clive D. Rodgers*, University of Oxford, England

SESSION C.4. INTERCOMPARISONS OF INVERSION METHODS
Chair: *E.D. Westwater*, NOAA/ERL Wave Propagation Laboratory

SESSION D. ERROR ANALYSIS AND FEASIBILITY STUDIES
Chair: *Vincent J. Falcone, Jr.*, Air Force Geophysics Laboratory

SESSION E. MULTIDIMENSIONAL METHODS
Chair: *John C. Gille*, National Center for Atmospheric Research

SESSION F. ARTIFICIAL INTELLIGENCE METHODS, PATTERN RECOGNITION, AND CLASSIFICATION
Chair: *William L. Smith*, University of Wisconsin-Madison

SESSION G. DATA COMPACTION AND MANAGEMENT
Chair: *H.K. Ramapriyan*, NASA Goddard Space Flight Center

SESSION H. DISPLAY SYSTEMS METHODOLOGY
Chair: *Owen E. Thompson*, University of Maryland

SESSION I. GLOBAL CHANGE MONITORING BY REMOTE SENSING
Chair: *John S. Theon*, NASA Headquarters

PREFACE

This volume contains the written versions of the papers presented to the interactive workshop on Advances in Remote Sensing Retrieval Methods, held in Williamsburg, Virginia, December 15 to 18, 1987. This workshop is the fourth in the series and was organized to provide a forum to discuss the present state of knowledge of retrieval methods in various remote sensing areas, to both identify important problems that require further research and to discuss new avenues of approach that may be applicable. Eighty-nine scientists from government laboratories, universities, and private industry in the United States, Canada, Australia, and the United Kingdom participated in the workshop at which 52 papers were presented. Forty-one of those papers and the discussions that followed them are contained herein.

As recognition becomes widespread that mankind is altering the environment by changing the composition of the atmosphere, the radiative properties of the atmosphere and its underlying surfaces, the hydrological cycle processes, etc., remote sensing assumes a new and more significant importance in providing information to study the atmosphere, the land, and the oceans on a global scale. These components and their complex interactions are now called the Earth system, and remote sensing of geophysical processes has gained new importance. Earth system science requires many observable parameters and it requires these measurements concurrently so that the interactions among them can be analyzed, understood, and modeled.

The workshop was intended to explore the connections and common avenues of approach among retrieval methods and broaden their applicability and utility. In an attempt to meet this goal, this book presents the latest methods of retrieving useful information from satellite observations to derive data on atmospheric temperature, atmospheric moisture, surface temperature, ocean surface wind, snow and ice cover, cloud amount and height, etc. These retrievals take into consideration scanning geometry and frequency of coverage, both individually and in combination.

The scope of the workshop included continuing research in areas covered in the first three workshops as well as topics not previously discussed, in particular: Error Analysis and Feasibility Studies, Display Systems Methodology, and Global Change Monitoring. The three earlier workshops were: Inversion Methods in Atmospheric Remote Sounding, 1976; Interpretation of Remotely Sensed Data, 1979; and Advances in Remote Sensing Retrieval Methods, 1984. All four of the workshops were conducted in Williamsburg, Virginia. The proceedings of the first and second were published by Academic Press in 1977 and 1980, and the third by A. Deepak Publishing in 1985.

The workshop program was divided into 12 sessions, each covering a specific topic and chaired by the following scientists: H.E. Fleming, Remote Sensing by Tomographic Techniques; E.J. Hurley, Role of Calibration in Retrievals; J.I.F. King, Mathematical Inversion Techniques; Grace Wahba, Maximum Entropy Techniques; C.D. Rodgers, Simultaneous Retrieval of Several Parameters; E.D. Westwater, Intercomparisons of Inversion Methods; V.J. Falcone, Error Analysis and Feasibility Studies; John Gille, Multidimensional Methods; W.L. Smith, Artificial Intelligence

Methods, Pattern Recognition, and Classification; H.K. Ramapriyan, Data Compaction and Management; O.E. Thompson, Display Systems Methodology; and J.S. Theon, Global Change Monitoring by Remote Sensing. Ample time was allowed for discussions following each paper; discussions were recorded, transcribed, and minimally postedited for inclusion in this volume.

To ensure proper representation of the major disciplines involved, a workshop program committee composed of the following scientists was set up: A. Deepak (Chairman), Science and Technology Corporation; M.T. Chahine, Jet Propulsion Laboratory; H.E. Fleming, NOAA/NESDIS; J.I.F. King, Air Force Geophysics Laboratory; J.S. Theon, NASA Headquarters; and O.E. Thompson, University of Maryland.

The editors wish to acknowledge the enthusiastic support and cooperation of the members of the Technical Program Committee, session chairmen, speakers, and participants for making this another very stimulating and valuable workshop. Special thanks are due Henry Fleming for his extra efforts in assisting in the preparation of an excellent technical program.

The workshop was cosponsored by the NASA Office of Space Sciences Application and the Air Force Geophysics Laboratory in cooperation with NASA Langley Research Center. It is our hope that the research findings presented and compiled in this volume will provide an important source of information for many researchers in the field.

Adarsh Deepak
Henry E. Fleming
John S. Theon

Vincent J. Abreu, University of Michigan, Space Physics Research Laboratory, 2455 Hayward, Ann Arbor, Michigan, 48109, USA

Thomas C. Adang, Det 1, HQ AWS (MAC), Pentagon, Washington, D.C. 20330, USA

Gail Anderson, AFGL/OPI, Hanscom AFB, Massachusetts 01731, USA

Lillian L. Barnes, NOAA/NCASC, Computer Division, DC63, Washington, D.C. 20233, USA

T. Dale Bess, NASA LaRC, Mail Stop 420, Hampton, Virginia 23665, USA

N.L. Bonavito, NASA/GSFC, Code 636, Greenbelt, Maryland 20771, USA

David E. Bowker, NASA/LaRC, Mail Stop 473, Hampton, Virginia 23665, USA

Robert Cahalan, NASA/GSFC, Code 613, Greenbelt, Maryland 20771, USA

Richard P. Cebula, ST Systems Corporation, 4400 Forbes Boulevard, Lanham, Maryland 20706, USA

Er-Woon Chiou, STX, 28 Research Drive, Hampton, Virginia 23666, USA

William P. Chu, NASA/LaRC, Mail Stop 475, Hampton, Virginia 23665, USA

Thomas V. Clarmann, Kernforschungszentrum Karlsruhe IMK, Postfach 3640, D-7500 Karlsruhe 1, FRG

Shepard A. Clough, AER, Inc., 840 Memorial Drive, Cambridge, Massachusetts 02139, USA

Brian J. Connor, NASA/LaRC, Mail Stop 401B, Hampton, Virginia 23665, USA

Karl W. Cox, SM Systems and Research Corporation, 8401 Corporate Drive, Suite 510, Landover, Maryland 20785, USA

David S. Crosby, U.S. Department of Commerce, NOAA/NESDIS, Code E/RA14 WWB, Room 810, Camp Springs, Maryland 20746, USA

Derek M. Cunnold, Georgia Tech, School of Geophysical Sciences, Atlanta, Georgia 30332, USA

John C. Curlander, Jet Propulsion Laboratory, Mail Stop 156-119, 4800 Oak Grove Drive, Pasadena, California 91109, USA

Richard E. Davis, NASA/LaRC, Mail Stop 474, Hampton, Virginia 23665, USA

Adarsh Deepak, Science and Technology Corporation, 101 Research Drive, Hampton, Virginia 23666, USA

Martin J. Donohoe, NASA/GSFC, Code 415, Greenbelt, Maryland, 20771, USA

John E. Dorband, NASA/GSFC, Code 635, Image Analysis Facility, Greenbelt, Maryland 20771, USA

Vincent J. Falcone, Jr., AFGL/LYS, Hanscom AFB, Massachusetts 01731, USA

Henry E. Fleming, U.S. Department of Commerce, NOAA/NESDIS, Code E/RA14 WWB, Room 810, Camp Springs, Maryland 20746, USA

Mark Forrester, The Mitre Corporation, 7525 Colshire Drive, McLean, Virginia 22102, USA

Paul F. Fougere, AFGL/LIS, Hanscom AFB, Massachusetts 01731, USA

Michael Franek, Science and Technology Corporation, 101 Research Drive, Hampton, Virginia 23666, USA

Chien-Cheng Fu, ST Systems Corporation, 4400 Forbes Boulevard, Lanham, Maryland 20706, USA

James J. Gallagher, U.S. Naval Underwater Systems Center, Code 3112, New London, Connecticut 06320, USA

Catherine Gautier, UCSD/SIO, Mail Code A-021, California Space Institute, La Jolla, California 92093, USA

Gerald Geernaert, Navy Center for Space Sciences, Naval Research Lab, Code 8314, Washington, D.C. 20375, USA

John C. Gille, NCAR, P.O. Box 3000, Boulder, Colorado 80307, USA

Ronald S. Gird, The Analytic Sciences Corporation, 1700 N. Moore Street, Suite 1800, Arlington, Virginia 22209, USA

Barbara B. Gormser, STX, 28 Research Drive, Hampton, Virginia 23666, USA

Andreas K. Goroch, NEPRF, Monterey, California 93940, USA

Christopher Grassotti, AER, Inc., 840 Memorial Drive, Cambridge, Massachusetts 02139, USA

Milton Halem, NASA/GSFC, Code 630, Greenbelt, Maryland 20771, USA

Peter Hamilton, SAIC, 4900 Water's Edge Drive, Suite 255, Raleigh, North Carolina 27606, USA

Daesoo Han, NASA/GSFC, Code 636, Greenbelt, Maryland 20771, USA

Michael J. Harris, SM Systems and Research Corporation, 8401 Corporate Drive, Suite 510, Landover, Maryland 20785, USA

Mike Hoke, AFGL/OPI, Hanscom AFB, Massachusetts 01731, USA

Warren A Hovis, Jr., TS InfoSystems, Inc., 4200 Forbes Boulevard, Lanham, Maryland 20706, USA

Allan Huang, University of Wisconsin-Madison, Space Science and Engineering Center, 1225 W. Dayton Street, Madison, Wisconsin 53706, USA

Richard Hucek, Research and Data Systems Corporation, 10300 Greenbelt Road, Lanham, Maryland 20706, USA

Ed J. Hurley, NASA/GSFC, Code 636, Greenbelt, Maryland 20771, USA

R.G. Isaacs, AER, Inc., 840 Memorial Drive, Cambridge, Massachusetts 02139, USA

John B. Jalickee, National Weather Service/NOAA, W10SDx2, 8060 13th Street, Silver Spring, Maryland 20910, USA

Douglas W. Johnson, Battelle Memorial Institute, 505 King Avenue, Columbus, Ohio 43201, USA

Arthur K. Jordan, Naval Research Laboratory, Space Science Division, Code 4110J, Washington, D.C. 20375, USA

Seung T. Kim, ST Systems Corporation, 4400 Forbes Boulevard, Lanham, Maryland 20706, USA

Jean I.F. King, AFGL/LY, Hanscom AFB, Massachusetts 01731, USA

Thomas J. Kleespies, AFGL/LYS, Hanscom AFB, Massachusetts 01731, USA

Steven J. Leon, Southeastern Massachusetts University, Department of Mathematics, North Dartmouth, Massachusetts 02747, USA

Robert M. Lewitt, University of Pennsylvania, Medical Image Processing Group, 419 Blockley Hall, Philadelphia, Pennsylvania 19104, USA

Kuo-Nan Liou, University of Utah, Department of Meteorology, Salt Lake City, Utah 84112, USA

Mong-Ming Lu, NASA/GSFC, Code 611, Sigma Data Services Corporation, Greenbelt, Maryland 20771, USA

Gary K. Maki, University of Idaho, College of Engineering, Microelectronics Research Center, Moscow, Idaho 83843, USA

Stephen A. Mango, Naval Resarch Laboratory, Code 5381 MA, Washington, D.C. 20375, USA

Robert McClatchey, AFGL, Hanscom AFB, Massachusetts 01731, USA

Norman J. McCormick, University of Washington, Department of Nuclear Engineering, BF–20, Seattle, Washington 98195, USA

Larry M. McMillin, U.S. Department of Commerce, NOAA/NESDIS, Code E/RA14 WWB, Camp Springs, Maryland 20746, USA

Juock Namkung, NASA/LaRC, Mail Stop 401A, Hampton, Virginia 23665, USA

Arthur Neuendorffer, NOAA/NESDIS, Physics Branch E RA14, WWB 810, Washington, D.C. 20233, USA

Ira G. Nolt, NASA/LaRC, Hampton, Virginia 23665, USA

Denise Pobedinsky, USACRDEC, SMCCR-DDT, E3330, Aberdeen Proving Ground, Maryland 21010, USA

H.K. Ramapriyan, NASA/GSFC, Code 636, Greenbelt, Maryland 20771, USA

Ichtiaque Rasool, NASA Headquarters, Code E, 600 Independence Avenue S.W., Washington, D.C. 20546, USA

Fred Reames, University of Wisconsin-Madison, 1225 W. Dayton Street, Madison, Wisconsin 53706, USA

Mary G. Reph, NASA/GSFC, Code 634, Greenbelt, Maryland 20771, USA

H.E. Revercomb, University of Wisconsin-Madison, Cooperative Institute for Meteorological Satellite Studies, 1225 W. Dayton Street, Madison, Wisconsin 53706, USA

Clive D. Rodgers, University of Oxford, Department of Atmospheric Physics, Clarendon Laboratory, Oxford, OX1 3PU, England

Judith A. Schroeder, NOAA/WPL, R/E/WP5, 325 Broadway, Boulder, Colorado 80303, USA

William Shaffer, Science Systems and Applications, Inc., 7401–B Forbes Boulevard, Seabrook, Maryland 20706, USA

Danny L. Sims, USAF, HQ AWS/DNXP, Scott AFB, Illinois 62225, USA

Mary Ann Smith, NASA/LaRC, Mail Stop 401, Hampton, Virginia 23665, USA

William L. Smith, University of Wisconsin-Madison, Cooperative Institute for Meteorological Satellite Studies, 1225 W. Dayton Street, Madison, Wisconsin 53706, USA

John Stout, Kirk-Mayer, Inc., Energy, P.O. Box 592, Livermore, California 94550, USA

James P. Strong, NASA/GSFC, Code 636, Greenbelt, Maryland 20771, USA

John Theon, NASA Headquarters, Code EE, 600 Independence Avenue S.W., Washington, D.C. 20546, USA

Owen E. Thompson, University of Maryland, Meteorology Program, Space Sciences Bulding, College Park, Maryland 20742, USA

Lloyd A. Treinish, NASA/GSFC, Code 634, Greenbelt, Maryland 20771, USA

Grace Wahba, University of Wisconsin-Madison, Department of Statistics, 1210 W. Dayton Street, Madison, Wisconsin 53706, USA

H. Andrew Wallio, NASA/LaRC, Mail Stop 401 A, Hampton, Virginia 23665, USA

Ed R. Westwater, NOAA/ERL, Wave Propagation Laboratory, 325 Broadway, Boulder, Colorado 80303, USA

Young Yee, Atmospheric Sciences Lab, White Sands Missile Range, New Mexico 88002, USA

Glenn K. Yue, NASA/LaRC, Mail Stop 475, Hampton, Virginia 23665, USA

Alexander S. Zachor, Atmospheric Radiation Consultants, Inc., 59 High Street, Acton, Massachusetts 01720, USA

Csaba K. Zoltani, Ballistic Research Laboratory, Attn: SLCBR–IB–A, Aberdeen Proving Ground, Maryland 21005, USA

THE NASA PERSPECTIVE ON REMOTE SENSING

John S. Theon
NASA Headquarters
Washington, D.C. 20546, USA

Three of mankind's five senses function remotely (if the olfactory sense can be considered remote). Thus our utilization of this approach is as old as the origin of our species. We use our personal remote sensing abilities in every aspect of our conscious lives. One of the oldest sciences based on remote sensing is astronomy. The very nature of astronomy requires that all the observations be made remotely. Our remote sensing capabilities have progressed considerably over the past few centuries with significant contributions by many people including Galileo and Newton. The telescope and the nature (spectrum) of light led the way to more sophisticated and more quantitative remote sensing techniques.

However, it was not until the beginning of the space age a mere 30 years ago that remote sensing became the important tool that it is today for observing the Earth. The combination of remote sensing capabilities together with the perspective and coverage provided by a spaceborne observing platform is a very potent one indeed. Spaceborne instrumentation may seem to be very expensive upon first glance, but it expands our capabilities to acquire observations over vast areas of the Earth quickly, repetitively, and reliably, thereby making it the most cost effective means for information gathering possible. Since the first meteorological satellite was orbited in 1960, we have made enormous progress in remote sensing of the Earth. As the technology has advanced, it has permitted more sensitive, more complex exploitation of the electromagnetic spectrum to obtain the information we seek about an ever-growing list of geophysical variables. Starting with qualitative observations of cloud patterns almost three decades ago, we now have developed capabilities to measure the solar output, the Earth's radiation budget, cloud cover, type

RSRM '87: ADVANCES IN
REMOTE SENSING RETRIEVAL METHODS
A. Deepak, H.E. Fleming, and J.S. Theon (Eds.)   xvii

and height, atmospheric temperature and moisture, land and sea surface temperatures, atmospheric winds and temperatures, precipitation, ocean color, ocean upper-layer circulation, land surface vegetation, atmospheric aerosol distributions, snow and ice cover, surface topography, geological formations, etc., to name a few. We have progressed to the point where we can remotely monitor most of the key parameters that are important in the integral of processes we define as the climate system.

As the population of the world continues to increase to unprecedented numbers, and as we consume ever increasing energy and other resources to raise our living standards, mankind has begun to modify the environment significantly. Deforestation, changing land use patterns, and changing agricultural practices are widespread. Our consumption of carbon fuels has increased the carbon dioxide content of the atmosphere significantly, and manmade compounds such as the freons have depleted the protective ozone layer in the stratosphere. These changes in our environment are observable, but we do not yet understand their full implications for the climate system. We do now recognize the interrelated nature of the climate system and we know that we cannot understand the system by isolating its components as we have traditionally done in the Earth sciences. We do need to understand the system so that we can account for all its significant parts both natural and manmade. This requires an interdisciplinary approach.

Advances in remote sensing are made possible not only by advances in instrumentation technology, but also by our ability to translate and interpret the measurements acquired by these instruments into useful geophysical parameters. Inasmuch as all remotely sensed data are acquired in terms of voltages or radiances related to voltages, they must be converted into parameters the Earth scientist can use. This procedure, called a retrieval or inversion, can be developed from first principles, from statistical correlations, or from something in between these extremes.

In its report to NASA, the Earth System Science Committee (ESSC) made a persuasive case that advances in remote sensing and the enormous gains in computing power in recent years, coupled with the great strides in our insights into the behavior of the important components of our climate system, have matured sufficiently to set the stage for a quantum leap forward in understanding that system. The ESSC proposed a long-term program to monitor, analyze, and understand the climate system, especially changes in the system that can occur on time scales of a decade to a half century. The committee members suggested that the potential exists for the development of comprehensive models of the climate system, based upon long-term observations, which can describe and even predict the behavior of the system. Natural climate changes do evolve over centuries or millenia. However, the concern is that man's ever increasing impact on the climate system might produce more rapid, more economically disruptive changes in climate. Certainly the geophysical records show that there have been climate changes in the past, but we must understand these as well as man's possible influence if we are to minimize man's impact on the climate system.

In response to the ESSC study, the NASA Earth Science and Application Program has developed plans for a comprehensive Earth Observing System (Eos), which will use as many as four large, polar-orbiting platforms carrying up to 20 sensors each. The space station would also carry instruments and small missions would be used to augment the polar platform observations. Led by NASA, the Eos plan relies upon participation of the National Science Foundation, the National Oceanic and Atmospheric Administration, and other federal agencies in the United States as well as major contributions by the European and Japanese space programs. With the development and implementation of this plan in the 1990s, our spaceborne remote sensing capabilities will become more complete than ever before. This is not to imply that space sensors alone can do the job. They will be complemented by a myriad of conventional and special observing systems.

The Eos Program will place the greatest demands ever upon the remote sensing retrieval community and it will also present the greatest opportunity for progress we have ever had. We are here today to learn about the state of the art in remote sensing retrievals. I look forward to a productive and interesting meeting.

THE AIR FORCE PERSPECTIVE IN REMOTE SENSING OF METEOROLOGICAL PARAMETERS

Robert A. McClatchey
Air Force Geophysics Laboratory
Hanscom Air Force Base
Bedford, Massachusetts 01731, USA

The Air Force requirements for the remote sensing of meteorological parameters is dictated by its global responsibilities and the focus on increasingly smaller scale atmospheric phenomena in the future. The satellite-based remote sensors are the platform of choice (indeed, virtually the only platform) for obtaining the initial data required by global and mesoscale numerical prediction models. Air Force and DOD requirements generally fall into the following categories:

a. Global numerical prediction models
b. Mesoscale numerical prediction models
c. Airport/Air-base support
d. Support to special weather-sensitive systems (e.g. communication systems, battlefield operations, etc.).

The required parameters include (in decreasing order of priority) cloud cover, temperature profiles, moisture profiles, winds, and a number of other parameters. Table 1 summarizes these requirements in some detail.

With respect to the matter of retrieval techniques, those people engaged in operational satellite-based remote sensing of the atmosphere have found it necessary to resort to the use of algorithms based on regression in order to obtain useful results. In principle it would appear that it should be possible to base a retrieval technique on principles of radiative transfer and accurate measurements of the upwelling radiation at appropriate frequencies (or angles). But, in the final analysis, it has generally been found necessary to resort to the use of some kind of climatological data base and regression technique in order to obtain useful results. I have a concern about such an approach. That concern is that the non-climatological, unusual, and potentially important meteorological situations may be missed because

RSRM '87: ADVANCES IN
REMOTE SENSING RETRIEVAL METHODS
A. Deepak, H.E. Fleming, and J.S. Theon (Eds.)   xxi

of the strong dependence of the technique on climatological statistics. In principle, it would be far more satisfying to develop a physical approach based on radiative transfer to optimize the meteorological information in the radiation measurements. I believe this remains a significant challenge for those of us in the satellite meteorology community.

Almost all of the observations to date (in more than 25 years of orbiting meteorological satellites) have depended on passive infrared or microwave systems for remote temperature sensing. Figure 1 presents the "weighting functions" for a typical moderate resolution infrared passive instrument and for a high resolution (monochromatic) infrared or microwave instrument. Although these curves are somewhat idealized and actual instrument weight functions may differ somewhat, the point is that an improvement by about a factor of 2 is all that we can hope to obtain compared with current moderate resolution infrared sensors and we cannot hope

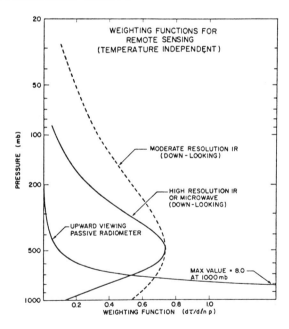

Figure 1: Weighting functions ($d\tau/d\ln p$) for a broad-band strong absorbing gas (looking down), for a "monochromatic" measurement in the space between strong absorption lines (looking down) and for the same "monochromatic" case (looking up from the ground).

for any additional improvement in the microwave. If we take the half-power points of the weighting function curves as representative of the vertical resolution of temperature sensing, it is easy to see that vertical resolution is of the order to 8-9 km and far greater than the requirements identified on Table 1. Even if we are successful in improving on this with high measurement accuracy and shrewd analytical techniques, it seems to me that we will never be able to address the requirements with passive sensors. These problems are even more difficult for moisture, clouds and wind.

I therefore draw the following conclusions:

a. Remote sensing of meteorological parameters is crucial for global forecasting objectives.
b. Improvements are possible with Passive Systems - high horizontal resolution can be attained and high spectral resolution has the potential of some significant improvement.
c. We should combine infrared and microwave sensors on the same platform to optimize the retrievals of meteorological data with passive systems.
d. Active systems (both lidar and radar) are the only approaches with the potential of more completely addressing the Air Force and DOD requirements. Active systems offer the hope of attaining the required vertical resolution and also the possibility of obtaining required parameters not accessible to passive systems.

REFERENCES

Aerospace Corporation: 1983, Defense Meteorological Satellite System, System Requirements Document; Vol. I., DMSS-100, 15 Dec. 1983.

Table 1

Summary of Atmospheric Observation Requirements of DWSP (Aerospace Corp., 1983)

Parameter	Area Covered	Horizontal Resolution	Mapping Accuracy (km)	Vertical Resolution	Measurement Range	Measurement Precision	Measurement Accuracy	Data Refresh Period	Timelines
Cloud Cover	Global	0.5KM	0.5km	±30m (<300m) ±300m (>3000m)	NS	NS		30min	15min
Moisture Profile	Global	100km	20km	±30m (<300m) ±300m (>3000m)	NS	±0.3mm	±1%	3hr	1hr
Temperature Profile	Global	100km	20km	±30m (<3000m) ±300m (3000-10000m) ±600m (>10000m)	210-310k	±1K	±1%	3hr	1hr
Visibility 0.4-0.7 μm	3500x 7500km	10km (<4500m)	2km	150m (<4500m) ±300m (4500-7500m)	0-30km	NS	±1km	1hr	1hr
Winds	Global	10km	2km	±30m (<3000m) ±300m (3000-10000m) ±600m (>10000m)	1-75m/a 0-3600°		±5%,<2m/s ±5°(dir.)	3hr	1hr
Clear Air Turbulence	Global	45km	10km	300m (<3000m) 600m (>3000m)	NS	NS	±1.5m/s	1hr	15min
Propagation Characteristics	Global	5km	10km	±100m	0.01-1.0km^{-1} 0-300N	NS	0.01km^{-1}	3hr	15min
Surface Pressure	Global	10km	10km	±20mb	930-1050mb	NS	±1mb	3hr	30min
Density	Global	25km	5km	1500m	0-2000g/m^2	NS	±5%	1hr	10min

xxiv

OVERVIEW OF INVERSION METHODS USED IN COMPUTED TOMOGRAPHY[1]

Robert M. Lewitt
Department of Radiology
University of Pennsylvania
Philadelphia, Pennsylvania 19104, USA

ABSTRACT

Methods for image reconstruction from projections constitute the foundations of computed tomography, and have proven useful in a wide variety of remote sensing applications in which the measured data have the form of line integrals (or strip integrals) of some physical property of the object under investigation. The purpose of this paper is to provide a brief overview of the inversion methods used in computed tomography, together with a list of representative references that may be used as a starting point for further investigation. Some relationships between the inversion methods used in medical applications of computed tomography and the retrieval methods used in atmospheric remote sounding are outlined.

1. INTRODUCTION

Many of the methods used for retrieval of atmospheric and oceanic parameters from remotely-sensed data share a common foundation with inversion methods used in other disciplines. The determination of the internal structure of a given object from measurements made outside the object is used extensively in diagnostic radiology and nuclear medicine to image human anatomy and physiological function. The determination of internal structure is also used in industrial imaging applications for the nondestructive evaluation of material specimens. The interiors of human or inanimate bodies may be probed by the external application of electric currents, magnetic fields, electromagnetic radiation, acoustic waves, etc., or by the external monitoring of such phenomena emanating from the interiors of the bodies. The methods of computed tomography are applicable when the measured data have the form of line integrals (or strip integrals) of some physical property of the object under investigation.

The end result of computed tomography (CT) is an image that represents the spatial distribution of a physical parameter (e.g., x-ray attenuation coefficient) within the object. The process by which

---

[1]The work of the author is supported by research grant HL 28438 from the National Institutes of Health, Department of Health and Human Services, U.S.A.

such an image is obtained from the measured data is known as image reconstruction from projections, since the data correspond to integrated projections (from a variety of angles of view) of the unknown spatial distribution within the object.

Methods for image reconstruction from projections have received considerable attention in the last two decades, in scientific fields ranging in scale from radio astronomy to electron microscopy, and in experimental situations representing all physical states of matter. The initial development of the subject proceeded almost independently in the various fields of application, with a considerable amount of duplication of effort and rediscovery of old results. As the subject has matured, a number of comprehensive books and survey articles have appeared, to the benefit of both newcomers and experienced practitioners alike.

The purpose of the present paper is to provide a brief overview of the inversion methods used in CT, together with a list of representative references that may be used as a starting point for further investigation of the literature. Most of the publications cited are books or survey articles that have appeared since 1980, and which provide introductions to the mathematical foundations of CT and to the research literature on image reconstruction from projections. Finally, we point out a few relationships between the inversion methods used in medical applications of CT and the retrieval methods used in atmospheric remote sounding.

2.   REPRESENTATIVE PUBLICATIONS

The following books, survey papers, conference proceedings and special issues of journals have been selected as providing an overview of CT (with a bias toward its medical applications), from the perspective of the 1980's. In spite of the best intentions of the author, such a list can never be considered complete; the author apologizes in advance for any deficiencies that may be noticed.

2.1 BOOKS

The theoretical foundations of image reconstruction from projections (also referred to as inversion of the Radon transform) and a wide range of applications are discussed in the books by Herman (1980), Deans (1983), and by Kak and Slaney (1988). Image reconstruction is discussed by Barrett and Swindell (1981), Macovski (1983), and Robb (1985) in the context of radiological imaging. Image restoration is a research field that is closely allied with image reconstruction, and both are treated by Rosenfeld and Kak (1982), and by Bates and McDonnell (1986). General image recovery problems, including image reconstruction from projections, are discussed by Stark (1987) and Huang (1984), and the book edited by Herman (1979) contains an older, though still valuable, collection of contributed articles on implementation and applications. Deeper mathematical analysis of the foundations of image reconstruction from projections may be found in the books of Natterer (1986) and Helgason (1980).

2.2 SURVEY PAPERS

Various approaches to image reconstruction from projections are discussed in the survey papers by Lewitt (1983) and by Censor (1983); see also Censor and Herman (1987), and Altschuler et al. (1981). Other surveys place greater emphasis on analytic inversion formulas (Barrett, 1984) and sampling considerations (Lindgren and Rattey, 1981), or on the mathematical context of image reconstruction (Louis and Natterer, 1983; Smith and Keinert, 1985; Herman and Tuy, 1987). Other approaches include statistical formulation of the problem (Vardi, Shepp and Kaufman, 1985), estimation of object model parameters (Bresler and Macovski, 1987), and inversion methods for curved-ray projection data (Andersen, 1987). X-ray CT is surveyed by Robb (1982a, 1982b), and a wide range of image reconstruction problems and applications are reviewed in the survey paper of Bates, Garden and Peters (1983).

2.3 CONFERENCE PROCEEDINGS AND JOURNAL SPECIAL ISSUES

Conference proceedings of an applied nature include those edited by Nalcioglu and Prewitt (1982) and by Nalcioglu et al. (1986), and proceedings with a mathematical orientation include those edited by Herman and Natterer (1981) and by Shepp (1983). Relevant special issues of journals include those edited by Kak (1981), by Herman (1983) and by Gordon (1985). Papers on various aspects of image reconstruction from projections appear frequently in several of the IEEE Transactions journals, especially the Transactions on Medical Imaging, the Transactions on Acoustics, Speech, and Signal Processing, and the Transactions on Nuclear Science. Also, the Journal of the Optical Society of America A: Optics and Image Science, and its related journal Applied Optics frequently contain relevant papers.

3. GENERAL APPROACHES TO FORMULATION OF ALGORITHMS FOR IMAGE RECONSTRUCTION

Algorithms for image reconstruction from projection data may be classified into two categories: transform methods and series-expansion methods. Tutorial articles on these general approaches and on the formulation of algorithms based on them may be found, for example, in the papers by Lewitt (1983) (on transform methods) and by Censor (1983) (on series-expansion methods).

The transform approach to image reconstruction from projections gives rise to a variety of algorithms, including those widely used in commercial CT scanners for medical imaging. The approach used to derive a transform method involves the following sequence of steps:

(1) formulate a mathematical model of the problem in which the known and unknown quantities are functions whose arguments come from a continuum of real numbers;

(2) solve for the unknown function by producing an inversion formula;

(3)   adapt the inversion formula for application to discrete and
      noisy data.

      In step (2), it is found that there are several formulas which
are theoretically equivalent solutions to the problem posed in step
(1). When each of these formulas is discretized (step (3)) it is
found that the algorithms which result do not perform identically on
real data, since different approximations have been introduced in
step (3).

      In contrast with step (1) above, the series-expansion approach
to image reconstruction begins with a mathematical model of the problem which relates a finite set of known numbers (the projection
data) to a finite set of unknown numbers, representing the image.
The image is expressed as a linear combination of basis functions,
in which the coefficients are initially unknown and will be determined from the data. This discrete formulation leads to a system of
equations whose solution is found numerically, as opposed to the analytical solution obtained in step (2) of the transform approach.
Examples of image basis functions include the simple square pixel,
the more sophisticated finite elements, and products of radial polynomials and angular harmonics.

      The system of equations resulting from the series-expansion
approach is usually very large (but sparse), and is most often
solved by an iterative method. However, noniterative methods may be
practical if the system matrix has a special structure. On the other
hand, the inversion formula obtained in step (2) of the transform
approach is most commonly noniterative, but iterative inversion formulas may also be derived, and then discretized to obtain practical
algorithms for discrete data; for an example, see Lewitt (1983). The
iterative reconstruction algorithms that are derived from either
approach are well-suited to imaging situations where the projection
data are incomplete or are of low precision, since they allow any
available *a priori* knowledge concerning the object to be taken into
account in the iterative refinement of the image.

   4.   SOME RELATIONSHIPS BETWEEN REMOTE SENSING RETRIEVAL METHODS
        AND MEDICAL IMAGE RECONSTRUCTION METHODS

      The books and survey articles that are cited at the end of this
paper, together with their references, contain among them citations
to many hundreds of publications on the theory and applications of
image reconstruction from projections. Similarly, the papers in this
volume, together with those in previous volumes (Deepak, 1980; Deepak et al., 1985) and their references, indicate the breadth and
depth of the literature on remote sensing retrieval methods. It
would be a major project to investigate in detail the potential
areas of overlap between these extensive sets of literature. The
scope of the discussion in this section is restricted to simply
pointing out a few of the relationships between the methods used in
the respective disciplines.

Image reconstruction from projections, and many of the more general problems of retrieval from remotely-sensed data, are examples of ill-posed problems, meaning that small perturbations in the data can lead to large perturbations in the solution (Tikhonov and Arsenin, 1977). The formulations of many retrieval problems lead to Fredholm integral equations of the first kind, which are notoriously ill-posed. For references to the extensive literature on this topic, see, for example, Natterer (1986), Rushforth (1987), Viera and Box (1985), Wahba (1985).

Some methods for image reconstruction from projections that we classify here as noniterative transform methods have been used to invert photometric measurements of atmospheric airglow from orbiting satellites (Solomon, Hayes and Abreu; 1984, 1985). With some idealization of the physical model, the problem of recovering the spatially-varying volume emission rate reduces to the problem of reconstructing a function from its line integrals. The method of solution (Solomon, Hayes and Abreu; 1984) involves discretization of an analytic inversion formula based on angular harmonic decomposition.

Some iterative methods based on the finite series-expansion approach to reconstruction from projections are related to iterative methods used in atmospheric retrieval problems. For example, an iterative method called the image space reconstruction algorithm (ISRA) (Daube-Witherspoon and Muehllehner, 1986) and which was developed in the context of positron emission tomography (PET), has since been found (De Pierro, 1987) to be related to the iterative algorithm of Chahine (1970) which is widely referenced in the geophysical literature (Chu, 1985).

The principles underlying the retrieval of vertical atmospheric temperature and constituent profiles from satellite radiance measurements were recognised many years ago: measurements may be made at a number of different zenith angles (King, 1958), or measurements may be made in a number of different spectral intervals (Kaplan, 1959). The geometry of a combination of angle and frequency scanning (Fleming, 1979), and the geometry of angle scanning itself, are both closely related to the geometry of medical CT scanning, as pointed out by Fleming (1982, 1985). Compared to standard medical CT, some of the limitations of satellite CT (such as restricted angular range) and the additional degrees of freedom in the data acquisition (such as multiple-frequency observations) are also found in microwave radar tomographic imaging (Farhat et al., 1984). However, the coherent signal detection possible in the latter case results in much greater independence of the data at the various frequencies than is possible with the usual radiance measurements.

The use of multiple frequencies for nadir sounding of the atmosphere enables the contributions from different heights to be retrieved from the radiance data. Similarly, in nuclear medicine, systems for time-of-flight positron emission tomography (TOFPET) collect data which enable the contributions from different parts of the line of integration to be separated, at least approximately. A system for TOFPET has much better spatial resolution perpendicular to the line of integration than it has along the line of integration. This is not necessarily the case in satellite tomography, where the retrieval operation may resolve temperatures along the line of integration at intervals less than the beam width, i.e., the instantaneous field of view of the satellite radiometer (Fleming, 1985). Nevertheless, some of the image reconstruction algorithms developed for TOFPET (see, for example, Politte and Snyder (1984)) might have relevance to some of the angle scanning schemes of satellite tomography.

Retrieval methods based on statistical estimation theory are well-established in atmospheric remote sensing (see, for example, Rodgers (1985), Rosenkranz et al. (1985), Wahba (1985)), and are presently receiving much attention in the medical imaging field. These include methods based on maximum-likelihood estimation (Lange et al., 1987; Snyder et al., 1987) and its generalizations (Levitan and Herman, 1987), and Bayesian approaches (Geman and Geman, 1984; Hart and Liang, 1987).

5. CONCLUSION

It is hoped that this brief overview will be of assistance to those who are involved in research and development of remote sensing retrieval methods, and who wish to become acquainted with the literature in other fields concerning the theoretical basis and practical applications of CT.

REFERENCES

Altschuler, M.D., Y. Censor, G.T. Herman, A. Lent, R.M. Lewitt, S.N. Srihari, H. Tuy and J.K. Udupa, 1981: Mathematical Aspects of Image Reconstruction from Projections. In L.N. Kanal and A. Rosenfeld (Eds.), _Progress in Pattern Recognition_, Vol.1, North-Holland, Amsterdam, 323-375.

Andersen, A.H., 1987: Tomography Transform and Inverse in Geometrical Optics, _Journal of the Optical Society of America A_, _4_, 1385-1395.

Barrett, H.H., and W. Swindell, 1981: _Radiological Imaging: Theory of Image Formation, Detection and Processing_, Academic Press, New York.

Barrett, H.H., 1984: The Radon Transform and its Applications. In E. Wolf (Ed.), _Progress in Optics XXI_, Elsevier, Amsterdam, 217-286.

Bates, R.H.T., K.L. Garden and T.M. Peters, 1983: Overview of Computerized Tomography with Emphasis on Future Developments, _Proceedings of the IEEE_, _71_, 356-372.

Bates, R.H.T., and M.J. McDonnell, 1986: _Image Restoration and Reconstruction_, Clarendon Press, Oxford.

Bresler, Y., and A. Macovski, 1987: Three-Dimensional Reconstruction from Projections with Incomplete and Noisy Data by Object Estimation, _IEEE Transactions on Acoustics, Speech, and Signal Processing_, _ASSP-35_, 1139-1152.

Censor, Y., 1983: Finite Series-Expansion Reconstruction Methods, Proceedings of the IEEE, 71, 409-419.

Censor, Y., and G.T. Herman, 1987: On Some Optimization Techniques in Image Reconstruction from Projections, Applied Numerical Mathematics, 3, 365-391.

Chahine, M.T., 1970: Inverse Problems in Radiative Transfer: Determination of Atmospheric Parameters, Journal of Atmospheric Sciences, 27, 960-967.

Chu, W.P., 1985: Convergence of Chahine's nonlinear relaxation inversion method used for limb viewing remote sensing, Applied Optics, 24, 445-447.

Daube-Witherspoon, M.E., and G. Muehllehner, 1986: An Iterative Image Space Reconstruction Algorithm Suitable for Volume ECT, IEEE Transactions on Medical Imaging, MI-5, 61-66.

Deans, S.R., 1983: The Radon Transform and Some of Its Applications, Wiley, New York.

Deepak, A. (Ed.), 1980: Remote Sensing of Atmospheres and Oceans, Academic Press, New York.

Deepak, A., H.E. Fleming, and M.T. Chahine (Eds.), 1985: Advances in Remote Sensing Retrieval Methods, A. Deepak Publishing, Hampton, Virginia.

De Pierro, A.R., 1987: On the Convergence of the Iterative Image Space Reconstruction Algorithm for Volume ECT, IEEE Transactions on Medical Imaging, MI-6, 174-175.

Farhat, N.H., C.L. Werner, and T.H. Chu, 1984: Prospects for three-dimensional projective and tomographic imaging radar network, Radio Science, 19, 1347-1355.

Fleming, H.E., 1979: Satellite Remote Sensing by a Combination of Angle and Frequency Scanning. In C.L. Wyman (Ed.), Space Optics, SPIE Proceedings 183, Society of Photo-Optical Instrumentation Engineers, Bellingham, Washington, 120-125.

Fleming, H.E., 1982: Satellite Remote Sensing by the Technique of Computed Tomography, Journal of Applied Meteorology, 21, 1538-1549.

Fleming, H.E., 1985: Temperature Retrievals via Satellite Tomography. In A. Deepak, H.E. Fleming and M.T. Chahine (Eds.), Advances in Remote Sensing Retrieval Methods, A. Deepak Publishing, Hampton, Virginia, 55-67.

Geman, S., and D. Geman, 1984: Stochastic Relaxation, Gibbs Distributions, and the Bayesian Restoration of Images, IEEE Transactions on Pattern Analysis and Machine Intelligence, PAMI-6, 721-741.

Gordon, R. (Ed.), 1985: Industrial Applications of Computed Tomography and NMR Imaging: an OSA Topical Meeting, Applied Optics, 24, 3948-4140.

Hart, H., and Z. Liang, 1987: Bayesian Image Processing in Two Dimensions, IEEE Transactions on Medical Imaging, MI-6, 201-208.

Helgason, S., 1980: The Radon Transform, Birkhäuser, Boston.

Herman, G.T. (Ed.), 1979: Image Reconstruction from Projections: Implementation and Applications, Springer-Verlag, Berlin.

Herman, G.T., 1980: Image Reconstruction from Projections: The Fundamentals of Computerized Tomography, Academic Press, New York.

Herman, G.T., and F. Natterer (Eds.), 1981: Mathematical Aspects of Computerized Tomography (Oberwolfach, West Germany, February

1980), Lecture Notes in Medical Informatics 8, Springer-Verlag, Berlin, West Germany.

Herman, G.T. (Ed.), 1983: Special Issue on Computerized Tomography, Proceedings of the IEEE, 71, 291-435.

Herman, G.T., and H.K. Tuy, 1987: Image Reconstruction from Projections: An Approach from Mathematical Analysis. In P.C. Sabatier (Ed.), Basic Methods of Tomography and Inverse Problems, Taylor & Francis, Philadelphia, Pennsylvania.

Huang, T.S. (Ed.), 1984: Image Reconstruction from Incomplete Observations, Advances in Computer Vision and Image Processing, Vol 1, JAI Press, Greenwich, Connecticut.

Kak, A.C. (Ed.), 1981: Special Issue on Computerized Medical Imaging, IEEE Transactions on Biomedical Engineering, BME-28, 49-234.

Kak, A.C., and M. Slaney, 1988: Principles of Computerized Tomographic Imaging, IEEE Press, New York.

Kaplan, L.D., 1959: Inference of Atmospheric Structure from Remote Radiation Measurements, Journal of the Optical Society of America, 49, 1004-1007.

King, J.I.F., 1958: The Radiative Heat Transfer of Planet Earth. In J.A. Van Allen (Ed.), Scientific Uses of Earth Satellites, University of Michigan Press.

Lange, K., M. Bahn, and R. Little, 1987: A Theoretical Study of Some Maximum Likelihood Algorithms for Emission and Transmission Tomography, IEEE Transactions on Medical Imaging, MI-6, 106-114.

Levitan, E., and G.T. Herman, 1987: A Maximum A Posteriori Probability Expectation Maximization Algorithm for Image Reconstruction in Emission Tomography, IEEE Transactions on Medical Imaging, MI-6, 185-192.

Lewitt, R.M., 1983: Reconstruction Algorithms: Transform Methods, Proceedings of the IEEE, 71, 390-408.

Lindgren, A.G., and P.A. Rattey, 1981: The Inverse Discrete Radon Transform with Applications to Tomographic Imaging Using Projection Data. In C. Marton (Ed.), Advances in Electronics and Electron Physics, Vol. 56, Academic Press, New York, 359-410.

Louis, A.K., and F. Natterer, 1983: Mathematical Problems of Computerized Tomography, Proceedings of the IEEE, 71, 379-389.

Macovski, A., 1983: Medical Imaging Systems, Prentice-Hall, Englewood Cliffs, New Jersey.

Nalcioglu. O., and J.M.S. Prewitt (Eds.), 1982: Proceedings of International Workshop on Physics and Engineering in Medical Imaging (Pacific Grove, California, March 1982), IEEE Computer Society Press, Silver Spring, Maryland.

Nalcioglu, O., Z.H. Cho and T.F. Budinger (Eds.), 1986: Proceedings of International Workshop on Physics and Engineering of Computerized Multidimensional Imaging and Processing (Newport Beach, California, April 1986), SPIE Proceedings 671, Society of Photo-Optical Instrumentation Engineers, Bellingham, Washington.

Natterer, F., 1986: The Mathematics of Computerized Tomography, Wiley, Chichester.

Politte, D.G., and D.L. Snyder, 1984: Results of a Comparative Study of a Reconstruction Procedure for Producing Improved Estimates of Radioactivity Distributions in Time-of-flight Emission Tomography, IEEE Transactions on Nuclear Science, NS-31, 614-619.

Robb, R.A., 1982a: X-ray Computed Tomography: From Basic Principles to Applications. In <u>Annual Review of Biophysics and Bioengineering</u>, Vol.11, Annual Reviews, Palo Alto, California, 177-201.

Robb, R.A., 1982b: X-ray Computed Tomography: An Engineering Synthesis of Multiscientific Principles, <u>CRC Crit. Rev. Biomed. Eng.</u>, 7, 265-333.

Robb, R.A. (Ed.), 1985: <u>Three-Dimensional Biomedical Imaging</u>, CRC Press, Boca Raton, Florida.

Rodgers, C.D., 1985: A Strategy for Optimal Profile Retrieval from Limb Sounders, In A. Deepak, H.E. Fleming and M.T. Chahine (Eds.), <u>Advances in Remote Sensing Retrieval Methods</u>, A. Deepak Publishing, Hampton, Virginia, 93-103.

Rosenfeld, A., and A.C. Kak, 1982: <u>Digital Picture Processing</u> (Second Edition), Academic Press, Orlando, Florida.

Rosenkranz, P.W., K.S. Nathan, and D.H. Staelin, 1985: Use of Two- and Three-Dimensional Spatial Filtering for Inversion of Radiometric Measurements. In A. Deepak, H.E. Fleming and M.T. Chahine (Eds.), <u>Advances in Remote Sensing Retrieval Methods</u>, A. Deepak Publishing, Hampton, Virginia, 373-382.

Rushforth, C.K., 1987: Signal Restoration, Functional Analysis, and Fredholm Integral Equations of the First Kind. In H. Stark (Ed.), <u>Image Recovery: Theory and Application</u>, Academic Press, Orlando, Florida, 1-27.

Shepp, L.A. (Ed.), 1983: <u>Computed Tomography</u> (AMS Short Course, Cincinnati, Ohio, January 1982), Proceedings of Symposia in Applied Mathematics 27, American Mathematical Society, Providence, Rhode Island.

Smith, K.T., and F. Keinert, 1985: Mathematical Foundations of Computed Tomography, <u>Applied Optics</u>, 24, 3950-3957.

Snyder, D.L., M.I. Miller, L.J. Thomas, and D.G. Politte, 1987: Noise and Edge Artifacts in Maximum-Likelihood Reconstructions for Emission Tomography, <u>IEEE Transactions on Medical Imaging</u>, MI-6, 228-238.

Solomon, S.C., P.B. Hays, and V.J. Abreu, 1984: Tomographic Inversion of Satellite Photometry, <u>Applied Optics</u>, 23, 3409-3414.

Solomon, S.C., P.B. Hays, and V.J. Abreu, 1985: Tomographic Inversion of Satellite Photometry. Part 2, <u>Applied Optics</u>, 24, 4134-4140.

Stark, H. (Ed.), 1987: <u>Image Recovery: Theory and Application</u>, Academic Press, Orlando, Florida.

Tikhonov, A.N., and V.Y. Arsenin, 1977: <u>Solutions of Ill-Posed Problems</u>, V.H. Winston and Sons, Washington, DC.

Vardi, Y., L.A. Shepp and L. Kaufman, 1985: A Statistical Model for Positron Emission Tomography, <u>Journal of the American Statistical Association</u>, 80, 8-37.

Viera, G., and M.A. Box, 1985: Information Content Analysis of Remote Sensing Experiments Using Eigenfunctions. In A. Deepak, H.E. Fleming and M.T. Chahine (Eds.), <u>Advances in Remote Sensing Retrieval Methods</u>, A. Deepak Publishing, Hampton, Virginia, 287-302.

Wahba, G., 1985: Variational Methods for Multidimensional Inverse Problems. In A. Deepak, H.E. Fleming and M.T. Chahine (Eds.), <u>Advances in Remote Sensing Retrieval Methods</u>, A. Deepak Publishing, Hampton, Virginia, 385-407.

DISCUSSION

Wahba: I guess I have a unique perspective on this because I worked a little bit in tomography before I got into meteorology. In fact, I went to a number of tomography meetings and I want to compliment your extremely interesting talk and I think that many of the techniques that have been developed will carry over and will be of much interest to meteorologists. I just want to make a few cautionary remarks on the kinds of techniques that have been developed and the points you make that are important for meteorology and ones that you should be careful about. The main difference between the problems that medical tomographers are doing and what meteorological tomographers are doing is that the medical tomographers design an instrument which generally has a lot of regularity so that you can use transforms easily and you have a lot of structure in the matrices that you have to deal with. Now meteorologists aren't so lucky in that the physical geometry is not so easily controlled. I mean, satellite data is not taken on a regular grid as a simple example. You don't have so much control over your windows as you do in the design of X-ray equipment. Some medical tomography problems have a lot of mathematical simplifications because you have transforms. There is a tremendous temptation to try to carry that over to meteorological tomography but I think that maybe you shouldn't yield to that temptation because generally your data is not that regular. The kind of work that has been done in medical tomography that will carry over, I think, is this tremendous body of methods for iterative solutions for very large linear systems that don't have too much structure. I think that people working in computerized tomography should, if they haven't already, become familiar with this tremendous, very impressive body of work. Also, this is my own personal view, but I think the continuous formulation is much more important in earth sciences, relatively speaking. I mean you could make a case for going both ways, discrete and continuous, in medical tomography, again because of this regularity. But we don't particularly have this regularity in meteorology and to try to discretize things into regular patterns too early in the earth sciences, examples I know of, could lead to loss of information. I think one should be very conscious of formulating a problem continuously. You remarked that you like to separate the formulation of the problem from the algorithm for solving it. Again, I think this is a very important point. One should formulate the problem to represent physical reality as accurately as you can, and usually this will be a continuous formulation. Then, you separately look for the most appropriate algorithm for solving the problem. Meteorologists should do this because the algorithm will depend on many things; the number of data points, the degree of ill posedness, known side conditions, etc. The algorithm is discrete, of course, but the later the discretization, the better.

King: When you use the box basis functions and you have rejection which goes across the diagonal, how do you handle it in that particular basis function formulation?

Lewitt: There are so many of them that we might think of the unknown distribution as just being a matrix of little boxes where there is nothing special about diagonal direction. That is not the only kind of basis function that one could use. You could use basis functions in polar coordinates, such as polynomials in the radial direction and angular harmonics in the angle direction. There are very many choices for basis functions. What you choose really depends on the underlying physics and the underlying mathematical model as to which kind of expansion of basis function can represent your unknown with the fewest number of parameters and with the most decoupled basis functions.

Hucek: I notice you didn't mention anything about error or noise that you might have in obtaining the measurement values, and I am just wondering: how do you determine the ultimate resolution of the particular patterns that you may be able to achieve, or when you look at a pattern that you know that all of the information is real?

Lewitt: That is a difficult problem. I think the way to deal with noise in a good solid way is to include it in the model which is done say in positron emission tomography. You make an explicit model which accounts for the Poisson nature of the radiation emission and absorption process and then formulate the problem as being one in which you are looking for the maximum likelihood solution based on that data. The problem is that your distribution may not be much like the one that you get from your maximum likelihood data but, at least, you have some well-founded methodology. Image evaluation and the estimation of image error is often a difficult problem. Was that the line of your question?

Hucek: Pretty much, though I was particularly interested in the discrete formalization where you expanded in basis functions. Is there some way you know to what resolution or to how many terms you can carry that expansion?

Lewitt: Within the discrete formulation you don't want to end up with an underdetermined system. You don't want to have a thousand pieces of data and be trying to determine 10,000 unknown parameters, unless you have some side information like positivity, the boundary of distribution, or some limits on the value, some extra information that you can feed into the retrieval process. When too many terms are included in the expansion, the solution is highly non-unique, like it would be if you had no side information with the numbers I quoted.

THE TOMOGRAPHIC INVERSION
OF SATELLITE PHOTOMETRIC DATA

Vincent J. Abreu and P.B. Hays
Space Physics Research Laboratory
Department of Atmospheric, Oceanic and Space Science
University of Michigan
Ann Arbor, Michigan 48109, USA

Stanley C. Solomon
High Altitude Observatory
National Center for Atmospheric Research
Boulder, Colorado 80307, USA

ABSTRACT

A tomographic technique developed to carry out the inversion of satellite photometric measurements is reviewed. The tomographic inversion is based on the method of Cormack (1963). The Cormack inversion is performed on limb brightness measurements in Fourier space; nadir measurements are later combined with the inverted data in order to improve the horizontal resolution of the inverted profiles.

1. INTRODUCTION

The photometric data to be inverted are those provided by the Visible Airglow Experiment (VAE) on board the Atmosphere Explorer (AE) satellites (Hays et al., 1973). This experiment was designed to measure thermospheric and auroral emission features and consisted of two filter-wheel photometers perpendicular to each other and located in the orbital plane of the spacecraft. Photometer channel one had a narrow field of view, 3/4 degree half-angle cone, for high spatial resolution, while channel two had a 3.0 degree half angle cone. Integration periods of 0.031 seconds for channel one and 0.125 seconds for channel two were matched to the spin rate of the satellite so that counts were summed over angular intervals roughly equal to the field of view when spinning. The translational and spinning motion of the satellite allowed the thermospheric emissions to be viewed from different angles. The measured quantity is the emission column brightness, B, which in units of Rayleighs is given by

$$B = 10^{-6} \int_{o}^{\infty} \eta \, (s) \, ds \qquad (1)$$

RSRM '87: ADVANCES IN
REMOTE SENSING RETRIEVAL METHODS
A. Deepak, H.E. Fleming, and J.S. Theon (Eds.)

13

where η is the volume emission rate in photons cm^{-3} s^{-1} and s is distance along the photometer line of sight. The line integrals measured by the VAE channel one have been used in a tomographic inversion to obtain the altitude profile and horizontal structure of the emission field along the satellite track (Solomon et al., 1984; 1985; 1987; 1988). Here we will review the inversion method and show results.

2. INVERSION TECHNIQUE

2.1 TOMOGRAPHIC INVERSION OF LIMB MEASUREMENTS

The inversion method developed is based on the work of A. M. Cormack (1963,1964). This method yields an analytic solution and unlike the Radon inversion, it uses only limb data. Figure 1 shows a representation of the inversion geometry. Here the emission field is represented by the functions $g(r,\theta)$. The line integrals are the functions $f(p,\phi)$, where p is the perpendicular distance from the origin and ϕ is the angle which that perpendicular makes with the reference direction. The inversion problem is to obtain $g(r,\theta)$ from measurements of $f(p,\phi)$. The function, $g(r,\theta)$ may be expanded in a Fourier series with respect to the angular variable θ. The expansion is expressed as the $G_n(r)$, a series of complex functions of r from n=0...∞. Similarly, the function $f(p,\phi)$ may be expanded in a Fourier series $F_n(p)$. Cormack (1963) derived an expression relating $G_n(r)$ to $F_n(p)$:

$$G_n(r) = -\frac{1}{\pi} \frac{d}{dr} \int_r^{r_s} \frac{r F_n(p) T_n(p/r)}{p(p^2 - r^2)^{\frac{1}{2}}} dp \qquad (2)$$

where r_s is the domain radius and T_n is the nth Chebyshev polynomial. The details concerning the numerical implementation of the technique are given by Solomon et al., (1984). The procedure for inverting the photometric brightness measurements can be summarized as follows:

1. Interpolate the brightness measurements into a 2-dimensional grid

2. Take the Fourier transform with respect to the angular (horizontal) variable

3. Invert each Fourier term using the Cormack equation

4. Filter out high-frequency terms

5. Take the inverse Fourier transform to obtain the 2-dimensional emission rate function.

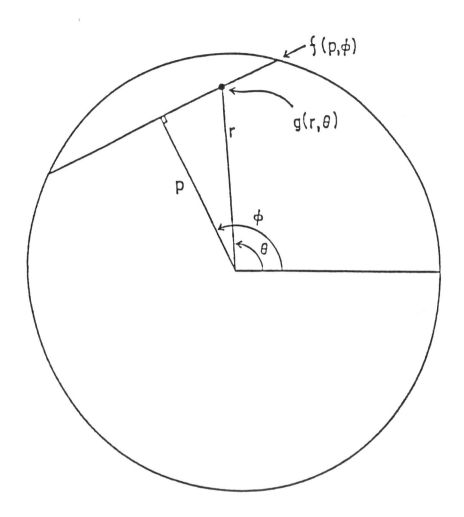

FIGURE 1. Geometry of the Cormack Inversion
Every line integral f is the sum of all
g(r, θ) specified by distance p and angle φ.

The Cormack inversion is conceptually sound and can be of value for the analysis of satellite photometry. In practice, however, there is an upper limit to the number of Fourier coefficients which may be retained in performing the inversion without encountering spurious oscillations. The limit depends on the instrumental characteristics, photometer counting statistics, temporal variations of the emission, and distance below the satellite orbit. Typical oscillatory behaviour for the VAE data begins after 20 to 35 terms, clearly limiting the horizontal resolution of the inverted data. When the AE satellites were in their spinning mode, the VAE photometers spent a large fraction of their measurement time pointing down at the earth. Consequently, nadir measurements are available and have been incorporated with the tomographically inverted data in order to increase the horizontal resolution.

2.2 INCORPORATION OF NADIR OBSERVATIONS

The nadir measurements, $b_m(\theta)$, cannot be directly incorporated because they are contaminated by light scattered by the lower atmosphere and the ground. During the night, and assuming that there are no other significant sources of illumination, the ground brightness, $B_g(\theta)$, depends on the atmospheric brightness, $b_a(\theta)$, the planetary albedo $\alpha(\theta)$, the height of the emitting layer z_a, and the scattering phase function. If it is assumed that the scattering is Lambertian, that the emitting layer has insignificant vertical extent and is a function only of θ, that a plane parallel atmosphere adequately approximates the actual case, and that the albedo is constant, then:

$$b_a(\theta) = b_m(\theta) - \int_0^{2\pi} b_m(\theta')\, w^{-1}(\theta - \theta')\, d\theta' \qquad (3)$$

where the one-dimensional inverse weighting function w^{-1} is defined:

$$w^{-1}(\rho) = \frac{2\alpha}{\pi} \int_0^\infty \frac{\exp(-k z_a)\cos(k\rho)}{1 + 2\alpha \exp(-k z_a)}\, dk \qquad (4)$$

This relationship was derived by Hays and Anger (1978) and extended to oblique angles by Abreu and Hays (1981). The details of how the nadir measurements are incorporated with the inverted data can be found in Solomon et al. (1985) and Solomon (1987). Here we present the highlights of the procedure:

1. Use nadir data and the emission profile resulting from the Cormack inversion to determine the ground albedo.

2. Take the Fourier transform of the nadir measurements and subtract the ground contribution.

3. Combine the normalized nadir high frequencies with the Cormack inversion low frequencies.

4. Take the inverse Fourier transform to obtain the 2-dimensional emission rate function.

Step 3 above is accomplished using transition weighting functions, which are constructed from error functions (Solomon et al., 1985).

3. INVERSION OF SIMULATED EMISSION FEATURES

The technique just presented has been evaluated using simulated data, which have been subjected to the analysis method. Figure 2a shows two fictitious emission functions in photons cm^{-3} s^{-1}, which have been made by multiplying a Chapman function with respect to altitude by a Gaussian with respect to angle along the satellite track. The feature on the left has a 30 km scale height, peak altitude at 150 km, and a half-width of 2.9 degrees, while the feature on the right has a 15 km scale height, 120 km peak altitude, and 4.0 degree half-width. The maximum volume emission rates were normalized to values of 160 and 120 photons cm^{-3} s^{-1} respectively. These functions are representative of the cross section of two auroral arcs.

The brightnesses that would be observed at the VAE data rate, are shown interpolated into a spatial grid in Fig. 2b. These data have been subjected to the procedure outlined in section 2.1. Figure 2c shows the inverted simulated emission function obtained by using the Cormack method. The effects of noise and field of view broadening have been included in the simulation. The inversion reproduces the form of the original features fairly well, and recovers the total integrated emission rate within 1%. The recovered features, however, are smeared horizontally and the vertical profiles are also slightly flattened. These effects are due to the truncation of the Fourier series and to the limited instrumental resolution with respect to tangent height. The inversion of the simulated feature on the left has been reduced at its peak to 60% of the original; the broader feature on the right has only been reduced to 85% of its peak value.

Figure 2d shows the inverted simulated emission function recovered following the procedure outlined in section 2.2. The Cormack inversion was performed through 25 terms, and the albedo used in calculating the ground emission was 0.8. The horizontal resolution is better, especially with respect to the narrower feature on the left. Increase in horizontal resolution also causes an improvement in the

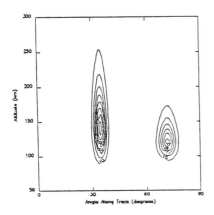

2a.   Contours of a fictitious emission
      function in photons cm^{-3} s^{-1}.

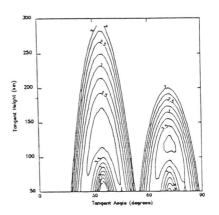

2b.   Simulated brightness measurements
           made at the VAE data rate, and
           interpolated onto a grid. Log$_{10}$
           Rayleighs are contoured with
           respect to tangent height h and
           tangent angle ϕ.

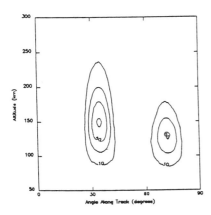

2c.   Inverted simulated emission function,
      photons cm^{-3} s^{-1}, including noise and
      field of view broadening.

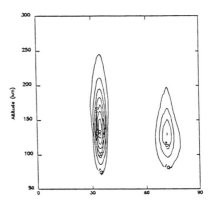

2d.   Volume emission rate contours
           (photons cm^{-3} s^{-1}) of recovered
           function using combination of
           limb scan and nadir data.

FIGURE 2.   Simulated Emission and Brightness Measurements

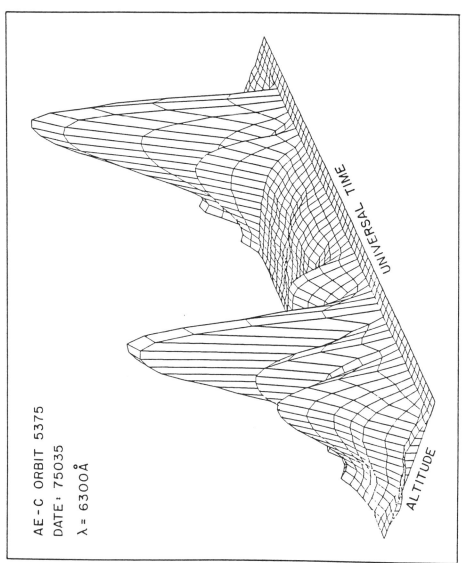

AE - C ORBIT 5375

DATE: 75035

λ = 6300Å

FIGURE 3 Example of a 6300A volume emission rate profile recovered along the satellite track.

recovery of the altitude profile. Neither the limb scan nor combined inversion can resolve the sharp altitude dependence at the peak as they are both limited by the VAE field of view and integration period.

4. DISCUSSION

The technique reviewed here has proven very useful in the recovery of auroral emission features. This method has been applied to studies of the auroral 6300Å, 5577Å, 3371Å and 4278Å emissions (Solomon, 1987; Solomon et al , 1988;). Figure 3 is an example of a 6300Å volume emission rate profile recovered along the satellite track for an orbit of the AE-C satellite. Comparisons of the inverted volume emission profiles with theoretical models have provided a unique way of fine tuning the models.

The next step in the development of this technique involves incorporating data taken looking into the earth at oblique angles. The mathematical formalism in these instances is more complex, but it deserves attention since a large fraction of the data is obtained while looking downward due to the spinning motion of the satellite .

ACKNOWLEDGMENT

This work was supported by NASA grant NAGW-496 to the Univerisity of Michigan.

REFERENCES

Abreu, V. J., and P. B. Hays, 1981: Parallax and Atmospheric Scattering Effects on the Inversion of Satellite Auroral Observations, Appl. Opt., 20, 2203.

Cormack, A. M., 1963: Representation of a function by its line integrals with some radiological applications, J. Appl. Phys., 34, 2722.

Cormack, A. M., 1964: Representation of a function by its line integrals with some radiological applications, II, J. Appl. Phys., 35, 2906.

Hays, P. B., G. Carignan, B. C. Kennedy, G. G. Shepherd, and J. C. G. Walker, 1973: The Visible Airglow Experiment on Atmosphere Explorer, Radio Sci., 8,369.

Hays, P. B. and C. D. Anger, 1978: Influence of Ground Scattering Effects on the Inversion of Satellite Auroral Observations, Appl. Opt., 17,1898.

Solomon, S. C., P. B. Hays, and V. J. Abreu, 1984: Tomographic Inversion of Satellite Photometry, Appl. Opt., 23, 3409.

Solomon, S. C., P. B. Hays, and V. J. Abreu, 1985: Tomographic inversion of satellite photometry. Part 2, Appl. Opt., 24, 4134.

Solomon, S. C., 1987: Tomographic Inversion of Auroral Emissions, PhD Thesis, University of Michigan.

Solomon, S.C., P. B. Hays, and V. J. Abreu: The auroral 6300Å emission: Observations and modelling, accepted by J. Geophys. Res., 1988.

DISCUSSION

Rodgers: The things you are looking at all seem to have a vertical structure. I was wondering if your approach will cope reasonably well with something which might have a slope to it?

Abreu: It will cope very well with situations in which the slope is negative-decreasing with increasing altitude. In the case where the source function increases with increasing altitude there is a problem.

McCormick: On your inversion of your data, did you have to assume a phase function? What was it?

Abreu: In the case presented here, we only corrected for the effects of a Lambertian ground. We have looked separately at the effects of a scattering atmosphere on the inversion and, of course, the situation is more complex. The scattering effects are smaller than the ground effects, especially in the aerosol region where the ground is covered by snow (albedo ≈ 0.8). In order to consider scattering, we used a phase function derived using Mie theory.

McCormick: How sensitive are your results to that phase function?

Abreu: Not extremely sensitive. We have done sensitivity studies with the phase function and the albedo because the albedo is also uncertain. The solution is more sensitive to albedo.

A STUDY OF SATELLITE EMISSION COMPUTED TOMOGRAPHY[1]

C. Grassotti, R.N. Hoffman, and R.G. Isaacs
Atmospheric and Environmental Research, Inc.
Cambridge, Massachusetts 02139, USA

ABSTRACT

We have performed a study to assess the capability of satellite emission computed tomography to infer the temperature structure of the atmosphere when realistic instrument geometry and noise characteristics are taken into consideration. Fleming (1982) demonstrated a significant improvement using the tomographic retrieval technique in an idealized simulation experiment. The present study rectifies a number of shortcomings of Fleming's (1982) experiment, following Fleming's (1985) suggestions. In particular, geometric effects of the instantaneous field of view of the sensor are properly accounted for in our study. In addition, realistic instrumental noise is included in our simulations. Finally, we have used realistic atmospheric cross sections and realistic geometry and simulation codes appropriate for the HIRS2 sensor. It is found that the tomographic approach is superior to the single angle approach in the cases studied when observational noise (instrument noise plus scene noise) is large (1.5 brightness temperature degrees in each channel). For smaller noise levels (0.75 degrees) the two approaches are comparable in the cases studied.

1. INTRODUCTION

Remote sensing of the atmosphere is presently achieved by the use of the multifrequency approach first put forward by Kaplan (1959). Kaplan's proposal to use satellites to remotely sense the atmosphere was in fact predated by King's (1958) suggestion to use multiple scan angles with a single frequency to achieve the same objective. While multichannel instruments have been operational for more than a quarter of a century, the multiangular approach has attracted little attention. Recently however, Fleming (1982, 1985) has proposed retrieving two- or three-dimensional atmospheric temperature fields from radiances observed in multiple channels at multiple angles. He has termed this approach satellite emission computed tomography, or simply computed tomography.

[1]This work was sponsored by the Air Force Systems Command, Air Force Geophysics Laboratory under contract F19628-86-C-0084.

Fleming's (1982) study of computed tomography showed considerable promise when applied using a simplified geometry and radiance calculation. However, Fleming (1985) later discussed the difficulties of interpreting his 1982 results in the context of an actual operational sensor and called for further simulation experiments to examine the problems which arise when realistic geometry is used. The present study was initiated in response to the suggestions of Fleming (1985), to which the reader is referred for further details of his methodology and results.

In this paper we shall first discuss the formulation of the problem and solution methodology as it pertains to our study. This includes a brief discussion of the forward problem used to generate radiance measurements, sensor geometry, and, lastly, the iterative algorithm used in the retrievals. Secondly, we review a few salient results from our sensitivity tests in which temperature retrievals were performed for two atmospheric cross sections which were constructed from a real atmospheric data set. Finally, we summarize our results and make recommendations for future work in this area. A more detailed discussion of this study may be found in the report compiled by Hoffman et al. (1987).

2. METHODOLOGY

Our approach employs Fleming's (1982) procedure to obtain tomographic temperature retrievals based on a radiance data set simulated for selected channels of the HIRS2 infrared component of the TIROS-N Operational Vertical Sounder (TOVS). However, the considerations of finite beamwidth and data independence discussed by Fleming (1985) are treated by constructing a realistic instrumental geometry applied to a real (albeit smooth) atmospheric data set.

2.1 RADIATIVE TRANSFER SIMULATION MODEL

Simulation of HIRS2 sensor brightness temperatures, i.e., the forward problem, for application in the present study are based on application of a realistic but highly simplified radiative transfer model. The form of the radiative transfer equation for computing clear column radiances follows the approach of Susskind et al. (1983). We simplify the equation by assuming surface emissivity to be constant, equal to unity and by neglecting the solar reflection term. Moreover, we avoid computationally intensive line-by-line transmission calculations by basing our forward calculations on the so-called "rapid" transmittance algorithm, which is essentially a band model approach specifically tuned for each of the HIRS2 channels. The algorithm which we used was developed at NASA Goddard Space Flight Center to provide accurate simulations of the HIRS2 radiances. A complete description is given by Susskind et al. (1983).

The rapid algorithm may be described symbolically by

$$R_{In} = F_I((\sec\theta)_n, \tilde{T}_{kn}, k=1,\ldots k_s)$$

where R_{In} is the radiance for channel I and path n, F_I is the calculated radiance for channel I, $(\sec\theta)_n$ is the secant of the zenith angle for path n (at the Earth's surface), $\tilde{T}_{kn}$ is the temperature at level k averaged over the IFOV for path n, and k_s is the level of the surface.

The forward problem F_I depends on several other parameters not explicitly noted. These include the profiles of H_2O, O_3, aerosol, the surface emissivity and surface temperature. However, for any particular cross section all these parameters excluding surface temperature have been held fixed at their climatological values both during simulation and retrieval. Surface temperature is assumed to be equal to the air temperature evaluated at the surface. Furthermore, we have neglected variations of atmospheric transmittance and linearized the Planck blackbody function. With these assumptions, the radiance in a particular channel at a particular zenith angle is a linear function of temperature along the path. The rapid algorithm requires temperature along the path at 66 pressure levels from 1050 to 1 mb distributed more or less evenly in log pressure. Consequently vertical basis functions, here taken to be empirical orthogonal functions (EOFs), must be used. Exactly the same forward problem has been used for both simulations and retrievals in this study.

2.2 GEOMETRIC CONSIDERATIONS

Realistic simulation of the problem geometry requires treating the instantaneous field of view (IFOV) beam of the instrument as a three-dimensional cone whose intersection with quasi-horizontal pressure surfaces is an ellipse. This formulation results in a transformation where the level temperatures $\tilde{T}$ are a linear combination of the gridded temperature with all of the geometric complications of the viewing geometry folded into the proportionality constants.

The basic approach is to weight the gridded temperatures T_k at level k over the appropriate instantaneous field-of-view (IFOV) such that for path n:

$$\tilde{T}_{kn} = \iint\limits_{IFOV} T_k \, da \Big/ \iint\limits_{IFOV} da$$

Assuming a realistic geometry based on the instrumental characteristics (spacecraft velocity, satellite scan angle and scan angle increment, etc.), this can be shown to give the desired temperature at level k for path n as:

$$\tilde{T}_{kn} = \sum_{i} \alpha_{ikn} T_{ik} \tag{1}$$

where i is the index along the subsatellite track. All the geometric complications of the satellite viewing geometry are thus included in the α_{ikn}. Note that the viewing paths and temperature grid spacing may be specified independently of one another.

2.3 INVERSION METHODOLOGIES

In satellite emission computed tomography the inversion method must solve for an entire atmospheric cross section. This is an advantage because the retrievals are more horizontally consistent and a disadvantage because larger problems must be solved. Fleming's method may be thought of as two procedures. The first procedure is the iteration which steps through the paths and frequencies and updates the atmospheric structure. The second procedure is the calculation of the increment for a single path and frequency. This latter procedure is basically a single step in a 1-D physical retrieval method.

Since only a fixed set of zenith angles will be used, $(\sec\theta)_s$, $s=1,\ldots N_s$, we may write the forward problem as

$$R_{In} = F_{Is_n}(\tilde{T}_{kn}) \tag{2}$$

showing explicitly that F depends on $\tilde{T}$ as a variable but on I and s_n only parametrically.

To allow for the general geometry we wish to consider, the level temperatures are a convolution over the grid point temperatures (cf Eq. (1)) which themselves are represented in terms of basis functions,

$$T_{ik} = \bar{T}_k + \sum_{r} \bar{B}_{kr}\beta_{ir}$$

Here, $\bar{B}_{kr}$ are the EOFs and the β_{ir} are the EOF coefficients. The β_{ir} are the retrievable quantities.

A satisfactory retrieval is characterized by

$$|\hat{R}_{In} - R_{In}| < \epsilon_I$$

for all I and n under consideration. Here $\hat{R}_{In}$ is the observation and ϵ_I is the noise level in channel I. We seek an approximation to $\hat{R}_{In} = R_{In}$ by linearizing Eq. (2) about a climate mean basic state. After use of the chain rule and several algebraic substitutions we obtain

$$(\hat{R}_{In} - \bar{R}_{Is_n}) = \sum_{ir} a_{ir}^{(In)} \beta_{ir}$$

where

$$a_{ir}^{(In)} = \sum_{k} \alpha_{ikn} \bar{B}_{kr} \frac{\overline{\partial R_{Is_n}}}{\partial \tilde{T}_k}$$

Temperature effects on the transmittances were ignored when calculating the $\partial R_{Is_n}/\partial \tilde{T}k$. This is consistent with the manner in which the radiances were simulated.

The row action method is essentially a relaxation technique which allows us to consider each path and frequency separately. Fleming's method is a variant of the method of Kaczmarz and others described by Censor (1981). In terms of our original variables the update is given by

$$\beta_{ir}^{(t+1)} = \beta_{ir}^{(t)} + \omega^{(t)} \delta^{(t)} D_{ir} a_{ir}^{(In)}$$

where $\omega^{(t)}$ is a relaxation parameter,

$$\delta^{(t)} = \hat{R}_{In} - \bar{R}_{Is_n} - \sum_{ir} a_{ir}^{(In)} \beta_{ir}^{(t)}$$

and

$$D_{ir} = \left| \sum_{I,n} a_{ir}^{(In)} \right|^{-1}$$

The update is not applied if $|\delta^{(t)}| < \epsilon_I$. The process converges when the rms radiance error in each channel falls below a constant (nominally one) times ϵ_I.

2.4 ATMOSPHERIC DATA

In order to model the horizontally and vertically inhomogeneous spatial structure of the realistic atmosphere, a suitable atmospheric cross section and atmospheric statistics are required. We have used the archived NMC analyses of temperature at 0000 GMT for the GWE SOP-1 period for this purpose. In particular, we used 8 February to 14 February 1979 in the North Atlantic region (20° to 50°W and 30° to 60°N) as a training set to develop our EOF functions. We then used portions of a N-S cross section along 35°W on 15 February 1979 as nature. The two cross sections we used in our retrieval experiments are 1200 km in length. The first, cross section X (Fig. 1a), runs from 58.4 to 47.6°N. The second, cross section R (Fig. 1b), joins two segments of our original cross section, from 49.4 to 44.0°N and from 38.4 to 33.2°N. In both cases and in both data simulation and

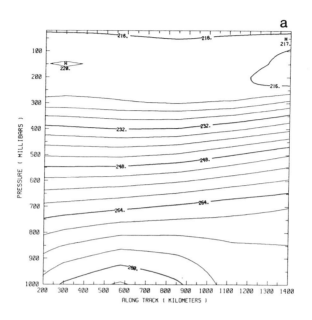

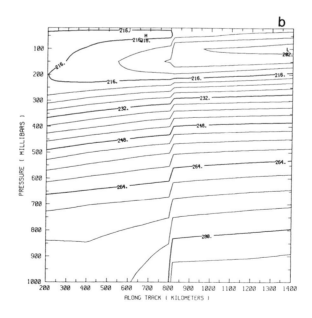

Figure 1.   North to south temperature cross sections (a) X and
(b) R used in sensitivity tests.

retrieval we used a 20 km horizontal grid. The original NMC data were interpolated to this grid.

3. SENSITIVITY TESTS

In this section we describe results of our computed tomography sensitivity experiments.   We contrast multiangular retrievals (ψ = 0°, ±15°, ±30°, ±45°, where ψ is the satellite viewing angle) with nadir only retrievals, for nominal (0.75 degree brightness temperature) and twice nominal (1.5 degrees brightness temperature) noise levels.   In these experiments 12 EOFs or equivalently unfiltered temperature profiles were used to simulate the radiances and 3 EOFs were used to retrieve the temperatures.   While the use of EOFs as basis functions is an implicit vertical filter the retrieval method includes no explicit or implicit horizontal filtering so that the retrieved temperature fields tended to fit the noise contained in the simulated radiance data.   Consequently, the retrieved temperature fields were all smoothed using a low pass filter in the horizontal. This had the effect of reducing the rms errors of all retrieved fields. Results are summarized in Table 1, which indicates that sensitivity tests were performed for 5 cases (A-E) for a total of 9 separate retrievals.

TABLE 1. RMS RETRIEVED TEMPERATURE ERRORS (K)
FOR DIFFERENT NOISE LEVELS, FOR MULTIANGULAR (MA)
AND NADIR ONLY, FILTERED AND UNFILTERED

Case	Cross Section	Noise Level	Unfiltered MA	Unfiltered Nadir	Filtered MA	Filtered Nadir
A	X	0.00	1.23	-	1.23	-
B	X	0.75	1.87	2.00	1.45	1.25
C	X	1.50	2.89	2.99	1.58	1.84
D	R	0.75	2.71	2.81	2.51	2.47
E	R	1.50	3.20	3.81	2.46	2.69

As an example Fig. 2 contains the retrieved temperature fields for cases C and E in which the noise level was set to 1.5K. Panels a and b are the retrieved fields for cross section X corresponding to the multiangular (MA) and nadir only experiments, respectively.   Panels c and d show the same for cross section R. All the retrieved fields qualitatively agree with the original temperature cross sections to which they correspond (Fig.1). However, the computed rms errors of these fields do reveal interesting differences.

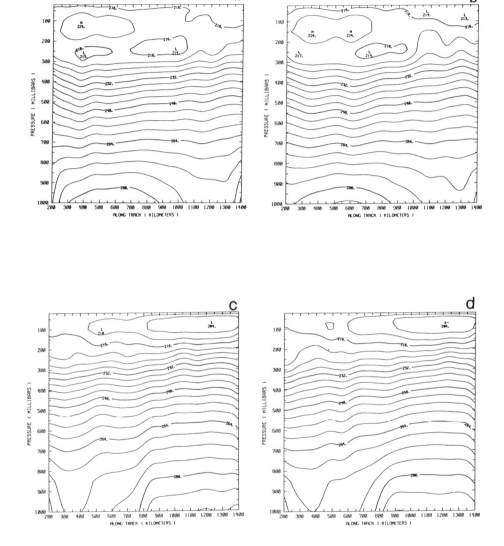

Figure 2.   Retrieved temperature fields (a) case C multiangle,
            (b) case C nadir only, (c) case E multiangle, (d) case
            E nadir only.

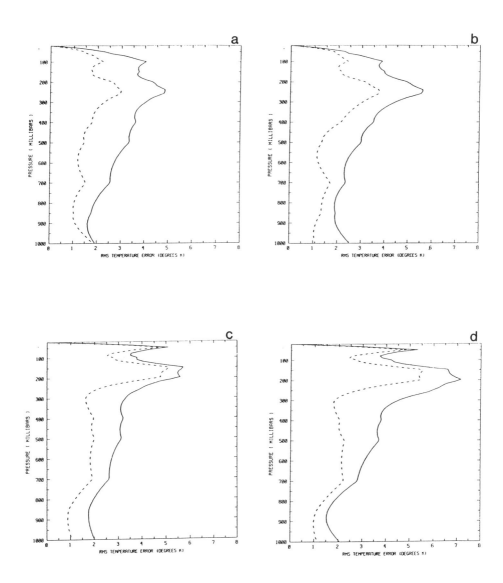

Figure 3.   RMS retrieved error for case C and case E.   See Fig. 2
            for panel definitions.   Solid and dashed curves
            correspond to unfiltered and filtered fields,
            respectively.

Figure 3 shows the vertical structure of the rms error for cases C (panels a and b) and E (panels c and d). In all panels the solid curve corresponds to the unfiltered retrieval while the dashed curve corresponds to the filtered retrieval. A common characteristic of the curves is the presence of error maxima near the tropopause and again near 700 mb. This feature was evident in nearly all rms error profiles. In part, it arises from the fact that we have retrieved only the first 3 of 12 vertical temperature EOFs.

For these higher noise cases it is apparent that, even after filtering, the rms temperature error for retrievals using multiple scan angles is lower than that for retrievals which used only a nadir view. This suggests that the additional information supplied by the off-nadir measurements may yield superior retrievals in high noise conditions.

The results presented here must be seen as preliminary primarily for two reasons: (1) A larger sample of cases should be examined and (2) The input temperature field should contain realistic medium to small scale features. With regard to (1), since each forward problem simulation is performed with a different set of randomly generated radiance errors, the observed radiances can vary noticeably from one simulation experiment to the next, even for the same input temperature field. For example, two different realizations of the multiangular experiment for X with 0.75K errors were conducted. In the first instance convergence was achieved in 23 iterations with pre- and post-filtered rms errors of 1.71 and 1.29, respectively. The second case had not quite reached convergence after 60 iterations and the overall rms errors were 1.87 (unfiltered) and 1.45 (filtered). This suggests that quite a few retrievals of an ensemble of temperature cross sections using the same noise level are needed to quantitatively assess the convergence properties and error characteristics of the retrieval technique.

4. CONCLUSIONS

An algorithm which accounts for all of the geometric details of realistic computed tomography has been developed and tested. The main results of sensitivity experiments indicate that:

- The multiangular retrievals perform better than the nadir retrievals at higher noise levels.
- Both multiangular and nadir retrievals captured a small scale temperature discontinuity. Horizontal filtering reduces the amplitude of this feature.
- The retrieval technique is relatively sensitive to the distribution of radiance errors which are generated randomly. This indicates the need for a number of retrieval experiments with the same noise level to determine retrieval accuracy.

In light of these results additional testing of the technique would be valuable. Such work should investigate retrieval sensitivity to the following parameters: input temperature field, instrumental scan pattern, IFOV size and shape, channel spectral resolution, number of channels, and number of basis functions used. Finally, a more sophisticated forward problem which takes into account the effect of clouds and moisture might be considered.

REFERENCES

Censor, Y., 1981: Row-action methods for huge and sparse systems and their applications. SIAM Rev., 23, 444-466.

Fleming, H. E., 1982: Satellite remote sensing by the technique of computed tomography. J. Appl. Meteorol., 21, 1538-1549.

Fleming, H. E., 1985: Temperature retrievals via satellite tomography. In Advances in Remote Sensing Retrieval Methods, edited by A. Deepak, H. E. Fleming, and M. T. Chahine, A. Deepak Publishing, Hampton, VA, pp. 55-69.

Hoffman, R. N., R. G. Isaacs, and C. Grassotti, 1987: A study of satellite emission computed tomography. AFGL-TR-87-0321, Air Force Geophysics Lab, Hanscom AFB, MA, 92 pp.

Kaplan, L. D., 1959: Inference of atmospheric structure from remote radiation measurements. J. Opt. Soc. Am., 49, 1004-1007.

King, J. I. F., 1958: The radiative heat transfer of planet earth. In Scientific Uses of Earth Satellites, J. A. Van Allen (ed.), University of Michigan Press.

Susskind, J., J. Rosenfield, and D. Reuter, 1983: An accurate radiative transfer model for use in the direct physical inversion of HIRS2 and MSU temperature sounding data. J. Geophys. Res., 88(C13), 8550-8568.

DISCUSSION

Wahba: Two questions. The first is, how did you choose the pass band of the filter in the horizontal? And the other is, if you had some outside information where the tropopause was and you enforced a minimum do you think that this would improve things?

Grassotti: As far as the pass band of the filter is concerned, we basically used a 3-point moving average filter which preferentially removed the smaller scales. I believe the response function was such that wavelengths of 160 km were reduced by roughly 80% and we found this to be satisfactory. Regarding your second point, a traditional problem in the retrieval of temperature has been getting that tropopause height, and I think that externally imposing a minimum would be useful. In fact, if you view the retrieval as a general optimization problem then one might apply a number of constraints upon the solution and minimize some particular function. For example, and this relates to the problem of noise in the horizontal, we might represent the horizontal variation of the temperature field in terms of Fourier basis functions and retrieve the coefficients subject to some a priori smoothness constraints. This would be done in addition to approximately fitting the radiances to within the noise, which we are already doing. In this case we would then be retrieving the coefficients of the Fourier-EOF basis functions of the temperature field.

Fleming: Did you have a particular motivation for stopping at three EOFs, or was it just a matter of the amount of work you wanted to do?

Grassotti: Well, it was a matter of time as much as anything. We found that the first three EOF's explained over 90% of the variance in nature so it seemed a logical point at which to truncate. But I think that we might want to extend these experiments using additional EOFs to obtain better results.

Fleming: The remaining 10% could be crucial.

Chiou: I would like to ask one short question about the EOFs that you construct. Did you construct them based on a particular period and particular area? What improvement will you get, or what kind of changes would you expect if you didn't consider that particular season and region?

Grassotti: We constructed the EOFs based on data from the week preceding the day on which we performed the retrievals and for a 30-degree by 30-degree latitude/longitude volume in the North Atlantic. So I don't know how it would affect our results if we used different EOFs. Our training set was fairly specific. Since we did choose the week preceding that day, we used the climatology established for the preceding week.

SATELLITE TOMOGRAPHIC REMOTE SENSING BY FREQUENCY SCANNING

Henry E. Fleming
Satellite Research Laboratory
NOAA/NESDIS
Washington, D.C. 20233, USA

Lillian L. Barnes
Computer Division
NOAA/NCASC
Washington, D.C. 20233, USA

ABSTRACT

Satellite tomographic remote sensing of the atmospheric temperature and constituent density structure, as reported in the literature,
is based either on angle scanning or on a combination of angle and
frequency scanning. However, there are some practical difficulties
associated with the geometric approach. A more fruitful solution to
the atmospheric remote sensing problem is to use only frequency scanning, but with the number of channels being in the hundreds, or even
thousands, instead of the usual 20 to 40 channels. In a certain sense
this approach is still a tomographic one because a fixed target in the
atmosphere is being sensed numerous times, not at different angles,
but at different frequencies. A numerical simulation study using 1825
channels is described. Retrieval results for various combinations of
channels are presented. It was found that a type of square root law
applies for which the temperature retrieval accuracy improves in proportion to the square root of the number of channels used. The most
important result is that when a sufficient number of channels is used,
an accuracy of 1°C (which is commonly accepted as being radiosonde
accuracy in the troposphere) can be achieved. An analysis of the
regions of the spectrum indicates which regions contribute most to
that accuracy.

1. INTRODUCTION

Satellite remote sensing by the technique of computed tomography
is accomplished as follows. Using satellite radiance measurements
taken along the orbital track at various angles, one can use computed
tomography to retrieve cross sections of the vertical atmospheric temperature and moisture structure. Measurements taken at various
frequencies can be used to supplement the angle scanning measurements.
Details for doing this and results are given in Fleming (1982). When
compared to conventional nadir sounding techniques, the retrieval
accuracies attained using computed tomography were far superior.

Unfortunately, the potential improvement in retrieval accuracy of
satellite tomography, based on theoretical considerations, very likely
cannot be realized in practice. When finite beam widths are consid-

RSRM '87: ADVANCES IN
REMOTE SENSING RETRIEVAL METHODS
A. Deepak, H.E. Fleming, and J.S. Theon (Eds.)   35

ered instead of geometric lines of sight, the grid boxes which make up
the sounding cross section are not geometrically independent. This
loss of geometric resolution can result in an almost complete negation
of the accuracy gains achieved over conventional sounding techniques.
Details can be found in Fleming (1985) and Grassotti et al. (1988).

The practical difficulties associated with the geometric approach
to computed tomography has led us in this paper to consider instead an
approach based solely on frequency scanning. While one can correctly
argue that frequency scanning is not a tomographic approach in the
strict sense of the word, it is in the following broader sense. Each
level of the atmosphere is being sensed numerous times, not at different angles, but at different frequencies. Just as new information is
acquired about a given atmospheric level every time it is viewed at a
different angle, so too is new information acquired about that level
when it is viewed at a different frequency. It is in this sense that
we say that frequency scanning is tomographic remote sensing.

The following questions now arise: (1) How does tomographic
remote sensing by frequency scanning differ from the type of atmospheric remote sensing that has been done for the past two decades,
and (2) What are the advantages of this approach? The answers are
that: (1) Very many more frequencies (channels) are used; they can
number in the thousands, and (2) There is a dramatic increase in
retrieval accuracy. These issues are now explored in depth.

2. FREQUENCY SCANNING METHODOLOGY

At present there are at least two viable instrumental approaches
to acquiring large numbers of channels simultaneously for purposes of
frequency scanning. The classical approach is to use an interferometer which, because of the Fellgett advantage and high energy
throughput, can yield very high spectral resolution power. A newer
approach is to use a grating array spectrometer (GAS). The essential
mechanisms of a GAS instrument are that the incoming energy is dispersed coarsely into orders in the y-direction, and in turn these
orders are finely dispersed in the x-direction. If the emergent, dou-
bly-dispersed energy is focused onto a detector array, one has high
spectral resolution channels equal in number to the total number of
detectors in the array.

It was pointed out that the distinctive feature of sounding by
tomographic frequency scanning is the number of channels involved. The
current operational sounding instruments have at most twenty-seven
channels; whereas, in frequency scanning thousands of channels might
be involved. To see the advantage of this, suppose we are retrieving
temperatures at 40 pressure levels, and suppose we have 2,000 channels. Then each level of the atmosphere is sensed, on average, by at
least 50 different channels (actually, by very many more than that,
because of the overlapping of the associated broad weighting functions). In other words, there are well in excess of 50 different
pieces of information (not totally independent, however) about the
temperature at each level of the atmosphere. This is bound to have a
major impact on the retrieval accuracy, as will be shown.

Fortunately, the impact of frequency scanning on the many retrieval algorithms in use is very limited. The only (but not trivial) detail that needs to be changed is the dimensionality of the problem. Normally, the number of pressure levels at which the atmospheric parameters are retrieved exceed the number of channels used. In tomographic frequency scanning the opposite is true; the number of channels greatly exceeds the number of pressure levels. This change from an underdetermined system of equations to an overdetermined system is the principle reason for the improvement in retrieval accuracy. However, the spectral variety, nonlinearity of the Planck function, and improved spectral resolution in the infrared part of the spectrum contribute toward improving the independence of the equations and, hence, also help to improve the retrieval accuracy.

To illustrate the kind of change in the retrieval algorithm that is required by the change in dimensionality, we consider the minimum variance simultaneous (MVS) retrieval method described in Fleming, et al. (1986b). The solution vector V is of the form

$$V = [T_1, \ldots, T_t, T_s, Q_1, \ldots, Q_q]^T \qquad (1)$$

where the vector elements T and Q represent temperature and water vapor mixing ratio, respectively, the superscript T indicates vector transpose, and the subscripts t and q represent, respectively, the number of temperature and mixing ratio elements in V, but the subscript s represents the surface. The number of elements in V is $m=t+q+1$.

The MVS solution is given by the formula

$$V = v + C (R - r) \qquad (2)$$

where R is the satellite-measured radiance vector with n elements (corresponding to the n channels), v and r are the initial solution and radiance vectors, respectively, and C is the retrieval matrix operator. In linearized form, the retrieval operator is

$$C = S A^T (A S A^T + N)^{-1} \qquad (3)$$

where the superscript -1 represents the matrix inverse, S is the mxm covariance matrix of the vector V, N is the nxn system noise covariance matrix, and A is an nxm matrix whose elements are products of the following three factors: weighting functions, Planck function linearizations, and numerical quadrature weights. For details see Fleming, et al. (1986b).

Notice that the composite inverse matrix in Eq. (3) is of overall dimension nxn, while S is of dimension mxm. For tomographic frequency scanning $n \gg m$; therefore, the computation of the inverse of that very large composite matrix will be very inefficient, and even could be ill-conditioned. The solution to this dilemma lies in the very important matrix identity

$$S\ A^T\ (A\ S\ A^T\ +\ N)^{-1} = (A^T\ N^{-1}\ A\ +\ S^{-1})^{-1}\ A^T\ N^{-1}\ . \qquad (4)$$

If we set C equal to the right-hand side of Eq. (4), all matrices that
need to be inverted, except N, are of the much smaller dimension mxm.
But even the nxn matrix N causes no difficulty because it usually
is either diagonal or has some other simple, tractable form.

It now is apparent what form of the MVS solution should be used
in the overdetermined case arising from frequency scanning. One continues to use the formula given by Eq. (2), but the right-hand side of
Eq. (4) is used for the retrieval operator C. Since the form of Eq.
(4) is very general, we automatically have available as special cases
additional frequency scanning retrieval algorithms, such as the minimum information solution. In addition, retrieval algorithms not covered by Eq. (4) can be derived in a manner analogous to that just
illustrated.

3. NUMERICAL SIMULATION STUDIES

The minimum variance simultaneous (MVS) retrieval method was used
to conduct several numerical simulation studies which compare
retrieval accuracies for various channel combinations and other parameter changes. Details of the simulation studies are as follows:

a. Forty atmospheric temperatures from 0.1 to 1000 mb, the surface temperature, and 15 water vapor mixing ratios from 300 to 1000
mb are retrieved; consequently, in Eq. (1) we have t=40, q=15, and
m=56.

b. Sample sets of temperature and moisture data to be retrieved,
and to be used for ground truth, were compiled for 3 latitudinal belts
and 2 seasons for a total of 6 sets. The sets are for the belts 0 to
30°N, 30 to 60°N, 60 to 90°N, and for the months of January and
August.

c. The physical model is the radiative transfer equation in which
the source function (i.e, the Planck function) and the kernel function
(i.e., the derivative of the atmospheric transmittance function) are
nonlinear functions of the parameters to be retrieved. Linearization
of the Planck function is by a linear Taylor expansion about temperature, while the kernel function is linearized by basing it on the
sample mean temperature/moisture profile of the set under study,
rather than on the individual ambient profiles.

d. A total of 1813 channels where selected from the middle infrared spectral region from 594.4 cm^{-1} to 2745.2 cm^{-1}. These channels
can be categorized as follows. There are 701 longwave CO_2 thermal
channels with a bandpass of 0.5 cm^{-1} each, 309 shortwave CO_2/N_2O
thermal channels with a bandpass of 2 cm^{-1}, and 803 water vapor channels with a bandpass of 1 cm^{-1}. For these bandpasses 1813 is the
maximum number of infrared channels available without bandpass overlap
or absorption by minor gases, such as ozone. In addition to the 1813
infrared channels, there are 12 microwave O_2 channels for a total of

1825 channels. The twelve microwave channels are used in all of the
simulation studies because in practice they always will be available
from the Advanced Microwave Sounding Unit (AMSU) instrument.

e. Our simulation study is equivalent to using cloud-free
radiances. Futhermore, unit surface emissivity is assumed in both the
infrared and microwave spectral regions. These assumptions, while not
totally realistic, are reasonable for the kinds of conclusions we are
trying to draw from the study. Also, most of the results are used
only in a comparative sense. On the other hand, if one were to
include a realistic model of the surface emissivities in the radiative
transfer equation, they could be determined as additional parameters
in the solution vector of Eq.(1) by the methods in Fleming (1988).

f. A noise equivalent delta-temperature of 0.2° C is assumed,
which is converted to noise equivalent delta-radiances that are
frequency dependent. This conversion is accomplished by using the
January, 30 to 60°N, sample mean temperature/moisture profile in the
forward calculation of the radiative transfer equation and converting
the resulting radiances into equivalent brightness temperatures, which
in turn are perturbed by plus and minus 0.1° C and converted back to
radiances. The difference between the purturbed radiances produces
the required noise equivalent delta radiances.

Since the purpose of the numerical simulation study is to determine the retrieval accuracy under various conditions, an explicit
expression for the retrieval error is required. This is accomplished
by calculating the covariance matrix of the vector that is the difference between the MVS solution vector and the ground truth vector. Then
the square roots of the diagonal elements of this covariance matrix
represent the desired retrieval rms error vector. Details can be
found in Fleming, et al. (1986a).

Denote the error covariance matrix just described by U, then in
terms of the operator C of Eq. (3),

$$U = S - C A S \tag{5}$$

where the right-hand expression of Eq. (4) is used for C, since n >> m
in tomographic frequency scanning. If the matrix identity

$$I - (X + Y)^{-1} X = (X + Y)^{-1} (X + Y) - (X + Y)^{-1} X$$

$$= (X + Y)^{-1} Y \tag{6}$$

is applied to Eq. (5), it reduces to the simpler form

$$U = (A^T N^{-1} A + S^{-1})^{-1} . \tag{7}$$

It follows that the vector E of retrieval rms errors is just

$$E = [diag(U)]^{1/2} . \tag{8}$$

These formulas were used to produce the results presented in the
next section. Note, however, that the covariance matrix S used in
Eqs. (5) and (7) is only an estimate of the true matrix because only a
finite set of less than 200 temperature/moisture profiles was available to calculate S. This, however, should not affect the conclusions
of this study.

4. RESULTS

4.1 TEMPERATURE RETRIEVAL ACCURACIES

The rms temperature error curves in Fig. 1, which were obtained
from Eqs. (7) and (8), illustrate the advantages of many-channel
frequency scanning. These curves are shown as a function of pressure
from 10 to 1000 mb, and the data points are evaluated at 30 pressure
values which are connected by line segments. Note that the data are
point values, not layer averages. The curve with triangles on the
extreme right-hand side of Fig. 1 is the square root of the original
variance. In other words, this curve represents the temperature rms
error that results from using the sample mean as the first approximation to the solution and can be thought of as that part of the temperature structure that remains to be retrieved. The sample set of 191
profiles for January, 30 to 60°N, is used in all illustrations of rms
temperature error because it has the largest original temperature
variance of all 6 sample sets.

The middle curve with squares in Fig. 1 is the rms temperature
retrieval error when the 27 sounding channels of the next generation
of operational sounders, namely, the High-resolution Infrared Radiation Sounder (HIRS) and the Advanced Microwave Sounder Unit (AMSU),
are used. Finally, the curve with circles on the extreme left-hand
side of Fig. 1 is the rms temperature retrieval error when the proposed 1825-channel frequency scanning instrument is used.

There are three important things to note in Fig. 1. First, both
instruments yield high retrieval accuracies, being (with only a few
exceptions) between 1° and 2°C for the HIRS and AMSU combination and
between 0.3° and 1°C for the 1825 channels. Second, The 1825 channels
yield considerably better results than do the 27 channels, as one
would expect. Third, it is significant that not only is the accuracy
of the circles curve considerably better than that of the squares
curve, but that the accuracy of the circles curve is better than the
1°C error in the troposphere that is commonly considered to be radiosonde accuracy. Achieving or exceeding radiosonde accuracy has always
been a goal of the temperature sounding community and the 1825-channel
sounder allows us for the first time to meet that goal.

4.2 WATER VAPOR RETRIEVAL ACCURACIES

The three curves in Fig. 2 are the same kinds of error curves as
in Fig. 1, except that they represent water vapor mixing ratio rms
errors instead of temperature rms errors. Also, the pressure ordinate
terminates at 300 mb instead of 100 mb, and the sample set for August,

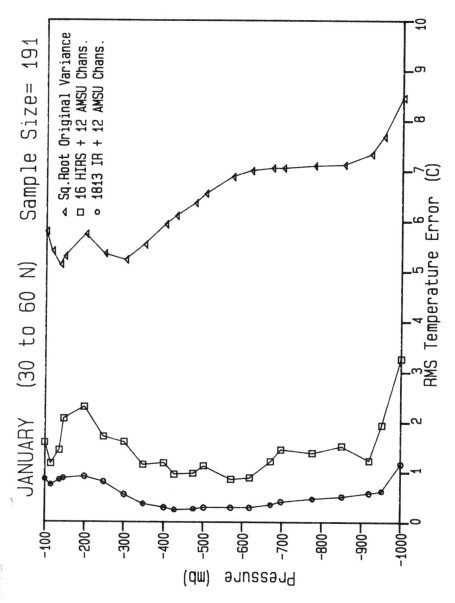

FIGURE 1.   Comparison of temperature retrieval accuracies.

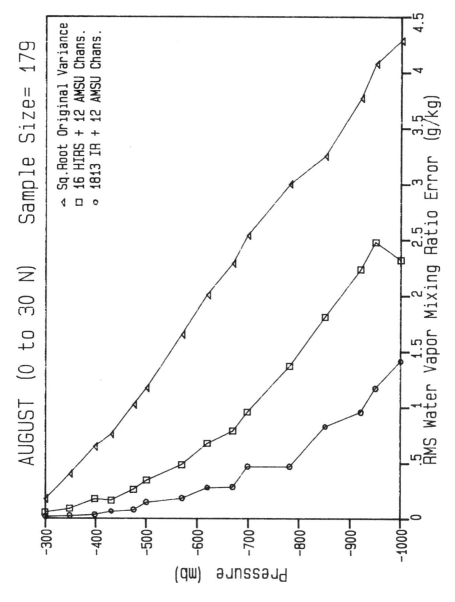

FIGURE 2. Comparison of water vapor mixing ratio retrieval accuracies.

0°to 30°N, is used instead of the January, 30°to 60°N, sample set because it has the greatest original water vapor variance of the 6 sets.

In Fig. 2 the percentage improvement in retrieval accuracy over
the original variance is not as pronounced as it is in Fig. 1, but
this is consistent with past experience with water vapor retrievals.
However, the relative improvement in accuracy of the circles curve
over that of the squares curve is about the same in Fig. 2 as it is in
Fig. 1. The power-law-like decay, with decreasing pressure, of the
vapor mixing ratio retrieval error to near zero above 300 mb is characteristic as well of the atmospheric water vapor mixing ratio itself.

4.3 CHANNEL OPTIMALITY

If one can achieve the high accuracy shown in Fig. 1 with 1825
arbitrarily-chosen channels, the question naturally arises as to
whether or not one can achieve the same accuracy with far fewer channels, provided they are chosen optimally. An answer to this question
is provided by Fig. 3. The circles curve on the left-hand side of the
figure is the same as that appearing in Fig. 1, except that in this
figure the abscissa has been magnified significantly by shortening the
rms temperature error range from ten degrees to only two degrees.

By a careful examination of the infrared spectrum, we selected a
subset of 407 channels (plus the 12 AMSU channels) from the original
set of 1813 channels that we judged to be essentially both minimal and
optimal. The retrieval results using these 419 channels are shown as
the triangles curve in Fig. 3. Since this error curve is systematically to the right of the one for the 1825 channels, it appears that
the number of channels used is the factor having overriding importance; the particular selection of channels is not nearly as important
a factor in determining retrieval accuracy.

To further test the hypothesis that the number of channels is of
primary importance, we used a second subset of 407 channels (plus the
12 AMSU channels). These channels were chosen to be midway in central
frequency between channel pairs of the previous subset of 407. Consequently, if the original subset is near optimal, this subset should be
far from optimal. Yet, the 419-channel nonoptimal results, as represented in Fig. 3 by the right-most squares curve, differ so little in
accuracy from the 419-channel optimal results that optimality appears
to be of little importance. In short, when minimizing retrieval error
it is the number of channels that matters most, not the frequency distribution of the channels (assuming their distribution is reasonable
in the first place). Of course, the statement just made applies only
to a large number of channels such as that employed by tomographic
frequency scanning. When a very limited number of channels is
involved, their central frequency distribution can become a significant factor in determining retrieval accuracy.

4.4 A SQUARE ROOT LAW

It is a well-known fact that the precision of an average of

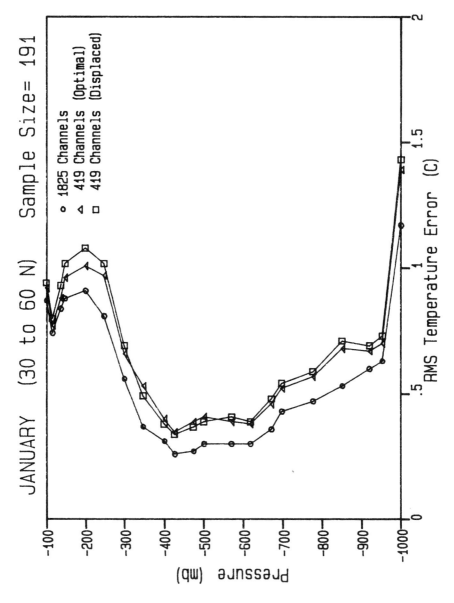

FIGURE 3.   Comparison of optimal, nonoptimal, and benchmark temperature
            retrievals.

measurements improves in proportion to the square root of the number
of independent measurements in that average. The results shown in Fig.
3 suggest that an analogous square root law applies to frequency scanning. The hypothesis is that the retrieval accuracy improves in proportion to the square root of the number of channels used in the
retrieval. The intent of Fig. 4 is to suggest that such a square root
law applies and is in fact the underlying rule for the improvement of
retrieval accuracy with the increase in the number of channels.

In Fig. 4 the curve with the open circles and the one with the
open triangles are repetitions of those in Fig. 3 and are the rms temperature error curves, respectively, for the 1825-channel and the optimal 419-channel retrievals. On the other hand, the retrieval error
curve with the closed circles in Fig. 4 is obtained by using the 419
optimal channels, but with their noise levels reduced by a factor of
0.48. This factor is the square root of the ratio 419/1825. In other
words, if a retrieval is made with 1825 channels set at their specified noise levels, and a second retrieval is made with 419 channels,
but with their noise levels reduced by a factor of 0.48 of their specified levels, then the two results should be identical if the square
root law holds. While the open- and closed-circle curves are not
identical in Fig. 4, they are so close together that, for all practical purposes, the truth of the square root hypothesis is confirmed.

One also can test the hypothesis by the converse argument, i.e.,
the 1825-channel retrieval can be made with the channel noise
increased by a factor of 2.09, which is the square root of the ratio
1825/419. This produces the retrieval error curve represented in Fig.
4 by closed triangles. When this curve is compared with the open
triangles curve (i.e., the one for 419 channels set at their specified
noise levels), the two curves are so close together that again, for
all practical purposes, the square root hypothesis is confirmed.

4.5 CONTRIBUTIONS OF THE THERMAL AND WATER VAPOR CHANNELS

The last thing we study is the separate contribution of the
thermal and water vapor channels to the temperature retrieval accuracy. As noted in Section 3, a total of 1010 infrared thermal channels and 803 infrared water vapor channels are available for the simulation study. These two sets of channels were used separately to do
temperature retrievals, with each set including the 12 AMSU microwave
channels.

In Fig. 5 the 1825-channel rms temperature error curve denoted by
circles is a repetition of the one shown in Fig. 4 and is our best
(baseline) result. On the other hand, the rms temperature retrieval
error curve denoted by the squares in Fig. 5 is for the 1022 thermal
channels, while the error curve denoted by the triangles is for the
815 water vapor channels. Two features of the curves in Fig. 5 are
immediately apparent. First, there is a considerable degradation in
accuracy in the thermal-channels curve which cannot be accounted for
solely by the reduction in the number of channels. Second, the water
vapor-channels curve is virtually coincident with the baseline

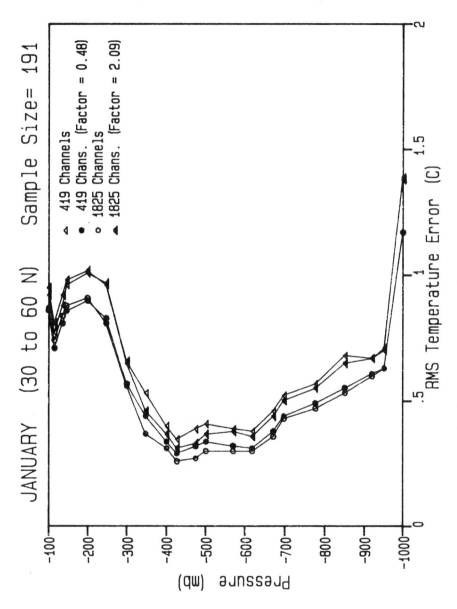

FIGURE 4.   Illustrations of the square-root law.

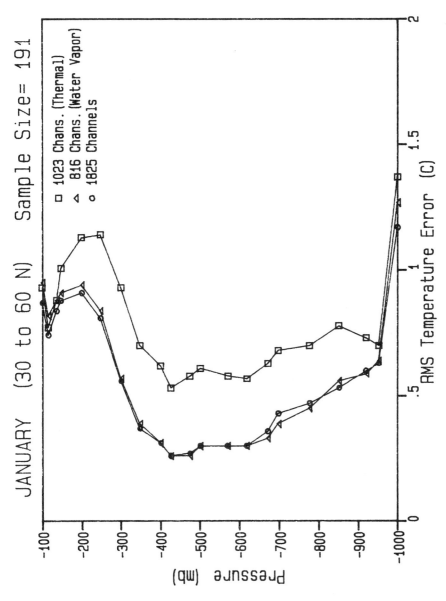

FIGURE 5.   The combined and separate contributions of the thermal and water vapor channels.

1825-channel curve. This second feature is now analyzed.

It is well known that the water vapor channels contribute to the
temperature retrievals through the Planck function in the radiative
transfer equation, but that does not explain the very strong contribution demonstrated in Fig. 5. That is accounted for by the following
three factors:

a. The band passes of the water vapor channels are so narrow relative to the width of the water vapor lines that wings of the lines
are being resolved. This results in narrow water vapor weighting
functions.

b. The water vapor weighting functions are narrowed further by
the power-law-like shape of the atmospheric water vapor profile in
that the low-altitude side of the weighting function is being squizzed
upward by the intense absorption of a relatively large quantity of
water vapor at high pressure, while much less absoption is taking
place at the high-altitude side of the weighting function because of
the much reduced absorber amount and pressure there.

c. The moderately strong nonlinearity of the Planck function with
respect to temperature in the middle infrared water vapor absorption
region of the spectrum, coupled with the atmospheric temperature lapse
rate, also narrows the weighting function in the linearized retrieval
problem.

These properties are explored in much greater detail in the paper
in this volume by Neuendorffer (1988). The implications that the
results of Fig. 5 have for the frequency scanning retrieval problem
are discussed in the next section.

5. CONCLUSIONS

Tomographic frequency scanning has been differentiated from the
usual sounding techniques in that tomographic frequency scanning uses
many more channels than does conventional remote sensing. This
results in a retrieval problem that requires solving an overdetermined
system of linear equations for frequency scanning as opposed to solving an underdetermined system in conventional sounding. The necessary mathematical details are given in Section 2.

Retrieval accuracies for various channel combinations and other
parameter changes were determined by numerical simulation studies,
whose details are given in Section 3. Results from the studies are
presented in the form of five figures. Moreover, these results are
consistent with those from the five other data sets that were run but
are not shown. It is concluded that radiosonde temperature accuracy
of 1°C can be achieved by using a sufficiently large number of
infrared channels, along with the 12 AMSU channels. The same conclusion can be drawn for the water vapor profiles.

Another conclusion to be drawn from the figures is that the

improvement in retrieval accuracy from tomographic frequency scanning over conventional methods is due almost exclusively to the incremental increase in information with each additional channel; not to the central frequency distribution of channels. In other words, channel optimization is not a contributing factor. Futhermore, the improvement in retrieval accuracy is essentially proportional to the square root of the increase in the number of channels used in the retrieval.

Finally, there is more temperature retrieval information in the water vapor channels than in the thermal channels. Figure 5 suggests that the thermal channels are not at all necessary and that in fact the 815 water vapor and microwave channels yield essentially the same information as do the 1825 channels. However, it is another question entirely as to whether or not all the potential information in the water vapor channels can be utilized in practice.

Because the water vapor mixing ratio profile must be known to high accuracy in order to use the water vapor channels in the temperature retrieval problem and because this profile is an unknown, one must successfully solve a nonlinear water vapor retrieval problem in order to do the temperature retrieval problem. An alternative to the sequential approach is to solve for temperature and water vapor simultaneously, as suggested in this paper. In either case, one is required to solve a difficult nonlinear problem. Therefore, it is unlikely that one can solve the temperature retrieval problem, using only the water vapor channels, and obtain the kind of accuracy shown in Fig. 5.

As a result of the analysis just presented, it is clear that one should use both thermal and water vapor channels in the practical solution of the temperature retrieval problem. This same analysis suggests that a simultaneous solution may have some definite advantages over a sequential approach when solving for temperature and water vapor.

On the other hand, the curve with the squares in Fig. 5 suggests that using the 1023 thermal channels alone to obtain temperature is also a reasonable approach, despite its small decrease in accuracy. That combination of channels still yields temperatures with radiosonde accuracy, except in the vicinity of the tropopause and at the surface. Futhermore, since that approach results in a considerably more linear problem, the probability of obtaining actual solution accuracies close to those given by the curve with the squares in Fig. 5 is quite high.

Irrespective of the combination of channels used, this paper has demonstrated the potential for obtaining temperature and moisture retrievals with radiosonde-like accuracies. However to reach that potential in actuality requires that one sucessfully solve a difficult nonlinear problem. Thus, the problem shifts from determining the kinds and number of channels to be used to that of our ability to solve the nonlinear water vapor problem.

ACKNOWLEDGMENTS

We would like to thank Arthur C. Neuendorffer for his guidance in
selecting the 407 optimal infrared channels and for several contributions resulting from discussions with him. We would also like to
thank Harold M. Woolf and William L. Smith for making their high spectral resolution transmittance data available to us.

REFERENCES

Fleming, H. E., 1982: Satellite Remote Sensing by the Technique of
    Computed Tomography, J. Appl. Meteor., 21, 1538-1549.

-----, 1985: Temperature Retrievals via Satellite Tomography. In A.
    Deepak, H. E. Fleming, and M. T. Chahine (Eds.), Advances in
    Remote Sensing Retrieval Methods, A. Deepak Publishing, Hampton,
    Virginia, 55-67.

-----, 1988: A Physical Retrieval Algorithm for Obtaining Microwave
    Soundings. In J. C. Fischer (Ed.), Passive Microwave Observing
    from Environmental Satellites: A Status Report Based on NOAA's
    June 1-4, 1987 Conference in Williamsburg, Virginia, NOAA Tech.
    Rep. NESDIS 35. [National Technical Information Service (NTIS),
    U. S. Department of Commerce, Sills Bldg., 5285 Port Royal Road,
    Springfield, VA 22161].

-----, D. S Crosby, and A. C. Neuendorffer, 1986a: Correction of
    Satellite Temperature Retrieval Errors Due to Errors in Atmos-
    pheric Transmittances, J. Climate Appl. Meteor., 25, 869-882.

-----, M. D. Goldberg, and D. S Crosby, 1986b: Minimum Variance
    Simultaneous Retrieval of Temperature and Water Vapor from
    Satellite Radiance Measurements, Preprints, Second Conf. on
    Satellite Meteorology/Remote Sensing and Appl., Williamsburg,
    Virginia, 20-23. [American Meteorological Society, 45 Beacon St.,
    Boston, MA 02108].

Grassotti, C., R. N. Hoffman, and R. G. Isaacs, 1988: A Study of
    Satellite Infrared Tomography. In this Volume.

Neuendorffer, A. C., 1988: The Feasibility of 1-Deg/1-km Tropospheric
    Retrievals. In this Volume.

DISCUSSION

McClatchey: I have two questions, so I thought I would ask them at once so I can get them in. The first one is, is it a matter of philosophy in terminology? I am puzzled by the title of your paper. I think of tomographic techniques as geometric scanning, and frequency scanning as something different—your title sounds like geometric approaches by frequency scanning. That is the first question. The second question is, can you clarify the accuracy and extent to which an error analysis is part of this? My question has to do with the accuracy of transmittances required to get the results you've shown here as well as the measurement accuracies required.

Fleming: As far as the use of terminology of tomography is concerned, I said it is only a tomographic-like approach. It is not tomography, you are right, because it has no geometric aspect and it has to have that to be tomography. I wanted the paper put in this session only to emphasize the big jump in number of channels from, at most, dozens to thousands. This makes the problems mathematically look like a tomographic one in that the system of equations becomes overdetermined. As far as the noise is concerned, the noise that I used is realistic. It is the level of noise that is being specified for these instruments, although there may be some system noise on top of it. The solution accuracy is rather insensitive to noise, and there is slow growth in error with increased noise. In the simulation I assumed perfect knowledge of transmittances, so that is something that would change when you get into the real world. That is why that water vapor result is as good as it is, but I don't think we can achieve those moisture accuracies in reality. The thermal problem is much easier than the water vapor problem. The purpose of these curves is just to show you what the potential is since this instrument hasn't been built. You are obviously going to get worse results in reality, but even a 50% degradation of these results is still impressive.

Rodgers: I think you will get some more insight into what is going on if you tried separating out the null space error from the measurement error. You've got them both thrown in there in your RMS error. It would be interesting to see how far down the null space error goes when you add all these channels.

Fleming: In addition to the information contained in the radiances, we have statistical information; so the null space error should be smaller than it would be otherwise.

Rodgers: Yes, your error covariance includes both kinds of error, the structure you can't see because of the width of the weighting functions plus the error due to noise. These are two separate components, that is what my paper is going to be about. It looks like in this case, it would be very instructive to separate out these two sources of error.

Falcone: Henry, I've got two questions also. You say this technique is in competition with the radiosondes. What do you intend to use the radiosondes for or this technique for? And number two, I've always been leary of a technique that used microwave and infrared together without separating the physical difference. Namely, you don't go through clouds with the infrared. The other thing is do you assume that the fields of view of both instruments are the same and they are coincident? Do you assume that?

Fleming: I'll answer the last question, first. Yes, we do assume coincidence. This is a perfect simulation. Secondly, if you use an approach in which you eliminate the cloud effects from the measurements before doing the retrieval, then there is no problem mixing infrared and microwave channels. Also, in the microwave channels you have to remove regions of precipitation and large amounts of cloud liquid water content. But assuming you can do those things as two separate steps before you do the retrieval, then you can combine channels indiscriminantly without any problem. There is one other problem, as you suggest. There may be differences in the fields of view, that will have to be corrected by an interpolation scheme or by averaging. After that you can throw all the channels together.

Falcone: What is the use of this data, is it NWP?

Fleming: Yes, that is the main use, but the data will be available to anyone.

Yue: I think we all face this same problem, that is, if you develop a more elaborate algorithm, or, in your case, you use a greater number of channels, in general, you can get a much better result. But the problem is sometimes you may spend a lot of computer time to just retrieve one profile. Sometimes we have a large quantity of data, and my question is how much more computer time do you use in order to get a much better result?

Fleming: Very little time because you have one big operation where you multiply a matrix by its own transpose in which the large dimension disappears forever. Then the remaining outside dimensions are simply numbers of pressure levels by number of pressure levels. It is not a big computer problem if you do it correctly.

INCREASING THE MATCHING RESOLUTION
OF HIERARCHICAL WARP STEREO ALGORITHMS

James P. Strong
NASA Goddard Space Flight Center
Information Analysis Facility
Greenbelt, Maryland 20771, USA

J. Anthony Gualtieri
NASA National Research Council
NASA Goddard Space Flight Center
Image Analysis Facility
Greenbelt, Maryland 20771, USA

ABSTRACT

Spatial resolution of elevation data obtained using hierarchical warp stereo techniques for matching corresponding pixels is often limited by noise in the two images and the size of the areas which contain recognizable features. This work describes a technique to increase spatial resolution of the matches by incorporating edge matching into the algorithm.

1. INTRODUCTION

In October 1984, the Shuttle Imaging Radar-B (SIR-B) experiment provided the first set of pseudo stereo Synthetic Aperture Radar (SAR) images from which elevations could be determined. In 1986, the French High Resolution Visible (HRV) sensor flown on the SPOT spacecraft provided completely overlapping stereo image pairs in the optical range. Future sensors such as the German Monocular Electroptical Stereo Scanner (MEOSS) to be flown on ERS-1 and the Optical Sensor (OPS) to be flown on J-ERS-1 will both provide a stereo viewing capability. With the potential availability of a vast number of stereo images from space there has been an increase in interest in developing automated techniques for performing stereo analyses. These analyses will provide elevation information for cartographers, the ability to remove geometric correction errors caused by smooth earth assumptions, and support geophysical and hydrological studies which require accurate regional terrain data.

Any automated algorithm for determining elevation from stereo image pairs must have the capability to match corresponding pixels in both images. The most common matching techniques match either edges or neighborhoods. Edge matching can produce accurate matches with few false matches or errors. However, matches are obtained only at edge points which make up only a small portion of the image. Neighborhood matching on the other hand can potentially provide matches at every pixel in the image. However, neighborhood matching is computationally expensive and image noise can limit the spatial resolution of the calculated elevation data. The advent of high speed massively parallel computers such as the Massively Parallel Processor (MPP) at the Goddard Space Flight Center and the commercially available Connection Machine has made the use of neighborhood matching at every pixel practical. The work reported in this paper shows how a combination of neighborhood matching and edge matching can be used to increase the resolution and accuracy of the computed elevations.

2. NEIGHBORHOOD MATCHING

Neighborhood matching can provide matches at every pixel in the image. Matching is performed by considering a neighborhood surrounding a given pixel in one image (i.e., a template) and moving the template within a search area in the stereopsis direction in the other image until a position is found which gives the best match. The size of the search area is a function of the maximum elevation to be expected. The difference in location between the center of the template neighborhood and the best match neighborhood is the disparity. The disparity values at all pixels form a two dimensional disparity function. This function is input to an imaging system model to compute the elevations for the image.

Many techniques have been reported in the literature for determining the best match (e.g. Gennery, 1980; Norvelle, 1981; Quam, 1984; Drumheller, 1986). If the contrast and brightness in both images is close, computing the sum of the absolute differences in the neighborhood is often sufficient. Since this is not generally the case, (especially in SAR images) a more robust technique is to calculate the normalized cross-correlation function between the neighborhoods. The position within the search area giving the maximum value is considered to be the location of the best match. A refinement to simply choosing the maximum is to fit a curve to a plot of the correlation function values (Quam, 1984). Locating the maximum of this curve can provide sub-pixel values for disparity. (For this paper, the maximum correlation value within the search area is called the "match score".) The spatial resolution of the disparity function is inversely proportional to the size of the neighborhood used for matching, i.e., smaller neighborhoods can detect higher frequency variations in elevation. One of the basic limitations to the spatial resolution of the disparity function and consequently the elevation data is caused by noise in the images. (SAR images are particularly noisy.) As neighborhoods get smaller, the noise has a greater influence on the cross correlation function and thus can produce false matches. In addition, when neighborhoods get too small, there are no recognizable features to match. For large neighborhood sizes, neighborhood matching is generally quite robust and can produce accurate disparity values even in low contrast areas.

3. THE STEREO MATCHING ALGORITHM

A major difficulty in matching neighborhoods is the local distortion caused by the different viewing angles. Local distortion occurs most severely in regions of rapidly changing terrain and creates a horizontally stretched or compressed area surrounding corresponding pixels in one image relative to the other. Thus the basic clue used to determine the elevation also makes the determination of that elevation more difficult.

Any technique for correcting local distortions must take into account the fact that the distortion function can have a broad band of spatial frequencies. For example, the distortion function for a mountain range would have low frequencies, but added to these would be high frequencies caused by rock formations making up the surface. When a human observer fuses two images seen through a viewer, the low-frequency information is used to obtain an initial fusion in which the eyes are brought into alignment (a technique used for automatic focusing of some cameras) and then high-frequency information brings out a detailed perception of depth. This suggests that a hierarchical approach for detecting matching pixels would be appropriate. With this approach, an initial match is performed on low-frequency information in an image and then increasingly higher frequencies are incorporated to obtain the final matching of corresponding pixels.

Even with no local distortion, errors caused by noise in the image and areas of low contrast will occur. One way of reducing errors in general is to provide redundancy by computing matches at nearly every pixel in the image. Then continuity constraints on the ground surface can be used to find local discontinuities in elevation.

3.1 HIERARCHICAL APPROACH

The matching algorithm developed for the MPP (Ramapriyan et al. 1986; Strong et al. 1987) is an example of what has been termed the Hierarchical Warp Stereo (HWS) technique. It is similar to the technique reported by Quam (1984). The algorithm is summerized in the following list of steps. Note that for the discussion, one image is considered as a reference image and the other the test image. All modifications are performed on the test image.

1. Remove linear distortions caused by the viewing system.

    This is particularly necessary with SAR images where incidence angle distortions in range are exaggerated. The function of this initial step is to reduce the search area to find matches.

2. Choose largest matching neighborhood or template size.

    For 512 x 512 images the m x m neighborhood for the first
    iteration has a value of m = 49 or 25. The larger is
    chosen if the images are particularly noisy. The size of
    the search area in each iteration depends on the expected
    residual distortion after distortions computed in the
    previous step are removed. The m x n search areas in this
    and subsequent steps are usually twice as wide as the
    matching neighborhood (i.e., n = 2m).

3. Match neighborhoods.

    The normalized mean and variance cross correlation function
    is used to match the neighborhoods. This function
    compensates for differences in contrast and brightness in
    the two images.

4. Update previous disparity function.

    The disparities computed at this iteration are added to
    those at the last iteration to give the net disparity. For
    the first iteration, the previous disparity is the linear
    distortion correction function used in step 1. The update
    is applied to all pixels. For later iterations, the update
    is performed only at the pixels where the match score is
    greater than a specified threshold, Tm.

5. Detect bad matches.

    Bad matches are detected at each iteration by marking
    pixels where the disparity function has a discontinuity
    exceeding a specified threshold Td (since disparities are
    computed at every pixel and the ground surface is assumed
    to be continuous, the disparity should also be
    continuous.) At each bad match pixel an octagonal
    neighborhood is placed in a "bad match mask". The
    "radius", Rd of the octagon is proportional to the
    discontinuity threshold value, Td. The overlapping
    octagons define bad match areas in the disparity function.

6. Remove bad disparity values.

    Acceptable disparity values are interpolated over the areas
    defined by the "bad match mask".

7. Smooth disparity function.

    To minimize any "overshoot" in the next step, the disparity
    function is smoothed by averaging over an s x s
    neighborhood approximately the size of the neighborhood
    used for matching.

8. Warp the test image to match the reference image.

    The disparity function is used as a distortion removal
    function to geometrically correct or warp the test image so
    that it more closely matches the reference image.

9. Choose next smaller size matching neighborhood.

    The next size is typically one half the present size.

10. Repeat steps 3 through 9 until bad matches determined in
    step 5 with Tm = 0 occur over most of the image. At this
    stage, there is no benefit in reducing the template size
    any further.

Figures 1 through 17 illustrate the stereo matching algorithm
implemented on the MPP applied to a pseudo stereo pair of synthetic
aperture radar images from the space shuttle SIR-B mission. Figures
1 and 2 were taken over a plateau region in northeast India at its
border with Bangladesh. They cover an area of approximately 120
square miles. Viewing Figures 1 and 2 through a stereo viewer shows
the plateau region and surrounding river valley clearly. Since
there are a large number of figures, their captions will be listed
below and will form the discussion of the results of the algorithm
applied to the Bangladesh images.

Figure 1. Reference image (REF) from an incidence angle of 45 deg.

Figure 2. Test image (TEST) from an incidence angle of 24 deg.

Figure 3. Updated disparity function computed in steps 3 and 4 with
           m = 25, n = 49, and Tm = 0. Dark (light) areas indicate
           that matching pixels in test image are to the left
           (right) of those in the reference image. Discontinuities
           in disparity function indicate locations of bad matches.

Figure 4. Match scores computed in step 3. Bright areas indicate
           high match score values. Typically, bad matches occur in
           dark areas.

Figure 5. Bad match areas computed in step 5 with Td = 7 and
           Rd = 7.

Figure 6. Interpolated disparity function computed in step 6.

Figure 7. Smoothed disparity function (D1) computed in step 7 with
           s = 25.

Figures 8 and 9. Reference image (REF) and warped version of TEST
                  (WT1) computed in step 8 using D1 as warping
                  function. Stereo viewing of REF and WT1 shows
                  considerable reduction in elevation.

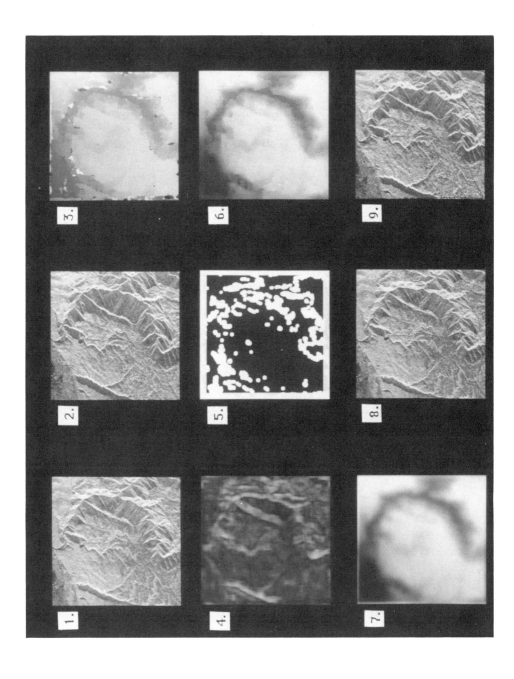

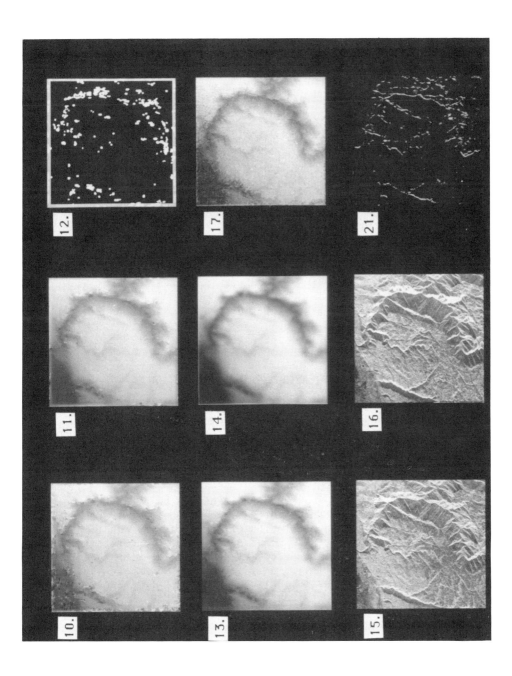

Figure 10.   Updated disparity function computed in steps 3 and 4
              with REF and WT1 as inputs, m = 13, n = 25, and
              Tm = 0. Large number of discontinuities indicates that
              Tm should be higher.

Figure 11.   Updated disparity function computed in steps 3 and 4
              with REF and WT1 as inputs, m = 13, n = 25, and Tm = 40
              giving an acceptable number of discontinuities. (Value
              of match score for perfect match is 255.)

Figure 12.   Bad match areas computed in step 5 with Td = 5 and
              Rd = 3.

Figure 13.   Interpolated disparity function computed in step 6.

Figure 14.   Smoothed disparity function (D2) computed in step 7 with
              s = 13.

Figures 15 and 16.   Reference image (REF) and warped version of TEST
              (WT2) computed in step 8 using D2 as warping
              function. Stereo viewing of REF and WT2 shows
              almost complete elimination of elevation.

Figure 17.   Updated disparity function computed in steps 3 and 4
              with REF and WT2 as inputs, m = 7, Tm = 0, and n = 13.
              Since discontinuities appear almost everywhere in the
              image, the algorithm terminates.

4. INCREASING SPATIAL RESOLUTION

Close examination of REF and WT2 shows some differences
especially at the edges of the bright regions corresponding to
valley walls. Figures 18 and 19 show an expanded area in the upper
right side of REF and WT2. Figure 20 shows the absolute differences
in brightness between REF and WT2 in the expanded area. Although
somewhat noisy (because the images are not identical, some bright
short elongated areas can be seen where there are incorrect values
of the disparity function. Since the expanded area is in a region
of large distortion, between the two images, the errors are due
primarily to the limited resolution of the disparity function at the
end of the second iteration. Human observers can accurately match
edges of bright regions in the two images. It is known that the
human visual system is particularly sensitive to edges. Thus to
improve the disparity resolution, the matching of edges has been
considered for all further iterations. The procedure implemented on
the MPP continues with the neighborhood matching technique but uses
smaller neighborhoods to detect edges.

4.1 CONTINUATION OF NEIGHBORHOOD MATCHING

Figure 21 shows the match scores obtained using a 7 x 7
neighborhood where the values less than Tm=90 have been set to

zero. One can see that pixels with high scores follow along the edges of bright regions in the reference image. Thus by setting Tm high enough, updates take place only along significant edges. Since it is assumed that the new disparities used to update the previous disparity function are accurate (from the high match score) any discontinuities in the updated disparity function are not bad matches. However the disparity function must be made continuous. Thus an interpolation is performed outward from the new disparity locations to form a continuous "fillet" between the new disparity values and the previous ones. The width of the fillet is typically equal to twice the template size m. Figures 22 and 23 show the expanded area in the reference image and a warped version of the test image after two additional iterations of the matching algorithm with m=7 and m=5. Figure 24 shows the corresponding difference image. One can see that many of the short elongated bright areas of Figure 20 have been eliminated indicating a more accurate match at these locations.

5. CONCLUSIONS

Hierarchical Warp Stereo methods employing normalized neighborhood correlation matching to determine disparities has proven to be a robust technique for producing disparity values at every pixel in the stereo pair. Implementation of this technique on a massively parallel computer has allowed the generation of disparity values over an entire image pair within practical computation times. (Less than one minute for 512 x 512 images.) However, image noise and lack of local information content can limit the spatial resolution obtainable with neighborhood matching techniques. The inclusion of edge detection and the matching of just the edges in further iterations of the HWS algorithm has allowed increased spatial resolution along significant edges within the images.

FIGURES

Figure 18 and 22.   Four times expansion of area in reference image (upper corner at row 100 and column 350 in original image).

Figure 19.   Corresponding area in WT2.

Figure 20.   Absolute differences in brightness between REF and WT2 in expanded area.

Figure 21.   Match scores for m = 7, n = 13, and Tm = 90.

Figure 23.   Corresponding area in warped version of test image after further iterations with n = 7, n = 13, and Tm = 90 and m = 5, n = 11, and Tm = 90

Figure 24.   Absolute differences in brightness between REF and warped test images of Figure 23.

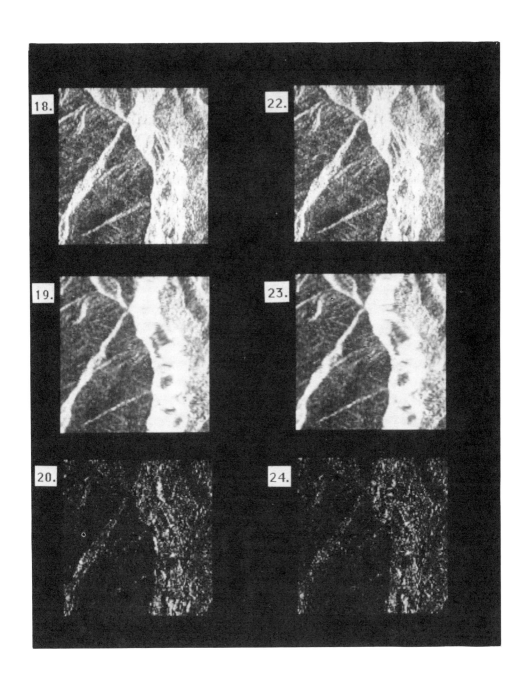

REFERENCES

D. B. Gennery, 1980: Modelling the Environment of an Exploring Vehicle by Means of Stereo Vision, Stanford Artificial Intelligence Laboratory, Memo AIM-339.

F. R. Norvelle, 1981: Interactive Digital Correlation Techniques for Automatic Compilation of Elevation Data, Engineering Topographic Laboratories Report No. ETL-0272.

L. H. Quam, 1984: Hierarchical Warp Stereo, Proceedings of the Image Understanding Workshop, New Orleans, DARPA, October.

H. K. Ramapriyan, J. P. Strong, Y. Hung, C. W. Murray, 1986: Automated Matching of Pairs of SIR-B Images for Elevation Mapping, IEEE Transactions on Geoscience and Remote Sensing, July.

M. Drumheller, 1986: Connection Machine Stereo Matching, Proceedings of the 5th National Conference on Artificial Intelligence, AAAI86.

J. P. Strong, H. K. Ramapriyan, 1987: Massively Parallel Correlation Techniques to Determine Local Differences in Pairs of Images, Proceedings of the Second International Conference on Supercomputing, May.

DISCUSSION

Johnson: In your description I noted that registration of the frames from one image to the other was only done in one direction. Is that actually the way you carried out the algorithm or did you go in the vertical direction as well?

Strong: If the images are well registered then you don't need to go in the vertical direction. However, in this synthetic aperture radar there was just a little bit of rotation, and, yes, we did go up and down by one or two, just to make sure.

Johnson: What were the look angles that you dealt with?

Strong: The incidence angles on these two images, one was 45 degrees and the other one was 25 degrees.

Johnson: Were the azimuths sufficiently different, or were they similar?

Strong: The azimuths were based on JPL data which they had already corrected to give equal increments in both images in azimuth as far as the fix resolution goes.

Johnson: No, that is not quite what I mean. You quoted the depression angle, I think. Were they coming from the same azimuth angle? Was the azimuth for the map the same for both of them?

Strong: Oh, yes.

Johnson: Okay, so it was two different orbits that were able to properly produce that.

Strong: Right, yes.

Johnson: One more thing. The data that you took from JPL was not corrected for layover, in particular, I assume.

Strong: The only thing they did was geometrically correct it to a more or less flat earth. They used ground range rather than slant range.

Neuendorffer: I just have one short comment. The reason you are probably in a tomography session is that an x-ray tomographer of 30 years ago would understand precisely what you are doing and they wouldn't understand anything about what anybody else talked about.

Forrester: I just wanted to make sure that when you are talking about the 25 by 25 size area, you are using the same resolution that you are using in the 13 by 13?

Strong: Yes. There are warp stereo techniques that will start out with a lower resolution image and use the same size neighborhood in different resolutions of the image. What we are doing is using the same resolution of the image but using different neighborhood sizes.

Forrester: The second question I have is whether the good enough fit of the correlation is solely dependent on the continuity of the surface? It certainly seems like a good enough match, I was just curious whether there is anything else.

Strong: I'm not quite sure. It just turns out that there are areas where we saw in the match scores that there was a low value, but we still got continuous surfaces. When we actually do the warping, comparing the reference and the warped test images is the only confirmation that we have that the algorithm is working correctly. The synthetic aperture radar images, at least with the pair that we had, are so noisy that if we just try to do a subtraction, they are still noisy. In fact, I have some pictures in the paper showing this. There is no easy measure of how well things are really matched up. It is only along the edges that you can really see whether things are matched up well or not.

Forrester: One more question, the sub-pixel location is not attempted primarily because of that displacement of edges?

Strong: When you do edge matching on images? All we were doing was extending the correlation techniques using smaller neighborhoods and only doing the matching where the match score was high. So, we weren't really doing edge matching there. We were just continuing the process as we normally would, except modifying it where the match score was high. If you do an edge-matching algorithm like the ones we've attempted, if you draw an edge in one image, it might follow the edge in the real edge okay, but in the other image it might be slightly different. You can see if you tried to match those edges up, you really wouldn't be matching the true edges in the two images. There are some papers in the literature that use what is called edge-focusing techniques which will start detecting low resolution edges and then home in on higher and higher resolution and get very close edges. That is what we want to try to explore.

Fleming: I have one more comment about tomography. There is another paper on tomography in the maximum entropy session because it is on both maximum entropy and tomography. This concludes Session A, thank you very much for your kind attention.

OPEN DISCUSSION

SESSION A. - HENRY FLEMING

I would like to go back to the first paper which was "Foundations of Computed Tomography" by Robert Lewitt and comment that his was essentially a methods paper. I thought he did an excellent job of presenting an overview of various methods. I could comment on the applicability of these particular methods to our problems but Grace Wahba did such a wonderful job commenting on that paper that I don't want to comment any further. The only other thing I might say about Lewitt's paper is that the people working in the medical tomography field have two advantages. First of all, they have more financial resources, so there are many more people doing this kind of work, particularly on the retrieval aspects. From a probabilistic point of view they are liable to be at the forefront of a lot of things just because of the amassing of resources. Furthermore, they can take these resources and put them in the single-minded direction of retrievals because they don't have cloud problems to solve, they don't have zenith angle problems, and they don't even have transmittance problems to solve, at least in X-ray tomography. They have a straightforward geometric problem and so they can put all their resources into just solving retrieval problems. Their only problem is a fan-beam problem which is equivalent to a field-of-view problem for us. I might add that Robert Lewitt's boss is Gabor Herman, who is also a recognized expert in computed tomography and who wrote a very nice book, much of which is relevant to our work. So, the importance of paper 1 is its nice compilation of retrieval of methodologies.

Now let me skip to paper A-5. First I would like to apologize to the authors for misreading their abstract and putting it in the wrong session. Stereographic techniques are definitely an approach that is geometric, but the paper was only about the end product and matching images. So the principal author was correct in suggesting that their paper really shouldn't have been in Session A. It should have been in a session on data handling. I thought it was a nice paper and good ideas. The chairperson of another session might want to comment further on it.

Let's go back to paper A-2. There are really very few good applications of tomography to the atmosphere. One of the outstanding ones is looking at the Aurora the way that was done in paper A-2. The basic technique involves a special case of the Radon equation, which I have never seen used before.

Paper A-3 is an outgrowth of something that I did in the previous workshop. The purpose of paper A-3 is essentially to settle the question of whether or not we really can do satellite tomography for the temperature sounding problem under totally realistic conditions. The authors have a good approach to the problem and did a nice piece of work. My only criticism of it is that they used three empirical functions to represent the solution, which may still leave the question unanswered. These three functions accounted for 90% of the total variance, but I think it is that remaining 10% that is going to make the difference in deciding whether or not the angle-scanning approach is going to do better than the nadir approach. So, I think the question is still open. My personal feeling, based on the work I have done on the problem, is that it is most likely not a viable method, but I wouldn't want to close the door on it until somebody actually does the numbers.

Paper A-4 is my paper and the only reason I wanted it in the tomographic section is that I was trying to emphasize the tomographic-like nature of frequency scanning when a large number of channels are involved. It is, of course, not true tomography because the geometric aspect is missing. But, for years, we've been talking about how many pieces of information we have and the problem always seemed to revolve around whether 8 channels are enough or are 10 channels enough or should we go to 12? The reason I wanted to put this talk in a tomographic context is that I don't think 10, 12, 20, or 50 channels are adequate. I think we have now learned that the point is not to talk about individual pieces of information because that is not the nature of the problem. The point is that if you have thousands of channels, particularly in the infrared region, you get a little incremental additional piece of information every time you add another channel. It is this collection of incremental pieces of information that is going to allow us to get to our goal of 1-degree rms temperature accuracy, which is the assumed accuracy of radiosondes. We now have the technology to get to that point and I am very excited about it. When talking about thousands of channels how do we get additional information from additional channels? We have three things working for us in the infrared region to prevent channels from becoming completely dependent. One is that the Planck function changes the nonlinearity with frequency over the spectrum we are talking about, which adds to the independence of the channels. Secondly, we have a very varied spectrum in the infrared region from the 15 micron band through the 4.3 band. And thirdly, the transmittances coming from that region of the spectrum are very temperature and moisture dependent which make them very nonlinear and add to the independence. So there is incremental information to be gained from each channel that is added. The increments are very small, nevertheless. Collectively you get a great deal of new information. That leads me to my earlier point about frequency scanning being tomographic-like in that when many channels are involved, one moves from an undetermined system of equations to a very overdetermined one.

This brings me to the last point that I want to make and that is that we now can see achieving our 1-degree temperature accuracy goal just over the horizon. Granted, we have only done it in simulation, so we've only shown its potential. Now we have to demonstrate it in the real world. When we get to the point of implementing the many channel approach, there are several problems that confront us immediately. One is that we have always used radiosonde data as ground truth. Also, we tune our retrieval algorithms to radiosondes. Now, if we are dealing with sounding instruments that have an inherent accuracy of 1 degree and we are tuning to ground truth that is no better than 1 degree, the question is: what is tuning what? I think the answer to what we have to do in this case lies in a comment that Clive Rodgers made during a coffee break. When somebody asked him "What do you think is the most important problem in temperature remote sensing?" and he replied "The forward problem." That is precisely the point. If we can do the forward problem that means we know all the physics we have to know to do the inverse problem, which is the limitation right now. If we want to take the physical approach, and I think we have to when we are talking about doing better than 1 degree, our problem for the future is really the forward problem. If we can do the forward problem, we can do the inverse problem.

Wahba: I would like to agree with everything that Henry said and note that he mentioned the nonlinearities are actually important in that we have to model them carefully in the forward problem. We are probably going to have to think about bigger and faster computing resources to deal with 12,000 data points in a most efficient manner when solving a nonlinear problem. So, I think that we should be aware of that if we don't want the computing power to be a limitation. When we really start getting serious about looking at data like his data and looking at the forward problem very carefully and the nonlinearities very carefully from a numerical point of view, then we might need bigger and better computer resources.

Clough: It is gratifying for those of us that have spent a better part of a lifetime working on the forward problem to observe the increased ability of that effort in this discipline. The ability to accurately calculate forward problems has not been as significant an issue in previous workshops as it has been in this one. The absolute accuracy of current measurements, in particular the spectral measurements, and the accuracy of the spectral radiance calculations, are, in general, better than 1°C equivalent brightness temperature. This is of the order if not less than the accuracy of the in-situ measurements used for validation. The other aspect of this session that I found particularly interesting was Clive Rodger's treatment of errors. This analysis may be applied directly to simulations using Bill Smith's simultaneous retrieval techniques which is a linearized formulation. There are no doubt cases where the linear approach may not be adequate, but nevertheless provides a useful initial step.

Rodgers: My comment is really a technical one. You were saying that with hundreds of thousands of channels we now have an overdetermined problem, I think this isn't true. It is still a mixed-determined problem. Some components with solutions are going to be overdetermined but some are not determined at all and are in the null space.

Fleming: Oh, yes. But the null space is shrinking each time we add channels.

Rodgers: I am not sure I believe that. I think there is a limit beyond which we are not going to be able to go. This business about using water vapor to improve the vertical resolution I think is really quite a major advance if that is where the improvement is coming from. But, it is hard to tell whether it will go any further than that. I suppose once upon a time, I didn't think we would go much beyond the resolution you can get from a monochromatic measurement in the wings of a CO_2 line.

Fleming: By transferring the problem to the water vapor region, it becomes a much more difficult problem because it is much more nonlinear but that is where the potential improvement lies.

Speaker: I think one of the fundamental problems in remote sensing, at least with infrared, is the problem with clouds. Again, even if you have 2000 measurements they are still affected to a considerable degree by clouds as we know that clouds occur about 50% of the time. So even though we may be able to derive extremely high resolution sounding perhaps during cloud pre-conditions the other half of the time you've got a serious problem. And I still think that requires a tremendous amount of work, coupling with other spectral regions, and perhaps active techniques as well.

Fleming: Oh yes. But the other side of the coin is that as we get more channels, the new technology is providing higher spatial resolution, which helps in solving the cloud problem.

Liou: I am not familiar with the tomographic method for remote sensing. I am wondering, what are the underlying physics associated with this particular method and, why is it so attractive for remote sensing? In terms of frequency or angular scan, what is the underlying fundamental that will give you more information if you increase more channels. You pointed out that the forward problem is important but what would be the underlying fundamental transfer theory which governs the information content from satellites?

Fleming: The information content comes from the fact that at the noise levels used for my simulations, the channels are not totally dependent. There is a small but positive incremental increase in information with each additional channel. Consequently, the collective impact of so many channels is large, the retrieval accuracy improves in proportion to the square root of the number of channels used in the retrieval.

Neuendorffer: First, I would like to reiterate that I think A-5 should stay in this session because that is the original meaning of tomography and it is fine for the definition of tomography to now include CAT scanning and a number of other things. But A-5 is tomography in the classical sense and it seems like it should belong in there.

Fleming: It is true, the technique of that paper is tomographic, but the subject matter of his particular talk was confined to matching images, and that is more of a data handling problem than one of geometry.

Isaacs: I think this meeting is particularly interesting with having people from all different research applications and getting a mix of ideas. New ideas are very important and I think we still need a lot of new ideas as far as doing inversions are concerned.

CALIBRATION OF SPACECRAFT AND AIRCRAFT SENSORS, VISIBLE AND NEAR INFRARED

Warren A. Hovis, Jr.
TS Infosystems, Inc.
Lanham, Maryland 20706, USA

ABSTRACT

One of the most serious problems that has affected remote sensing in the visible and near-infrared is the lack of calibration sources traceable to NBS standards for many classes of sensors and lack of calibration in flight. Large aperture sensors, such as the Landsat Thematic Mapper and Multispectral Scanner, the Nimbus 7 Coastal Zone Color Scanner (CZCS) and the NOAA Advanced Very High Resolution Radiometer are calibrated with a source that is partially traceable to NBS standards but NBS does not produce a large area source of diffuse radiance or verify the accuracy of the sources in use.

Sensors flown by NASA and NOAA are calibrated before launch but in-flight performance is either not measured at all or, if measured, is measured in a way that does not include all of the optical elements. SPOT is a notable exception having calibration from both a light source and the sun that includes all elements of the optics. Experience with the CZCS has demonstrated that a calibration source that does not include all of the optical elements can only show that degradation is not occurring in the internal components of the instrument, but cannot provide any information about degradation in the exposed optics. The CZCS has shown a marked loss of sensitivity, since launch, and the only way of recovering calibration is through simultaneous measurements from aircraft or from the surface. Such measurements are underway using high altitude aircraft and surface measurements.

Both types of measurements, surface and aircraft, suffer from inaccuracies due to lack of exact simultaneity, area of coverage, and uncertainty about the atmospheric contribution above the surface or aircraft. Complete, on board calibration is essential if sensor data is to be used for quantitative applications such as radiation budget and ocean color measurements.

1. INTRODUCTION

Calibration of space and aircraft sensors, in the visible and near-infrared, is limited in accuracy by the requirement that the source be at a high temperature in order to simulate the radiance that the sensor will see observing reflected solar energy. Such sources decay with time of operation due to evaporation of metal from the filament and deposition on the inside of the envelope. The two US standards are the NBS irradiance and radiance standards. The radiance standard is a ribbon filament lamp with a small area of the filament

RSRM '87: ADVANCES IN
REMOTE SENSING RETRIEVAL METHODS
A. Deepak, H.E. Fleming, and J.S. Theon (Eds.)

calibrated. It has neither the brightness nor the area of emission to calibrate a large aperture sensor such as the MSS or Thematic Mapper. The irradiance standard produces a given spectral irradiance in units such as watts/square meter, micrometer, at a specified distance from the lamp, usually about 50 cm.

The irradiance standard has sufficient brightness but the output is diverging from the small filament and could fill the area of a large collector but not the solid angle. Such a calibration would be invalid since the entire optical aperture must be filled for an accurate calibration. An ad hoc method has been used for calibration of sensors such as the AVHRR, MSS, Thematic Mapper, CZCS, and the GOES VISSR. This method uses an internally illuminated calibration sphere (Hovis and Knoll, 1983), but this technique is not verified by the NBS nor is NBS prepared to offer any substitute. Other techniques such as illuminating flat or curved surfaces were tried but uniformity was found to be unsatisfactory. This leaves the remote sensing community on its own until some standard is defined by the NBS or a foreign nations standards bureau.

The figure of 5% error in the sphere calibration has often been quoted, mostly out of context. McCulloch, McLean, and Mohr (1969), give estimates of the error at various wavelengths, but the actual error is not known since there is no standard for comparison. The relative calibration is in the 1% range. In the CZCS program, Rayleigh backscatter corrections using the NASA standard solar spectral irradiance and Rayleigh optical depths due to Frohlich and Shaw (1980) produced unusable a results, but when the data of Neckel and Labs (1981) were used the results were quite satisfactory. Since most of the signal seen by the CZCS is atmospheric in origin, 80% or more in the blue, a small error in either term would have produced results in disagreement with simultaneous ship measurements. This provides some confidence in the calibration since Labs and Neckel did not use a sphere in their calibration.

Another check was made during the assembly of the Nimbus 7 spacecraft. John Hickey of the Eppley Laboratories compared the CZCS sphere with the calibration of the Earth Radiation Budget (ERB) instrument. Although the ERB is a wide angle radiometer and the sphere is designed for use with scanners of large aperture and small viewing angle, he achieved agreement to 3% in the visible and near-infrared.

None of these comparisons is as assuring as would be a standard produced by the NBS and with a recommended procedure for use from that agency. The NBS has maintained that the space program is the only program that needs such a standard and, as such, should pay for the development of such a standard. To date, NASA and NOAA have not agreed to support such a development despite numerous calls for such support.

2. LABORATORY CALIBRATION

Laboratory calibration is accomplished by placing the sphere in a position to overfill the aperture of the sensor, just as the earth as an extended target will overfill the aperture. Care must be taken that the non-emitting surface of the sphere does not enter into the calibration due to reflected light entering the sphere. Draping with diffuse black coverings is essential. In the laboratory, lamps ranging in number from 12 to 16 are turned on or off sequentially to produce the various energy levels. In the red, the saturation level of the sensor may be reached. In this case, the output of the number of lamps that do not cause saturation is calculated as a percentage of the saturation radiance and the gain of the amplifier set accordingly. For instance, if 6 lamps produce 87% of saturation the amplifier is set to 87% of the maximum output.

Although the technique of calibration is important, the validity of calibration of wide spectral band channels remains questionable. Wide band channels on sensors such as the AVHRR, MSS, Thematic Mapper and SPOT are defined by interference filters. Such filters do not produce a symmetric, well-shaped response. Figure 1, shows the spectral response of the first two channels of AVHRR 9 compared to the output of a typical sphere. Both response functions show a clear bias, Channel 1 toward the long wavelength side and vice versa for Channel 2. Consider Channel 1. If the slope of the sphere output were opposite in direction, the same amount of radiance between the spectral limits would produce a lower signal level.

The GOES VISSR uses eight photomultiplier tubes to scan eight lines for each rotation of the spacecraft. The short wavelength response is controlled by a long pass filter, but the long wavelength cutoff is simply the rolloff insensitivity of the photomultiplier. Figure 2, shows the extremes and middle response curve for a GOES VISSR. As can be seen, there is significant variation in the red cutoff. The data is processed by averaging the detector output over a large, uniform area and assuming that each detector saw the same sum in radiance. The output is then normalized against the median and a correction applied to the other seven outputs. This destriping is adjusted about every two weeks as striping varies and, in some cases, the median tube changes. When this destriping adjustment is made previous calibrations, with destriped data, are no longer valid. Data can be collected without destriping but this is only done on request, for a short time, due to objections by users.

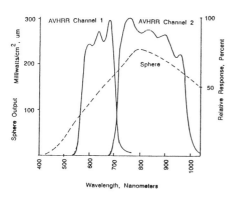

FIGURE 1

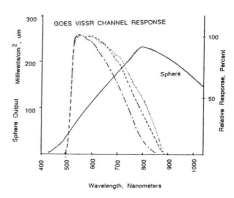

FIGURE 2

A problem that may also occur, especially in sensors using photomultipliers or photodiodes, is spectrally dependent decay in sensitivity. Detectors with photocatodes have been found to decay, in most cases, from the red end of the response. If this occurs a loss in sensitivity would occur and present onboard or vicarious calibration techniques have no way of discriminating between an overall loss of sensitivity and a spectrally dependent loss.

The bias problem occurs in both the MSS and Thematic Mapper of Landsat. Figure 3, shows the spectral response of a Thematic Mapper with the sphere output overlaid. Again, most of the channels have a clear bias toward one end of the spectrum or the other. The MSS used photomultipliers for the first three channels so the comment about spectrally dependent decay also applies to that sensor.

The Nimbus 7 CZCS had narrow symmetric spectral bands of 20 nm each for the bands at 443, 520, 550, and 670 nm. The bands were defined by a spectrometer with the detectors in the focal plane. This allowed for high efficiency in optical throughput and symmetric band shape. This type of band shape is less sensitive to the shape of the reflected spectrum sensed by the band and the shape of the calibration source than those defined with interference filters. It is, however, difficult to use with wide spectral bands, especially when multiple detectors are to be used for each band. The shape of the bands will be shown in the section dealing with post launch vicarious calibration. The CZCS detectors, in the first five bands, are silicon and less susceptible to decay than photomultipliers.

3. INFLIGHT CALIBRATION

Calibration of sensors in orbit has been poorly done, except in the SPOT case, if done at all. As previously mentioned, none of the NOAA imaging channels on the AVHRR or the VISSR have any provision for calibration in orbit. Attempts have been made to calibrate the sensors vicariously, using ground and aircraft measurements, but this is a poor substitute for a complete in-orbit system calibration. Problems with vicarious calibration will be discussed in the section dealing with that area. The imager, under construction, for the new three axis stabilized GOES also contains no onboard calibration for the reflected solar channel even through the data will be quantized to ten bits per word. There is a plan for calibration using star images. This will be discussed under vicarious calibration.

Sensors such as the MSS, Thematic Mapper and CZCS have an onboard calibration system, but in each case the calibration lamp is located inside the sensor and does not illuminate the entire optical system. The CZCS experience has shown that the major degradation is occurring in the more exposed fore optics. Figure 4, shows a schematic representation of the calibration scheme of the MSS and CZCS. The only major difference is that the MSS calibration is inserted by reflection from a rotating mirror and the CZCS calibration through a hole in one of the mirrors. The only information supplied by the CZCS calibration is that the degradation is not in the detectors, amplifiers or interior optics.

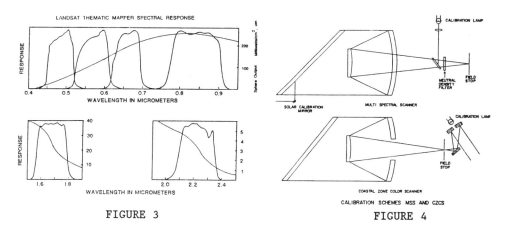

FIGURE 3   FIGURE 4

One of the two lamps in the CZCS was operated each time the sensor was turned on. This proved to be a mistake since the lamp showed degradation that was clearly in the lamp since the second lamp showed no degradation. Figure 5, shows the pattern of lamp degradation observed for the first lamp. The pattern is not the uniform, exponential decay that one might expect indicating that some action other than filament evaporation was occurring, possibly filament motion. In any event it does not appear wise or necessary to operate the lamps so frequently. Any catastrophic change in sensor sensitivity would be recognized in the data so there is no need for calibration on a more frequent basis than every two weeks to a month.

As previously mentioned, the SPOT imagers each contain a calibration system using both the sun and a quartz-halogen lamp. The sun is observed by a 48-fiber bundle that transmits the collected radiance to the area of a collimator where the sunlight, or the output of the lamp is fed through the entire system from a Lummers Cube.

4. VICARIOUS CALIBRATION

Vicarious calibration is that carried out by measuring the radiant output of some source not contained on the spacecraft. Examples are observations from the ground, from aircraft, and viewing stars. The ground and aircraft observations both suffer from the problem of unmeasured atmospheric backscatter in the path to the sensor on the spacecraft. For surface measurements the entire atmosphere intervenes and for aircraft measurements some smaller fraction, depending on altitude. Rayleigh backscatter can be calculated with presently available tools, but aerosol backscatter cannot. Multi-frequency LIDAR sensors may allow measurement of aerosol concentration and calculation of backscatter, but at this time no system is available that can cover the entire spectral range necessary for such calculation.

For the CZCS, when it became obvious that degradation was occurring, a system was developed to fly on a U2 or ER2 aircraft at 65,000- to 70,000-feet altitude. A small, double monochromator was designed and built and fitted to a mount that allowed the spectrometer field of view to be pointed off to either side of the aircraft and to change pointing angle once during a flight to allow for measurements with two satellite sensors in one flight. The reason for selecting a double monochromator, in lieu of attempting to duplicate the imager spectral bands, was that several sensors could be calibrated with a monochromator and it is very difficult to exactly match the fixed spectral bands of many imagers. See Figure 1 for an example.

The aircraft was flown on a direct line between the spacecraft and a target area and simultaneous measurements made. The data was then reduced to calibrated spectra and the radiance within the spectral response integrated (Figure 6). It was still necessary to calculate the Rayleigh backscatter above the aircraft and add this to the measured radiance to get the radiance exiting the top of the atmosphere toward the spacecraft.

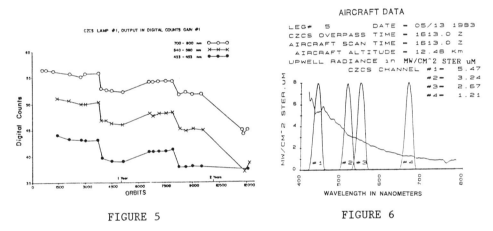

FIGURE 5   FIGURE 6

This technique eliminates much of the uncertainty in atmospheric backscatter calculation but still leaves the problem of exactly matching fields of view. Several surface measurements were made in Europe and the United States in an attempt to recalibrate the CZCS. The discrepancies were large, as might be expected.

5. CONCLUSIONS

There are many uncertainties in calibration in the visible and near-infrared starting with the fact that there is no NBS standard of diffuse radiance. As accuracy in radiance measurements increases with advanced sensors a strong effort should be made to fund NBS to produce such a standard. Onboard calibration must include all elements of the optical system to be of any value. Vicarious calibration is a poor substitute for an accurate prelaunch calibration and observation of changes with a full onboard calibration system.

REFERENCES

Frohlich, C., and Glenn E. Shaw, 1980: New Determination of Rayleigh Scattering in the Terrestrial Atmosphere, <u>Applied Optics</u>, <u>19</u>, 1773-1775.

Hovis, W. A., and J. S. Knoll, 1983: Characteristics of an Internally Illuminated Integrating Sphere, <u>Applied Optics</u>, <u>22</u>, 4004-4008.

McCulloch, A. W., J. T. McLean, and E. I. Mohr, 1969: Evaluation and Calibration of Some Energy Sources for the Visible and Near Infrared Regions of the Electromagnetic Spectrum, NASA X-622-69-195, 29 pgs.

Neckel, H., and D. Labs, 1981: Improved Data of Solar Spectral Irradiance, <u>Solar Physics</u>, <u>74</u>, 231-249.

DISCUSSION

Gautier: You have been pretty critical about calibration and I agree with you. Can you make some comments on how you would go about calibrating a visible instrument like AVHRR, for instance, or the future AVHRR?

Hovis: The future AVHRR is called AMRIR and I should mention that it does require in the specification an onboard calibration, but it does not specify the nature of the calibration. It is left up to the proposer. I don't think it is practical to attempt to fill the entire mirror diameter which will be the order of 20 centimeters. I would use a system somewhat like the French, but certainly fill each element of the system. That is, illuminate each element of the system. I am willing to accept the fact that if a large scan mirror is degrading it is probably degrading uniformly across the surface. We have a calibration system on the U2 ocean color imager that does this. I think that such a system, at least, is needed on any instrument that is intended to be a quantitative instrument. Again, I can't tell the NBS how to make a standard, but I think pressure should be placed upon them to do so.

Speaker: With regard to doing an absolute solar calibration, could you not work from the model solar atmosphere which astronomers produce to obtain the radiance the spectrum over a large range of wavelengths? These include a great deal of information from various sources but it is packed into the form of an outgoing spectrum?

Hovis: Outgoing spectrum? I'm not familiar with it. I showed you the three major recently accepted measurements that have been made. I don't see how you can do it except measure it. By the way, all of those were made inside the atmosphere either from mountaintop or from an airplane.

Speaker: Yes, this is working backwards from the Sun, from knowing the temperature distribution in the solar photosphere.

Hovis: I would like to see a comparison with the measurements made, but I still think we should utilize the shuttle capabilities. It can be done and, by the way, let me do a little commercial here, once you've bought the instrument, it costs only $10,000 per shuttle flight to fly on the Get-Away Special (GAS-program). Pretty cheap.

ON THE CALIBRATION OF THE SOLAR BACKSCATTER ULTRAVIOLET (SBUV) INSTRUMENTS FOR LONG-TERM OZONE MONITORING

R.P. Cebula
ST Systems Corporation
Lanham, Maryland 20706, USA

P.K. Bhartia
Science Applications Research
Lanham, Maryland 20706, USA

ABSTRACT

Proper separation of sensor sensitivity change from atmospheric variation is central to the success of the long-term stratospheric ozone monitoring program. Based on the success of earlier NASA instruments, the SBUV/2-series instruments are now being flown on NOAA operational satellites for long-term ozone monitoring during the mid 1980's to mid 1990's. For these instruments, the fundamental quantity in determining the ozone profile and the total ozone overburden is the scene reflectance or albedo. Because the retrieval requires only a relative measurement, many of the effects of instrument degradation cancel out. Hence, the instrument sensitivity for the ozone measurement primarily depends on the reflectance properties of the onboard diffuser plate that is employed during solar measurement but not during the terrestrial measurement. The SBUV/2 instruments are equipped with a means of operationally monitoring diffuser plate reflectivity changes. Unfortunately, the onboard diffuser calibration system on the first SBUV/2 flight unit has not provided data of the level of accuracy required. This emphasizes the need for independent means of monitoring long-term diffuser reflectance changes. The method that was developed for monitoring long-term changes in the Nimbus-7 SBUV instrument sensitivity is discussed, with particular attention to its strengths and weaknesses. The application of this method to the SBUV/2-series and future ozone monitoring instruments is addressed in the context of using the method as a backup to the use of on-board calibration systems.

1. INTRODUCTION

As evidenced by the attention being given in both the scientific and popular press, the importance of maintaining a program to provide accurate, global scale, long-term

RSRM '87: ADVANCES IN
REMOTE SENSING RETRIEVAL METHODS
A. Deepak, H.E. Fleming, and J.S. Theon (Eds.)   79

monitoring of the ozone layer is widely recognized. Of particular concern is the possibility that anthropogenic activities, specifically the release of chlorofluorocarbons (CFCs), may lead to ozone depletion. Based on the success of the Nimbus-4 Backscatter Ultraviolet (BUV) and Nimbus-7 Solar Backscatter Ultraviolet (SBUV) instruments, the instrument now in use for long-term stratospheric ozone monitoring is the Solar Backscatter Ultraviolet spectral radiometer Model 2 (SBUV/2). The first SBUV/2 instrument was launched on the NOAA-9 spacecraft in late 1984 and, like SBUV which was launched in 1978, is still operational. At least three more SBUV/2-type instruments are planned for launch on future TIROS satellites in the 1988 to mid-1990's time frame. Ozone monitoring through the end of the next decade will be provided by NOAA's Global Ozone Monitoring Radiometer (GOMR), which will include an instrument that uses the BUV technique to derive total ozone.

The SBUV instruments measure the total ozone overburden and the vertical ozone profile up to an altitude of 50-55 km. In their primary mode of operation the instruments, which are based on the double monochromator design described in detail by Heath et al. (1975), provide measurements in twelve $1 - nm$ wide wavelength bands spanning the region from 250 to 340 nm of the radiance backscattered from the Earth-atmosphere system. Periodic measurements of the incident solar irradiance are made by deploying a diffuser plate into the field of view to diffusely reflect sunlight into the instrument. The ratio of backscattered radiance for π units of solar flux to the solar irradiance, hereafter called the albedo, is inverted to derive ozone.

2. SENSITIVITY OF LONG-TERM OZONE DETERMINATION TO CALIBRATION ERRORS

Precise instrument characterization is central to the success of the long-term ozone monitoring program. There are two aspects to this problem: 1) absolute calibration of individual instruments, and 2) characterization of sensitivity change during the lifetime of each instrument. The significant difficulties involved in the first activity shall not be discussed herein. Knowledge of each instrument's sensitivity change in orbit must be maintained to a level such that the uncertainty introduced into the instrument's data base is less than the expected ozone change. Meeting this requirement is especially difficult for UV instruments because onboard radiometric calibration sources have not been available. Fortunately, the fundamental quantity in determining ozone via the BUV technique, the scene reflectance or albedo, is the ratio of the backscattered terrestrial radiance to incident solar irradiance. Hence, many effects of instrument degradation cancel out and in the first approximation maintenance of the albedo calibration for sensor change reduces to a correction for change in the reflecting properties of the diffuser plate which is employed only during the solar measurement. Uncertainties in the long-term reflectance calibration are manifest as altitude dependent uncertainties in derived ozone. Figure 1 shows the uncertainty introduced into the mid–latitude ozone amount versus pressure for an assumed ±2% uncertainty in the reflectance calibration. A ±4% uncertainty is introduced into the 1 mb ozone retrieval; the uncertainty is smaller at other pressure levels. The total ozone amount is based on reflectance measurements at pairs of closely-spaced wavelengths. For the total ozone determination, a 1% wavelength independent error in the calibration introduces an approximate 0.3% error into the retrieval. Similarly, a 1% relative error in calibration between the pair wavelengths introduces roughly a 1% error into the total ozone retrieval.

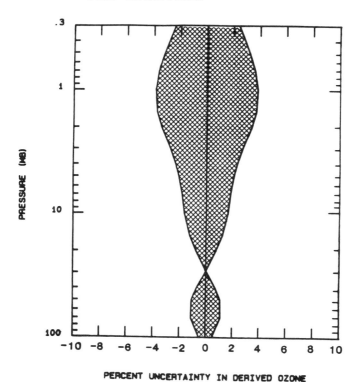

Figure 1. Impact of a ±2% uncertainty in the reflectance calibration on the mid latitude ozone retrieval. (After Cebula et al., 1988)

3. SBUV/2 DIFFUSER REFLECTIVITY MONITORING

Due to the critical dependence of the accuracy of the long-term ozone measurement on knowledge of diffuser characteristics, one of the principal hardware additions to the SBUV/2-series instruments is the inclusion of a system capable of measuring diffuser plate reflectivity changes. A frosted low pressure mercury lamp that is attached to the inner surface of the instrument aperture door is viewed directly when the door is closed. Alternately, the aperture door and the solar diffuser plate are repositioned so that mercury light reflected off the solar diffuser plate enters the instrument. The ratio of diffuser view signal to direct view signal provides a measurement of the relative reflectivity of the solar diffuser plate. The ratio of the relative reflectivities from any two individual diffuser calibration sequences provides a measurement of the amount of diffuser change during the intervening period.

The diffuser reflectivity measurement is, in principle, independent of long-term changes in lamp output and it is only required that the lamp output remain stable over the approximate 25-minute period needed for a single calibration sequence. However, the experience from SBUV/2 Flight unit Model 1 (FM#1) was that longer term lamp behavior can also affect the reflectivity determination. This may arise from the fact that, when viewed directly, the lamp does not uniformly fill the entrance aperture even though the lamp is viewed through a small quartz transmission diffuser which is located immediately in front of the lamp. Therefore, variation in the mercury arc position can cause a fluctuation in the recorded direct view signal. Spatial variations have less impact on the the diffuser view signal because they tend to be washed out when the lamp output is diffusely reflected off the solar diffuser plate and the signal more uniformly illuminates the entrance aperture. Short-term variations can increase the noise in a single relative reflectivity determination. More significant, however, would be any quasi-permanent arc repositioning; such a lamp behavior would be interpreted as an apparent change in diffuser reflectivity. Additionally, polymers outgassed from spacecraft and/or instrument components could be photolyzed onto the lamp transmission diffuser by the lamp emissions, direct backscattered terrestrial radiation, and solar radiation scattered off the solar diffuser (the former having the greatest probability). This could alter the brightness distribution of the lamp diffuser. The experience on FM#1 is that after approximately three months of operation, data from the onboard diffuser monitoring system suggested that the relative diffuser reflectivity began to degrade significantly.

Corresponding to this apparent degradation, the noise in an individual relative reflectivity measurement increased by a factor of 5 to 10. However, direct measurements of the solar flux which, in addition to diffuser degradation, include instrument optics and PMT gain degradation, indicated only small reductions in instrument throughput. Hence, data from the onboard system were being corrupted, possibly due to the systematic effects similar to those just enumerated.

Independent laboratory tests performed using two separate SBUV/2 instruments verified the sensitivity of the instrument throughputs to radiance nonuniformities on the surface of the mercury lamp transmission diffuser and the existence of such nonuniformities. The onboard diffuser reflectivity monitoring system for use on follow-up SBUV/2 instruments has therefore been modified to provide a spatially uniform mercury lamp radiance and to lessen the possibility of photolysis of contaminants onto the lamp's transmission diffuser. Given the failure of the FM#1 onboard system, an alternative approach must be used to maintain the long-term albedo calibration of this instrument. This failure also emphasizes the need for independent means of monitoring SBUV/2 diffuser changes. One approach is to use the method that was developed for SBUV on Nimbus-7. This method is discussed below.

4. NIMBUS-7 SBUV CALIBRATION METHOD

The Nimbus-7 SBUV instrument lacks an on-board diffuser reflectivity monitoring system, therefore an indirect method (Park and Heath, 1985; Cebula et al., 1988) is used to assess instrument sensitivity changes and to maintain the calibration for long-term ozone monitoring. Based on the distinct negative correlation between the relative solar irradiance output and the accumulated time the SBUV solar diffuser plate has been exposed to the sun, a model has been developed to estimate the amount of diffuser plate reflectivity degradation at any time. First, a wavelength dependent diffuser degradation parameter, r_λ, is defined as the fractional change in diffuser reflectivity at time t, $R_\lambda(t)$, with accumulated diffuser exposure to the sun, $E(t)$:

$$r_\lambda = \frac{-1}{R_\lambda(t)} \frac{\partial R_\lambda(t)}{\partial E(t)} \tag{1}$$

The SBUV model contains three assumptions: 1) the diffuser degradation parameter is time independent, 2) the diffuser degrades only during solar exposure, and 3) the diffuser's angular reflecting (goniometric) characteristic is time independent. With the introduction of the model assumptions, the partial derivative in Eq. (1) becomes a readily integrated total derivative giving

$$R_\lambda(t) = R_\lambda(t_o) \exp[-r_\lambda \cdot E(t)] \tag{2}$$

where R (t_o) is the diffuser reflectivity at launch. Hence the reflectivity of the SBUV solar diffuser is modeled to degrade exponentially with solar exposure. This degradation may be physically understood as the solarization of surface contaminants by the solar ultraviolet flux. Because the rate of solarization is proportional to the unsolarized area (as well as solar UV spectral intensity, and contaminant composition and density), the diffuser reflectivity decreases exponentially with the accumulated exposure. This process is analogous to the darkening of the emulsion on a photographic plate.

The diffuser degradation rate parameter was defined in Eq. (1) without reference to a particular model. Given the three model assumptions, the diffuser reflectivity at any time is given by Eq. (2). The accuracy of the long-term diffuser characterization depends on the accuracy of the diffuser model and the accuracy to which the degradation rate parameters can be determined. The task is then 1) to derive the wavelength dependent diffuser degradation rate parameters, and 2) assess the validity of the model. This process, described in detail by Cebula et al. (1988), is excerpted here. In regard to the first part of the task, when fitting the model to the SBUV solar data, changes in the instrument other than the diffuser, as well as real solar irradiance change had to be taken into account. Changes in the photomultiplier tube (PMT) gain were removed by monitoring the ratio of PMT output to the output of a reference photodiode and assuming that the photodiode has constant gain. Purely for the convenience of fitting the diffuser degradation model to the solar data, the instrument optics change was represented by an exponential decay with elapsed time. Finally, solar irradiance change was based on an empirical solar model (Heath and Schlesinger, 1986). Multiple linear regression fits of the diffuser degradation model to the SBUV solar data, made viable by the distinction between accumulated solar exposure time and elapsed time, were then used to determine r_λ. The fitting intervals were chosen to provide a clear distinction between exposure dependent and time dependent degradations. The lengths of the fitting intervals were kept as short as possible in order to minimize the impact of any errors of the forms assumed for nondiffuser changes on the r_λ determination.

While the fitting results showed excellent agreement between the SBUV solar data and the diffuser degradation model prediction, this is not sufficient to prove that the model can be used to correct for diffuser reflectivity change. Errors may enter into the diffuser degradation correction from six separate error sources: 1) formal statistical uncertainties (random errors) in the diffuser degradation parameters which arise from fitting a regression model to experimental data containing noise, 2) non-random errors in the degradation parameters from sources such as error in the PMT and solar change corrections over the course of the fitting intervals, 3) temporal variation in the diffuser's goniometric characteristic, 4) temporal variation in the r_λ-values, 5) diffuser degradation not caused by solar exposure, and 6) instrument optics changes caused by solar exposure.

Of the six potential error sources listed in the preceding paragraph, the last two are the most critical in maintaining the SBUV calibration for long-term ozone monitoring. These potential error sources arise first from the explicit model assumption that the diffuser degrades only during solar exposure, and second, the assumption (for the purpose of deriving the diffuser degradation parameters) that instrument components other than the diffuser do not degrade with solar exposure. If the first assumption is in error, the calculated diffuser degradation correction will be too small relative to the true degradation. This would cause SBUV-reported long-term ozone values to be low relative to true ozone values. If the second assumption is in error the opposite will occur, and SBUV-reported long-term ozone amount would be too high relative to the true ozone amount.

Several methods have been developed to assess the accuracy of the SBUV diffuser degradation model, in particular, the validity of the above two assumptions. In regard to the assumption that the diffuser is the only SBUV component which degrades with solar exposure, if this assumption is incorrect, then the resulting errors would be manifest as recognizable, exposure correlated features in the long-term SBUV albedos. Statistical tests indicate that the diffuser degradation-corrected albedo data are free of such features, supporting the validity of the assumption.

Validating the model assumption that the diffuser degrades only during solar exposure has proven to be more difficult, for only at the longest SBUV ozone wavelength does there currently exist an unambiguous validation method. Examination of the long-term $340-nm$ albedo provides the most comprehensive test of the accuracy of the SBUV long-term calibration for ozone determination. The ozone absorption cross section is very small at 340 nm so this channel is insensitive to any possible ozone

trend.   Assuming that the average annual cloudiness is
time independent (an assumption support by data from the
Nimbus-7 Earth Radiation Budget experiment), the 340−nm
albedo measured by SBUV should show no trend if the long-
term instrument calibration is correct.   As shown in
Figure 2, over the first eight years of instrument
operation, the 340 − nm albedo is trend free at
approximately the 2% level, and shows only a small short-
term increase in the Fall of 1982, coincident with the
injection of aerosols into the stratosphere due to the
eruption of El Chichon.   This suggests that at 340 nm the
SBUV calibration for long-term ozone monitoring is
accurate to within approximately ±2% over the first eight
years of instrument operation.

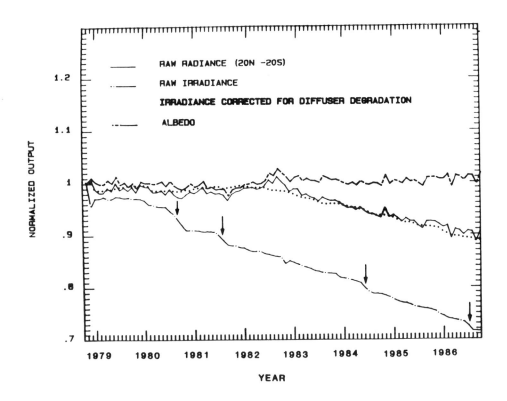

Figure 2.   SBUV 340-nm tropical albedo for the first eight
            years of operation.   The albedo is trend free
            at the ±2% level.   Over a 17% diffuser
            degradation correction is required at this
            wavelength.   The arrows indicate r-value
            measurement periods.

While the method developed to maintain the SBUV calibration for long-term ozone monitoring is well validated at 340 nm, techniques used to date to validate the calibration at the shorter wavelengths are subject to appreciable uncertainties. Unfortunately these are the wavelengths used to determine upper stratospheric ozone amounts. Therefore, validation of SBUV upper level long-term ozone trends ultimately relies on intercomparison with the shorter length data records of coincident, independent instruments.

The uncertainty in the long-term SBUV calibration at the shorter wavelengths arises from the long data record this instrument has provided. The long-term calibration method described above seems to have done extremely well for at least the first four years of instrument operation, during which the SBUV diffuser received about 450 hours of solar exposure. Individual SBUV/2 instruments are scheduled to be launched on approximate 2-year centers, with the older instrument being deactivated shortly after the new instrument achieves operational status. Assuming one r-value measurement each year, only about 200 hours of diffuser solar exposure is expected over the 2-year lifetime of an individual SBUV/2 instrument. The experience of the Nimbus-7 SBUV instrument suggests that the above method can provide an accurate diffuser degradation correction during this period.

5. SUMMARY

Accurate instrument calibration is central to the success of any long-term monitoring program. Given the success of earlier instruments, the long-term ozone monitoring program under the operational control of NOAA utilizes an updated version of the Nimbus-7 SBUV instrument, denoted by SBUV/2. The new instruments include an onboard system to monitor changes in the solar diffuser plate reflectance. However, data from the first SBUV/2 instrument (FM#1) indicate that the onboard monitoring system on this instrument is not operating to the required level of accuracy. The system for use on later SBUV/2 flight units has been modified to circumvent FM#1's failure mode. Given the failure of FM#1's onboard system, diffuser degradation on this instrument needs to be monitored with other means, such as the indirect technique developed for the Nimbus-7 SBUV instrument. Similarly, the application of this technique is recommended as a backup on all future SBUV/2 flight units, at least until such time that the accuracy of their on-board calibration systems are proven to satisfy the long-term ozone calibration accuracy requirements.

REFERENCES

Cebula, R. P., H. Park, and D. F. Heath, 1988: Characterization of the Nimbus-7 SBUV Radiometer for the Long-Term Monitoring of Stratospheric Ozone, J. Atm. Oceanic Tech., 5, 215-227.

Frederick, J. E., R. P. Cebula, and D. F. Heath, 1986: Instrument Characterization for the Detection of Long-Term Changes in Stratospheric Ozone: An Analysis of the SBUV/2 Radiometer, J. Atm. Oceanic Tech., 3, 472-480.

Heath, D. F., and B. M. Schlesinger, 1986: The Mg 280-nm Doublet as a Monitor of Changes in Solar Ultraviolet Irradiance, J. Geophys. Res., 91, 8672-8682.

Heath, D. F., A. J. Krueger, H. A. Roeder, and B. D. Henderson, 1975: The Solar Backscatter Ultraviolet and Total Ozone Mapping Spectrometer (SBUV/TOMS) for NIMBUS G, Opt. Eng., 14, 323-331.

Park, H., and D. F. Heath, 1985: Nimbus 7 SBUV/TOMS Calibration for the Ozone Measurement. In C. S. Zerefos and A. Ghazi (Eds.), Atmospheric Ozone, Proceeding of the Quadrennial Ozone Symposium Held in Halkidiki, Greece, 3-7 September 1984, D. Reidel, Hingham, MA, 412-416.

DISCUSSION

Hurley: Any questions? I would like to make a comment about this particular presentation. We hope that lessons learned on Nimbus-7 and SBUV and SBUV/2 will help us in developing systems for visible and near-IR scanners as well.

RADIOMETRIC CALIBRATION OF IR INTERFEROMETERS: EXPERIENCE FROM THE HIGH-RESOLUTION INTERFEROMETER SOUNDER (HIS) AIRCRAFT INSTRUMENT

H.E. Revercomb, H. Buijs[1], *H.B. Howell*[2], *R.O. Knuteson, D.D. LaPorte*[3],
W.L. Smith, L.A. Sromovsky, and H.W. Woolf[2]

Space Science and Engineering Center
Madison, Wisconsin 53706, USA

ABSTRACT

An accurately calibrated Fourier transform spectrometer has been developed to measure the upwelling infrared emission of the earth from high-altitude NASA research aircraft as part of the HIS program to improve the vertical resolution of temperature and humidity retrievals. The HIS instrument has demonstrated that the radiometric accuracy goals for high-resolution sounding (1°C absolute and 0.1°C RMS reproducibility) can be achieved. Accurate radiometric calibration over the full spectral range was demonstrated on the ground using a third reference blackbody. The unknown temperature of the third blackbody is determined routinely to within 0.2 K of its measured temperature. Achieving this level of accuracy in one of the three spectral bands required developing a technique for cancelling the effects of an anomalous phase response in that band. Additional errors could be present in flight, but comparisons with aircraft altitude temperatures and water surface temperatures are generally within about 1 K. An accuracy of < 0.5 °C is possible, except for regions of low brightness temperature in the 4.3 micron CO_2 band. Comparisons of HIS earth-emitted spectra with line-by-line calculations using the AFGL FASCODE demonstrate the high radiometric accuracy of measured high spectral resolution features.

1. INTRODUCTION

The aircraft model High-resolution Interferometer Sounder, designed for the NASA U-2 research aircraft, has demonstrated the scientific value of radiometrically-precise high spectral resolution emission measurements (Smith et al., 1987a, 1987b). The instrument has been flown reliably on over 40 flights including two major NASA field experiments. The HIS participated, with many other atmospheric sensing instruments, in the COmbined Huntsville Meteorological EXperiment (COHMEX) for studing severe storms and the First ISCCP

---

[1] BOMEM Inc., Ville de Vanier, Quebec, Canada G1M2Y2

[2] NOAA/NESDIS Systems Design and Appl. Branch, Madison WI 53706

[3] Santa Barbara Research Center, Goleta, CA 93117

RSRM '87: ADVANCES IN
REMOTE SENSING RETRIEVAL METHODS
A. Deepak, H.E. Fleming, and J.S. Theon (Eds.)
89

Regional Experiment (FIRE) for studing the effect of cirrus clouds on climate. The unique ability of this instrument to measure accurately the emission spectrum from a flexible, high-altitude platform with a large complement of other instrumentation should make it an important resource for many types of experiments for several years. The success of the aircraft instrument has led to a current effort to develop a spacecraft instrument which will provide improved operational soundings from geosynchronous orbit (Smith et al., 1983, 1984) by as early as 1995.

    As background for the calibration discussion, the primary parameters of the HIS aircraft instrument are summarized in Table 1. The specific implementation for the aircraft bears little resemblance to a spacecraft instrument, but the principles are the same. Also, the instrument noise performance, which along with the calibration defines the radiometric performance, is shown in Fig. 1 taken from Revercomb et al. (1987). The detector noise level derived from the variance of blackbody brightness temperatures in flight is very similar for both the hot and cold blackbodies. The noise performance of the short wavelength Band III is not as good as that of the other two bands, with an extra contribution between 2150 and 2350 cm^{-1} from pickup in the analog electronics. However, the RMS noise for most of the spectral range of bands I and II is very low, generally less than 0.2°C for a single 6-second interferometer scan. The additional noise arising from sample-position errors for the HIS varies with the aircraft configuration. It is caused by undamped vibrations from the aircraft engine and varies from substantially smaller to somewhat larger than the detector noise. This source of noise is small on the ground and would be negligible in a spacecraft instrument.

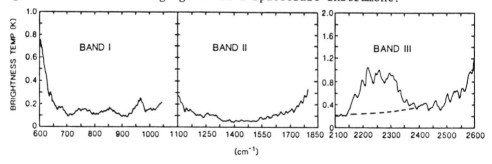

Figure 1.   Detector noise for the three HIS spectral bands.

2. RADIOMETRIC CALIBRATION AND VERIFICATION

    The basic approach for determining absolute radiances from the HIS nadir-viewing interferometer is the same as that used for filter radiometers and has been used successfully for other interferometric applications. (Hanel et al., 1980, 1972, 1971, 1970; LaPorte and Howitt, 1982). The detectors and electronics are designed to yield an output which is linear in the incident radiance for all wavenumbers in the optical passband of the instrument. Two blackbody reference

TABLE 1. CHARACTERISTICS OF THE HIS AIRCRAFT INSTRUMENT
(after Revercomb et al, 1987, 1988)

Spectral range (cm-1)[*]:
	Band I	590-1070
	Band II	1040-1930
	Band III	2070-2750

Field of view diameter (mr):
	Telescope	100
	Interferometer	30

Blackbody Reference sources:
	Emissivity	>0.998
	Aperture diameter (cm)	1.5
	Temperature stability (K)	±0.1
	Temperatures (K)	240, 300

Auto-aligned Interferometer: modified BOMEM BBDA2.1

Beamsplitter:
	Substrate	KCl
	Coatings (1/4 λ at 3.3 μm)	$Ge+Sb_2S_3$

Maximum delay (double sided)-current configuration (cm)
	Band I (hardware limit is ±2.0)	±1.8
	Bands II & III (limited by data system)	+1.2,-0.8

Michelson mirror optical scan rate (cm/s):   0.6-1.0

Aperture stop (at interferometer exit window):
	Diameter (cm)	4.1
	Central obscuration area fraction	0.17
	Area (cm^2)	10.8

Area-solid angle product (cm^2-sr):   0.0076

Detectors:
	Type	Ar doped Si
	Diameter (cm)	0.16
	Temperature (K)	6

Nominal instrument temperature (K):   260

* The ranges shown are design ranges. The current bandpass filters
were chosen from available stock filters, and will be changed as new
filters are acquired.

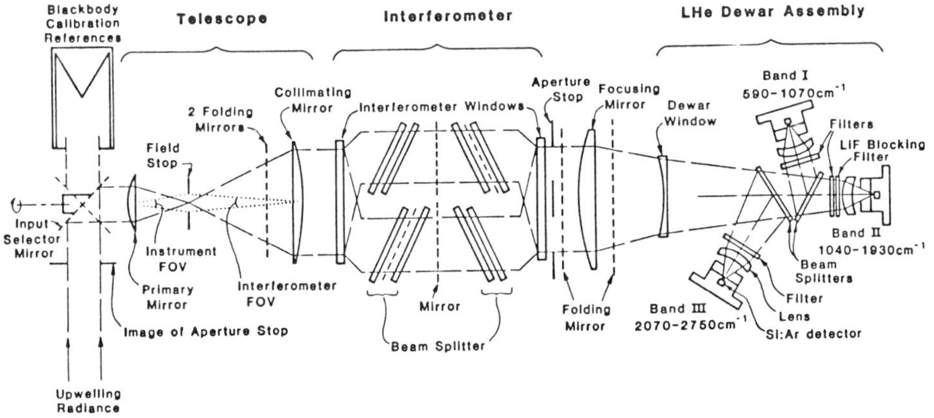

Figure 2. Schematic of HIS optics. Primary, collimating and focusing
mirrors are shown as lenses (after Revercomb et al., 1988).

sources are viewed to determine the slope and offset which define the
linear instrument response at each wavenumber.

In the HIS U-2 instrument, calibration observations of the two
on-board reference blackbodies are made every two minutes. There are
4 double-sided optical-path scans of each reference source for every
12 scans of the earth. As shown in Fig. 2 which summarizes the
optical configuration, the blackbodies are viewed by rotating the
telescope field-of-view from below the aircraft to inside a blackbody
aperture using a 45° plane mirror. There are no uncalibrated optical
surfaces, since the earth is viewed through an open aperture in the
pod which provides an aerodynamic shell.

The small size of the optical beam at the blackbody positions
makes the design of accurate radiation standards relatively easy. The
reference blackbodies are thermoelectrically-controlled, blackened,
copper cavities (Fig. 3). The insulated copper walls of the blackbody
cavities give good temperature uniformity, and because of the cavity
effect, the normal emissivity is very close to one (Table 1). The
temperatures are sensed with accurately calibrated platinum resistance
thermometers (PRTs) embedded in the base of each cavity (during
testing, a second PRT in the side of the cavity was used to verify
adequate temperature uniformity).

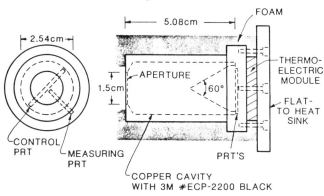

Figure 3. Blackbody reference cavity design.

One important, additional requirement when applying a two-point
calibration with blackbody references to an interferometer, as opposed
to an instrument measuring spectra directly, is that the instrument
responsivity should be independent of optical delay (or that any delay
dependences should be accurately known). Avoiding sources of delay
dependent response was a major objective in designing the HIS
instrument. To accomplish this, care was taken in the optical stop
design and alignment to prevent the effective aperture stop size from
changing with motion of the Michelson mirror. The best location for
the aperture stop, which is focused on the detectors, was found to be
at the exit window of the interferometer module (see Fig. 2).
Further, the field-of-view (FOV) of the interferometer is restricted
to 30 mr to limit self-apodization.

Now, turning to the mathematical expressions for the calibration, we present the new formulation for calibration which correctly accounts for the anomalous phase response observed in the HIS Band I (Revercomb et al., 1988). Assuming linearity as expressed above, the output interferogram F can be expressed in terms of the incident spectral radiance L_ν as follows, using a continuous representation:

$$F(x) = \frac{1}{2} \int_{-\infty}^{\infty} C_\nu \, e^{i2\pi\nu x} \, d\nu \qquad (1)$$

where the uncalibrated complex spectrum $(C_\nu \equiv C_{-\nu})$ is given by

$$C_\nu = r_\nu (L_\nu + L_\nu^o e^{i\phi^o(\nu)}) \, e^{i\phi(\nu)} \qquad (2)$$

and where x is optical path difference (delay), ν is wavenumber, $\phi(\nu)$ is the normal phase response to external radiation, $\phi^o(\nu)$ is the anomalous phase response from instrument emission, r_ν is the responsivity of the instrument, and L_ν^o is the offset from instrument emission (referred to input).

Equation (2) expresses the linear relationship between the uncalibrated spectrum and spectral radiance. The two unknowns to be determined from the two calibration observations are the responsivity and the offset radiance. The offset radiance defined here is the radiance which, if introduced at the input of the instrument, would give the same contribution as the actual emission from various parts of the optical train.

The phase characterizes the combined optical and electrical dispersion of the instrument. Note that both a normal and an anomalous phase are explicitly represented here. This is to allow for the possibility, encountered with the HIS band I, that the phase for radiance from the source is different from the phase for background emissions.

It is clear from Eq. (2) that the anomalous phase contribution can be eliminated along with the instrument radiance offset by differencing complex spectra from different sources. The difference spectra are identical to the difference spectra which would result if there were no anomalous phase contribution. The equations for the difference spectra are

$$C_\nu - C_{c\nu} = r_\nu [L_\nu - B_\nu(T_c)] \, e^{i\phi(\nu)} \qquad (3)$$

$$C_{h\nu} - C_{c\nu} = r_\nu [B_\nu(T_h) - B_\nu(T_c)] \, e^{i\phi(\nu)} \qquad (4)$$

where B_ν is the Planck blackbody radiance, and subscripts h and c label the quantities associated with the hot and cold blackbody. [Note that, for simplicity, the blackbodies are assumed to have unit emittance here. To account for actual emittances ϵ, the Planck radiances should be replaced with $\epsilon B + (1-\epsilon) B(T_a)$ where T_a is the ambient temperature.]

The new expression for the responsivity, which follows immediately from Eq. (4) by taking the magnitude of both sides, is

$$r_\nu = |C_{h\nu} - C_{c\nu}| / [B_\nu(T_h) - B_\nu(T_c)] \tag{5}$$

The offset which follows directly by substituting the responsivity into Eq. (2) is

$$L_\nu^o \, e^{i\phi^o(\nu)} = C_{h\nu} \, e^{-i\phi(\nu)} / r_\nu - B_\nu(T_c) \tag{6}$$

Note that the offset from instrument emission given by Eq. (6) is a complex function. In the shorter wavelength bands of the HIS where the beamsplitter is largely free of absorption, the anomalous phase is essentially zero, making the offset real. However, in Band I the HIS beamsplitter absorbs as shown in Fig. 4. In addition to reducing efficiency, absorption also alters the phase response. The resulting emission from the ambient temperature beamsplitter creates coherent beams in each leg of the interferometer, with the effective point of wavefront division in a plane different from the plane for reflection. This is the source of the anomalous phase.

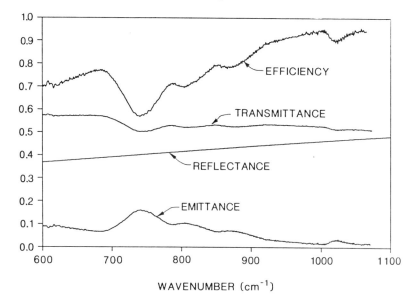

Figure 4. Characteristics of HIS beamsplitter for Band I.

Finally, the basic calibration expression which follows by taking the ratio of Eq. (3) to Eq. (4) is

$$L_\nu = Re[(C_\nu - C_{c\nu})/(C_{h\nu} - C_{c\nu})][B_\nu(T_h) - B_\nu(T_c)] + B_\nu(T_c) \tag{7}$$

For ideal spectra with no noise, this expression for the calibrated radiance would be real, since the phases of the ratioed difference spectra are the same. This cancellation of the phases avoids the square root of two noise amplification, which can be associated with taking the magnitude of spectra with none-zero phase. Because the phase of the ratio of difference spectra is zero to within the noise, the calibrated spectrum can equally well be defined in terms of the real part of the ratio (as shown), or in terms of the magnitude of the ratio.

The ground calibration tests to verify the above procedure consisted of measuring the radiance from a blackbody at approximately 280 K using calibration blackbodies at 300 K and 77 K. The results of one of these tests are shown in Fig. 5. The spectra for each of the

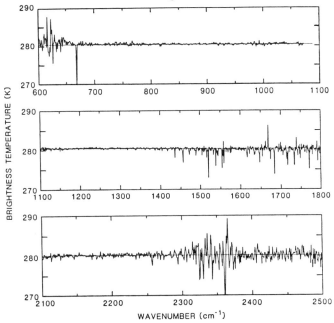

Figure 5. Laboratory calibration with source at 280.2 K (lines shown).

three HIS spectral bands are presented as brightness temperatures to make any errors stand out as a deviation from the measured blackbody temperature of 280.2 K. The deviations are almost exclusively caused by noise, in part due to operation in moist room air. The large noise between 600 and 650 cm^{-1} is caused by low optical throughput in this region during this test, and has since been improved. The noise spikes from 1450 to 1800 cm-1 are due to water vapor absorption between the interferometer and the detectors and are not present at flight altitude (Fig. 1). Room air CO_2 absorption is responsible for the increased noise amplitude between 2300 and 2400 cm^{-1}. The true calibration errors from this test are only about 0.1 to 0.2 K.

3. EARTH EMISSION SPECTRA

To illustrate some of the features of the raw data, the complex uncalibrated spectra from a flight on 17 June 1986 are shown in Figs. 6 and 7 for Bands I and II respectively. Both the magnitude and phase are determined directly from a complex Fourier transformation of the measured two-sided interferogram. It is apparent from the magnitude spectra that the cold blackbody temperature of 245 K for these measurements is not optimum, because there are many areas where the earth spectrum is smaller than that for the cold blackbody. The objective was to run the cold blackbody at about 220 K, but we could not dissipate enough heat from the thermoelectric cooler. The extrapolation necessary for low earth radiances can lead to somewhat larger calibration errors, but the effect is not large with the high emissivity sources used on HIS.

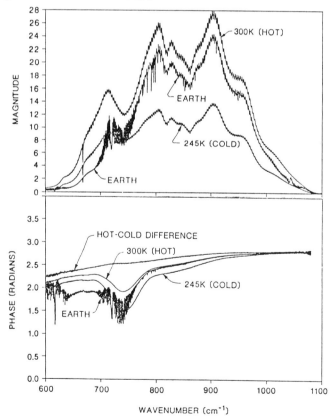

Figure 6. Magnitude and phase of Band I uncalibrated spectrum.

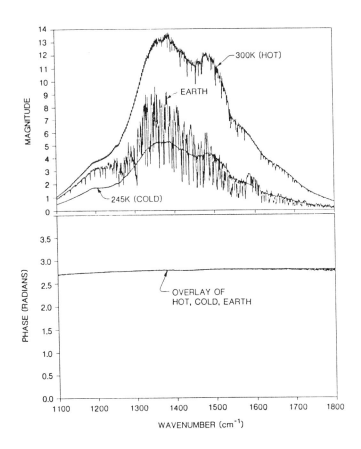

Figure 7. Magnitude and phase of Band II uncalibrated spectra.

The magnitude spectra have various features which need explanation. The general Gaussian shape of the magnitude spectra is caused by the numerical filtering which is performed in the instrument digital electronics (a hardware convolution is performed for signal-to-noise preserving sample volume reduction by factors of 14, 8, 8 in the 3 spectral bands). The sinusoidal components superimposed on the magnitude spectra are channeled spectra caused by the parallel surfaces of the arsenic-doped silicon detectors. Because the channeled spectra are very stable, they do not affect the calibrated spectra. The magnitude spectrum of the hot blackbody for Band II also shows some features for wavenumbers above 1350 cm^{-1} from the small water content at altitude, and band I shows the 667 cm^{-1} CO_2 Q-branch. These features are also stable and do not cause errors in the calibrated spectra.

The phase spectra for bands I and II differ markedly. For band II the phases are nearly linear and are source independent, the behavior expected with an ideal beamsplitter having zero dispersion and with an electrical response having a pure time delay. Band I phases, on the other hand, show significant deviations from linearity, and the phase for the earth view even has high resolution structure. As mentioned earier, these peculiar characteristics are caused by emission from the beamsplitter in band I. The high resolution structure occurs when the contribution from the earth with a well behaved phase and that from the instrument with an anomolous (but reasonably smooth) phase are combined to give a single phase.

To demonstrate the radiometeric integrity of the high-resolution features of spectra from the aircraft instrument, we now give some examples (Figs. 8-10) of comparisons between calibrated earth emission spectra and spectra from line-by-line calculations. All three examples are from a single Band I spectrum made with a six-second interferometer scan on 15 June 1986 over Northern Tennessee, the first day of COHMEX. The apodized resolution is about 0.5 cm^{-1}. The only selection criterion was a clear scene with reasonably close proximity to radiosonde measurements.

The calculations, which use the AFGL FASCODE version 2 (Clough et al., 1986) and the 1986 HITRAN database line tape (Rothman et al., 1987), have only recently been completed. Now, the reduction of the full resolution FASCODE spectrum to the HIS resolution is done without any loss of accuracy. The finite field-of-view of the interferometer (Table 1) is introduced into the calculated spectra as part of the resolution reduction process. The temperature and water vapor profiles are from the average of two radiosondes about 180 km apart in the special network. The six minor constituents included besides water vapor default to the midlatitude summer model.

The generally excellent agreement of the line structure and of the water vapor continuum as well is evident in Fig. 8. There are some interesting real differences which vary slowly with wavenumber and are presumably due to surface emissivity variations or to trace constituents not yet included in the calculations. Fig. 9 shows the short wavelength side of the 15 μm CO_2 band used for temperature sounding. The differences are less than about three degrees at most wavenumbers. The large feature near 720 cm^{-1} is a known deficiency of FASCODE caused by line mixing. Much of the residual difference is of the type which might be attributable to the failure of two radiosondes to accurately characterize the atmosphere.

Figure 10 shows the detailed nature of the differences for CO_2 lines from the same spectral region. The agreement in shape and wavenumber alignment of the spectra is phenomenal. This is especially remarkable since the HIS wavenumber calibration here is based on the known laser wavenumber and interferometer field-of-view, without adjustment. The difference is close to an offset minus a small fraction of the spectrum. This type of difference can largely be accounted for by adjusting the temperature and lapse rate of the atmosphere, although small calibration adjustments (to the radiosonde or to the HIS) might also be indicated.

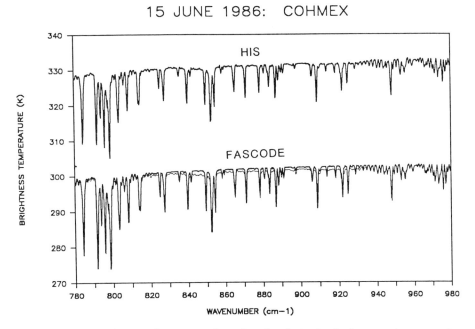

Figure 8. Comparison of measured and calculated window region spectra.

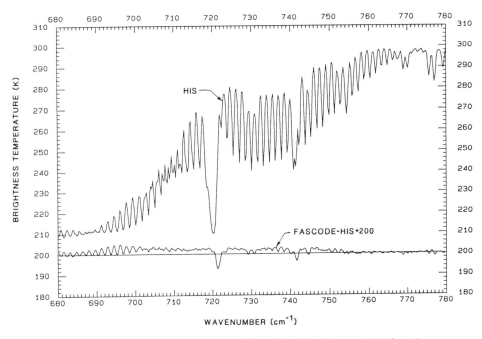

Figure 9. HIS and FASCODE compared in the 15 micron CO_2 band.

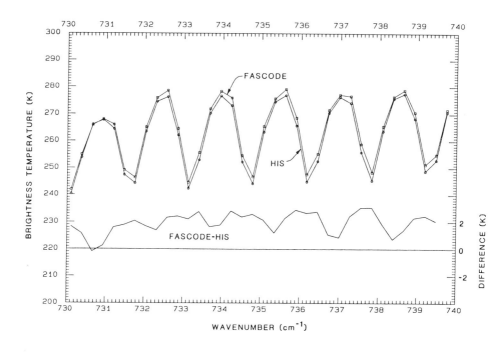

Figure 10. Closeup of HIS and FASCODE CO_2 lines (0.5 cm^{-1} resolution).

4. CONCLUSIONS

The high radiometric precision and accuracy of the HIS aircraft instrument clearly demonstrate that there are no fundamental problems in achieving excellent radiometric calibration with an interferometer. In fact, the interferometer has the great asset that its wavenumber calibration and resolution function are defined by a small number of parameters which can easily be known accurately.

ACKNOWLEDGMENTS

The authors thank the members of the instrument team at the Space Science and Engineering Center, BOMEM, Inc., the Santa Barbara Research Center, and the University of Denver, whose care in fabricating and aligning the HIS instrument made accurate radiometric calibration realizable. Thanks also to Frank Murcray for many helpful discussions. The HIS program is jointly funded by NOAA contract NA-84-DGC-00095 and NASA contract NAS5-27608.

REFERENCES

Clough, S.A., F.X.Kneizys, E.P.Shettle, G.P.Anderson, 1986: Atmospheric Radiance and Transmittance: FASCOD2, Sixth Conference on Atmospheric Radiation, American Meteorological Society, Boston, Massachusetts, 141.

Hanel, R.A., B Schlachman, F.D. Clark, C.H. Prokesh, J.B. Taylor, W.M. Wilson, and L. Chaney, 1970: The Nimbus III Michelson Interferometer, Applied Optics, 9, 1767.

Hanel, R.A., B. Schlachman, D. Rodgers, and D. Vanous, 1971: Nimbus 4 Michelson Interferometer, Applied Optics, 10, 1376.

Hanel, R.A., B. Schlachman, E. Breihan, R. Bywaters, F. Chapman, M Rhodes, D. Rodgers and D. Vanous, 1972: Mariner 9 Michelson Interferometer, Applied Optics, 11, 2625.

Hanel, R.A., D. Crosby, L. Herath, D. Vanous, D. Collins, H. Creswick, C. Harris, and D. Rhodes, 1980: Infrared Spectrometer for Voyager, Applied Optics, 19, 1391.

LaPorte, D.D. and R. Howitt, 1982: Ambient Temperature Absolute Radiometry using Fourier Transform Spectrometers, SPIE, 364.

Revercomb, H.E., D.D. LaPorte, W.L. Smith, H. Buijs, D.G. Murcray, F.J. Murcray, and L.A. Sromovsky, 1987: High-altitude Aircraft Measurements of Upwelling IR Radiance: Prelude to FTIR from Geosynchronous Satellite, Mikrochimica Acta, Springer-Verlag, Wien.

Revercomb, H.E., H. Buijs, H.B. Howell, D.D. LaPorte, W.L. Smith, and L.A. Sromovsky, 1988: Radiometric Calibration of IR Fourier Transform Spectrometers: Solution to a Problem with the High-Resolution Interferometer Sounder (HIS), Applied Optics, in press.

Rothman, L.S., R.R. Gamache, A. Goldman, L.R.Brown, R.A. Toth, H.M. Pickett, R.L. Poynter, J.-M. Flaud, C. Camy-Peryret, A. Barbe, N. Husson, C.P. Rinsland, and M.A.H. Smith, 1987: The HITRAN Database: 1986 Edition, Applied Optics, 26, 4058.

Smith, W.L., H.E. Revercomb, H.B. Howell, and H.M. Woolf, 1983: HIS - A Satellite Instrument to Observe Temperature and Moisture Profiles with High Vertical Resolution, Fifth Conference on Atmospheric Radiation, American Meteorological Society, Boston, Mass.

Smith, W.L., H.E. Revercomb, H.B. Howell, and H. M. Woolf, 1984: Recent Advances in Satellite Remote Sounding, International Radiation Symposium '84: Current Problems in Atmospheric Radiation, Edited by Giorgio Fiocco, A. Deepak Publishing, Hampton, Virginia, p.388.

Smith, W.L., H.E. Revercomb, H.M. Woolf, H.B. Howell, D.D. LaPorte, and K. Kageyama, 1987a: Improved Geostationary Satellite Soundings for the Mesoscale Weather Analysis/Forecast Operations, Proc. Symp. Mesoscale Analysis and Forecasting, Vancouver, Canada, 17-19 August.

Smith, W.L., H.M. Woolf, H.B. Howell, H.-L. Huang and H.E. Revercomb, 1987b: The Simultaneous Retrieval of Atmospheric Temperature and Water Vapor Profiles - Applications to Measurements with the High spectral Resolution Interferometer Sounder (HIS), this issue.

DISCUSSION

McClatchey: In your calibration curves that you showed part way through your briefing, where you wrote off the fact that you referred to data spikes as noise associated with water vapor and CO_2, I don't understand where that is coming from. Is that coming from a path in the instrument, or the laboratory, and is it really irrelevant?

Revercomb: Yes, it is irrelevant. This instrument is open to the atmosphere with the exception of the interferometer box itself. There is a long path from the telescope into the interferometer and from the interferometer to the detector. The interferometer itself is evacuated for flight to protect the beam splitter during ascent and descent. That path on the ground is full of CO_2 and water. In flight it is not. As a matter of fact, I have a number of examples of uncalibrated spectra taken at altitudes that I wanted to show. The CO_2 and water spikes present in laboratory spectra are absent. At altitude there are some small features of CO_2 but they are constant so they don't affect the calibration.

Falcone: This isn't really a question with your technique, it was just that in your instrument you have a helium dewar. An instrument that is going to fly in space is not going to have a helium dewar is it?

Revercomb: No. All of our space studies have been with passively cooled HgCdTe and InSb detectors. I should mention that we are currently doing a study for NOAA to modify the sounding instrument on the last two models GOES I-L replacing the filter wheel with an interferometer. It looks very good. I think there is going to be every attempt to proceed in that direction.

A CALIBRATION MODEL FOR NIMBUS-7 SMMR

Daesoo Han
NASA Goddard Space Flight Center
Greenbelt, Maryland 20771, USA

S.T. Kim, C.C. Fu, and D.S. MacMillan
ST Systems Corporation
Hyattsville, Maryland 20784, USA

ABSTRACT

A calibration model for the Nimbus-7 SMMR is studied. This model not only removes major drawbacks of the current calibration model but also helps us understand the performance degradation of the aging instrument. The current Nimbus-7 SMMR calibration algorithm was derived without considering the interference effect between the two orthogonally polarized signals merging at a ferrite polarization selector switch. The resulting calibrated brightness temperatures, considered as a function of scan angle ϕ, are not symmetric around $\phi=0$. This problem was resolved empirically by introducing "offset angles". However, neither the origin of the offset angles nor the manner in which the two orthogonal components are mixed has been fully understood. Thus, it was not possible to derive "absolutely calibrated" brightness temperatures.

The new calibration model proposed in this paper incorporates all the leakage factors associated with the ferrite switches along the signal paths. The resulting calibration equations, though much more complicated than those of the current model, clarify how the orthogonal components of surface brightness are coupled at radiometers. As a consequence, the origin of the offset angles is clearly identified as arising from the interference between the two orthogonal signals at the polarization selector switch. It is also shown that the current calibration algorithm can be derived from the new one if the restrictions imposed upon the current model are restored. In addition, the feasibility of "absolute calibration" using in-orbit data is discussed.

1. INTRODUCTION

The Nimbus-7 Scanning Multichannel Microwave Radiometer (SMMR) consists of six conventional Dicke-type radiometers: two, operating at 37 GHz, measuring simultaneously the horizontal and vertical polarizations; the other four, operating at 6.6, 10.7, 18, and 21 GHz, measuring the two orthogonal polarizations alternately during a scan. Thus the SMMR provides ten data channels, corresponding to five dual-polarized frequencies. A description of the SMMR instrument was given by Gloersen and Barath (1977). The basic calibration scheme of the SMMR utilizes a two-point reference signal

RSRM '87: ADVANCES IN
REMOTE SENSING RETRIEVAL METHODS
A. Deepak, H.E. Fleming, and J.S. Theon (Eds.)

103

system consisting of an ambient RF termination and a horn antenna viewing deep space.

All six radiometers share a single offset parabolic reflector. While the reflector scans 25° to the left and right of the satellite flight direction, the multi-frequency feed horn (MFFH) remains fixed. This configuration of the fixed MFFH relative to the scanning reflector introduces a polarization mixing in the measured signal, which would be given by Eq. (1) below, if the instrument were perfect:

$$\begin{pmatrix} T_x \\ T_y \end{pmatrix} = \begin{pmatrix} \cos^2\phi & \sin^2\phi \\ \sin^2\phi & \cos^2\phi \end{pmatrix} \begin{pmatrix} H \\ V \end{pmatrix}, \qquad (1)$$

where T_x and T_y are measured radiances for horizontal and vertical channels, respectively, H and V are horizontal and vertical components of the surface radiance, respectively, and ϕ is the scan angle. In Eq. (1), T_x and T_y are symmetric about the scan angle $\phi=0$. However, as noticed by Gloersen et al. (1980) and Gloersen (1983), the measured radiances are not symmetric about scan angle $\phi=0$, but the whole shape is shifted as shown in Fig. 1. The cause of the shift was not well understood at that time. So, to compensate for the shift, "offset angles", δ_x and δ_y were empirically introduced in the following manner:

$$\begin{pmatrix} T_x \\ T_y \end{pmatrix} = \begin{pmatrix} \cos^2(\phi + \delta_x) & \sin^2(\phi + \delta_x) \\ \sin^2(\phi + \delta_y) & \cos^2(\phi + \delta_y) \end{pmatrix} \begin{pmatrix} H \\ V \end{pmatrix}, \qquad (2)$$

where the values of δ_x and δ_y are determined from the measured data using regression technique.

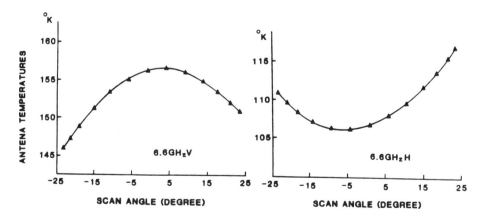

FIGURE 1.   Averaged antenna temperatures for 6.6 GHz horizontal and vertical channels versus scan angle. For averaging, one month of November, 1978 data over Pacific ocean are used.

As mentioned earlier, the radiometers operating at lower four frequencies utilize polarization selector switches for measuring one of the two polarizations. There are leakages in a polarization selector switch, which is about 1 % of the total input power. In Section 2, it is shown that the leakages, though small, cause interference between the signal to be measured and the one to be suppressed, thus generating a term which is not symmetric about φ=0. There is also additional mixing between the two polarizations that arises from the leakages at other switches. In Section 3, it is shown that the current calibration equations can be derived from the new ones if switch leakages are ignored. In the last section, methods to analyze inflight data to obtain "absolutely calibrated" radiances are briefly discussed.

2. CALIBRATION EQUATIONS

A schematic diagram of the RF components for a typical radiometer, which provides two orthogonal polarization channels operating at a common frequency, is shown in Fig. 2. In addition to the familiar modulator (or Dicke) switch, there are three latching ferrite switches: the sky/ambient switch (c) which selects a signal from either a sky horn or an ambient load; the polarization selector switch (p) which selects either a horizontally or vertically polarized signal from the multi-frequency feed horn; the cal/sig switch (s) which selects a signal from either switch c or p. These three switches, when suitably set, provide four distinct signal paths to the modulator switch input port marked I in Fig. 2; two

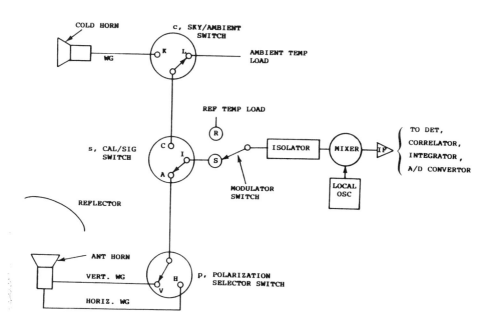

FIGURE 2. Schematic diagram of RF components of the Nimbus-7 SMMR.

paths for cold and warm reference signals and another two paths for
horizontally and vertically polarized radiances.

In the following we will derive a relationship between the
target temperature and the power at port I by piecing together the
input-output relationships for individual RF components along each
signal path. By way of introducing notations, we write down generic
input-output relationships for two- and three-port devices. For a
two-port device (with input port A and output port B), the output
power, T_B is given by

$$T_B = \alpha T_A + (1-\alpha)t, \tag{3}$$

where T_A is the input power, α is the transmissivity of the device,
and t is the physical temperature . For a ferrite switch x (=p,c,s)
with input ports A and B and output port C, the output power, T_C,
when port A is selected, is assumed to be given by

$$T_C = \alpha_{xAC}T_A + \beta_{xBC}T_B + (1-\alpha_{xAC})t, \tag{4}$$

where T_A and T_B are input powers at ports A and B, respectively,
α_{xAC} is the transmissivity from A to C, β_{xBC} is the leakage factor
from B to C, and t is the device's physical temperature.

We start with microwave radiation emitted from the earth's
surface, which may be expressed in terms of electric field as

$$\vec{E} = \begin{pmatrix} E_x \exp(iwt) \\ E_y \exp(iwt + i\Omega) \end{pmatrix}, \tag{5}$$

where E_x and E_y are the horizontal and vertical components of the
electric field defined on the coordinate system connecting the earth
surface and the antenna (Fig. 3),w is the microwave frequency, and Ω
is the phase difference which is assumed to have any value between 0
and 2π with equal probability. The electric field, $\vec{E}$ is reflected
by the reflector to the feed horn. Since the feed horn coordinates
are rotated relative to the scanning reflector by an amount of scan
angle ϕ, the two orthogonal components E_{1x} and E_{1y} selected by the
horn are given by

$$\vec{E}_1 = \begin{pmatrix} \cos\phi & \sin\phi \\ -\sin\phi & \cos\phi \end{pmatrix} \vec{E}. \tag{6}$$

The two orthogonal components are routed separately to switch p via
waveguides or coaxial cables, undergoing attenuation by a factor of
g_x and g_y, respectively. Though prelaunch test results suggested
that the switch leakages from an unselected input port to the output
port should be negligible, the effective leakages may not be
negligible, when mismatched components are attached to the ports.
We denote by k_{1x} the magnitude of the blocked portion of the
horizontal polarization component, and by k_{2x} the leaked portion

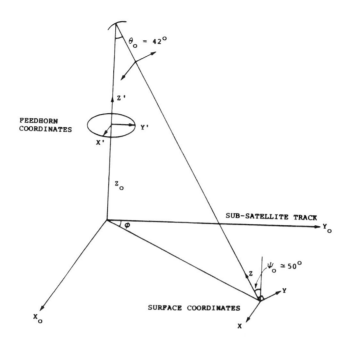

FIGURE 3.   Coordinates of surface and antenna feed horn. The scan
            angle ϕ, as shown, is positive. Also shown are the
            antenna look angle θ_o, measured from nadir, and the angle
            of incidence at the surface which is about 50°.

from the vertical polarization added to the horizontal component;
and similarly for k_{1y} and k_{2y}. We note that these quantities are
related to the transmissivities and leakage factors of switch p as
follows:

$$\alpha_{pHA} = (1 - k_{1x})^2,$$

$$\alpha_{pVA} = (1 - k_{1y})^2,$$

$$\beta_{pHA} = k_{2y}^2,$$

$$\beta_{pVA} = k_{2x}^2.$$

Immediately after a polarization is selected, the electric field, E_2
would be given by

$$\vec{E}_2 = \begin{pmatrix} 1 - k_{1x} & k_{2x}\exp(i\theta_x) \\ k_{2y}\exp(i\theta_y) & 1 - k_{1y} \end{pmatrix} \begin{pmatrix} g_x & 0 \\ 0 & g_y \end{pmatrix} \vec{E}_1, \qquad (7)$$

where θ_x and θ_y are phase differences and $\vec{E}_1$ is given by Eq. (6).
The power at the output port of switch p, solely due to surface
radiation, is given by

$$T_x = < (\text{Re } E_{2x})^2 >_{t.\Omega}, \qquad\qquad (8.a)$$

$$T_y = < (\text{Re } E_{2y})^2 >_{t.\Omega}, \qquad\qquad (8.b)$$

where $< >_{t.\Omega}$ denotes the averaging operation over t and Ω. From Eqs. (5) through (8), one obtains

$$T_x = A_x[H \cos^2(\phi+\delta_x) + V \sin^2(\phi+\delta_x)] + B_x(H + V), \qquad (9.a)$$

$$T_y = A_y[H \sin^2(\phi+\delta_y) + V \cos^2(\phi+\delta_y)] + B_y(H + V), \qquad (9.b)$$

where

$$H = E_x^2,$$

$$V = E_y^2,$$

$$A_x = \{[g_x^2(1-k_{1x})^2 + g_y^2k_{2x}^2]^2 - [2g_xg_y(1-k_{1x})k_{2x}\sin\theta_x]^2\}^{1/2},$$

$$B_x = 0.5 [g_x^2(1-k_{1x})^2 + g_y^2k_{2x}^2 - A_x],$$

$$A_y = \{[g_y^2(1-k_{1y})^2 + g_x^2k_{2y}^2]^2 - [2g_yg_x(1-k_{1y})k_{2y}\sin\theta_y]^2\}^{1/2},$$

$$B_y = 0.5 [g_y^2(1-k_{1y})^2 + g_x^2k_{2y} - A_y],$$

$$\tan(2\delta_x) = 2g_xg_y(1-k_{1x})k_{2x}\cos\theta_x/[g_x^2(1-k_{1x})^2 - g_y^2k_{2x}^2],$$

$$\tan(2\delta_y) = 2g_yg_x(1-k_{1y})k_{2y}\cos\theta_y/[g_y^2(1-k_{1y})^2 - g_x^2k_{2y}^2].$$

We denote by $T_{A.H}$ and $T_{A.V}$ the powers at port A when switch p selects port H and V, respectively. Then $T_{A.H}$ ($T_{A.V}$) is the sum of T_x (T_y) and $Q_{A.H}$ ($Q_{A.V}$) which are noise powers added incoherently along the respective signal paths from the feed horn to port A:

$$T_{A.H} = T_x + Q_{A.H}, \qquad\qquad (10.a)$$

$$T_{A.V} = T_y + Q_{A.V}, \qquad\qquad (10.b)$$

where $Q_{A.H}$ and $Q_{A.V}$, computed using Eqs. (3) and (4), are given below:

$$Q_{A.H} = \alpha_{pHA}[\alpha_{awH}(1-\alpha_{ahH})t_{ah} + (1-\alpha_{awH})t_{awH}]$$
$$+ \beta_{pVA}[\alpha_{awV}(1-\alpha_{ahV})t_{ah} + (1-\alpha_{awV})t_{awV}] + (1-\alpha_{pHA})t_h,$$

$$Q_{A.V} = \alpha_{pVA}[\alpha_{awV}(1-\alpha_{ahV})t_{ah} + (1-\alpha_{awV})t_{awV}]$$
$$+ \beta_{pHA}[\alpha_{awH}(1-\alpha_{ahH})t_{ah} + (1-\alpha_{awH})t_{awH}] + (1-\alpha_{pVA})t_h,$$

where t_{ah}, t_{awH}, t_{awV}, and t_h are physical temperatures of antenna horn, horizontal wave guide, vertical wave guide, and switch assembly, respectively, and $\alpha_{ahH(V)}$, $\alpha_{awH(V)}$ are transmissivities of the antenna horn and wave guide for the horizontally (vertically) polarized signal path. From Eqs. (9) and (10), one obtains

$$\begin{pmatrix} H \\ v \end{pmatrix} = R \begin{pmatrix} T_{A.H} - Q_{A.H} \\ T_{A.v} - Q_{A.v} \end{pmatrix} , \qquad (11)$$

where R, which we may term the generalized polarization rotation matrix, is given by

$$R = \begin{pmatrix} A_x \cos^2(\phi+\delta_x) + B_x & A_x \sin^2(\phi+\delta_x) + B_x \\ A_y \sin^2(\phi+\delta_y) + By & A_y \cos^2(\phi+\delta_y) + B_y \end{pmatrix}^{-1} \qquad (12)$$

Equation (11), which expresses the surface brightness in terms of the powers at port A for two antenna paths, is one of our two basic sets of calibration equations.

In the remainder of this section, we derive the second basic set of equations. The power at port K of switch c is given by

$$T_K = \alpha_{cw}[\alpha_{ch}T_c + (1-\alpha_{ch})t_{ch}] + (1-\alpha_{cw})t_{cw}, \qquad (13)$$

where α_{ch} and α_{cw} are the transmissivities of the cold horn and wave guide, respectively, and t_{ch} and t_{cw} are the physical temperatures of the cold horn and wave guide, respectively.

We denote by $T_{I.H}$, $T_{I.v}$, $T_{I.c}$, $T_{I.w}$ the powers at port I for the following four signal paths designated by H, V, C, and W, each of which is defined by a set of switch settings of switches c, p, and s as shown in Table 1.

TABLE 1. DEFINITION OF PATHS VIA SWITCH SETTINGS

Path	Switch Settings
H	p on H, c on L, s on A
V	p on V, c on L, s on A
C	p on H, c on K, s on C
W	p on V, c on L, s on C

Applying the input-output relationship of Eq. (4) to each switch along a chosen path, one obtains

$$T_{I.H} = \alpha_{sAI}T_{A.H} + \beta_{sCI}(t_h+\beta_{cKC}T_K) + (1-\alpha_{sAI})t_h, \qquad (14)$$

$$T_{I.v} = \alpha_{sAI}T_{A.v} + \beta_{sCI}(t_h+\beta_{cKC}T_K) + (1-\alpha_{sAI})t_h, \qquad (15)$$

$$T_{I.c} = \alpha_{sCI}(\alpha_{cKC}T_K + \beta_{cLC}t_h + (1-\alpha_{cKC})t_h)$$
$$+ \beta_{sAI}T_{A.H} + (1-\alpha_{sCI})t_h, \qquad (16)$$

$$T_{I.w} = \alpha_{sCI} (t_h + \beta_{cKC}T_K) + \beta_{sAI}T_{A.v}$$
$$+ (1-\alpha_{sCI})t_h. \qquad (17)$$

The integrate and dump output of the radiometer, denoted by C, is given in terms of the input power, T_I at port I, as follows:

$$C = G (T_I - t_h) + O, \tag{18}$$

where G is the radiometer gain and O an offset added to make C increase with T_I. Both G and O are assumed to remain constant from one end of scan to the other, and are effectively eliminated from calibration equations through introduction of normalized counts, N_H and N_v defined below:

$$N_H \equiv (C_{A.H} - C_w)/(C_C - C_w), \tag{19.a}$$

$$N_v \equiv (C_{A.v} - C_w)/(C_C - C_w), \tag{19.b}$$

where $C_{A.H}$, $C_{A.v}$, C_C, and C_w are the radiometer outputs corresponding to the four paths H, V, C, and W, respectively. Substituting Eq. (18) into Eqs. (19.a) and (19.b), one obtains

$$N_H = (T_{I.H} - T_{I.w})/(T_{I.c} - T_{I.w}), \tag{20.a}$$

$$N_v = (T_{I.v} - T_{I.w})/(T_{I.c} - T_{I.w}). \tag{20.b}$$

Substituting Eqs (14) through (17) into Eqs. (20.a) and (20.b), we obtain the second basic set of calibration equations given below:

$$T_{A.H} = \{[-(\alpha_{sAI} - \beta_{sAI})D - \beta_{sAI}C]N_H + \beta_{sAI}(C-D)N_v + \alpha_{sAI}C\}/\Gamma, \tag{21.a}$$

$$T_{A.v} = \{(-\alpha_{sAI}D + \beta_{sAI}C)N_v - (\beta_{sAI}C)N_H + \alpha_{sAI}C\}/\Gamma, \tag{21.b}$$

where

$$C = (\alpha_{sCI} - \beta_{sCI})\beta_{cKc}T_K + (\alpha_{sAI} - \beta_{sCI})t_h,$$

$$D = - \alpha_{sCI}(\alpha_{cKc} - \beta_{cKc})T_K + \alpha_{sCI}(\alpha_{cKc} - \beta_{cLc})t_h,$$

$$\Gamma = (\alpha_{sAI} - \beta_{sAI})[\alpha_{sAI} + \beta_{sAI}(N_v - N_H)].$$

Using measured quantities, N_H and N_v, one can compute from Eq. (21), $T_{A.H}$ and $T_{A.v}$, and then substitute them into Eq. (11) to obtain surface radiances. Thus, Eqs. (11) and (21) constitute a new set of calibration equations for the Nimbus-7 SMMR.

3.   CURRENT CALIBRATION EQUATIONS

The calibration equations currently in use for the Nimbus-7 SMMR data production can be derived from Eqs. (11) and (21) if we reinstate the assumption that switch leakages are negligible.

Setting all the leakage factors appearing in Eqs. (21.a) and (21.b) to zero, we obtain simplified expressions for $T_{A.H}$ and $T_{A.v}$:

$$T_{A.H} = t_h + \alpha_0(T_K - t_h)N_H, \qquad (22.a)$$

$$T_{A.v} = t_h + \alpha_0(T_K - t_h)N_v, \qquad (22.b)$$

where $\alpha_0 = \alpha_{sCI}\alpha_{cKC}/\alpha_{sAI}$, and T_K is given by Eq. (13).

Polarization rotation, which accounts for not only the antenna scan but also the polarization mixing at switch p, is effected empirically by introducing "offset angles", δ_x and δ_y into the polarization rotation matrix given below:

$$R' = \begin{pmatrix} \cos^2(\phi+\delta_x) & \sin^2(\phi+\delta_x) \\ \sin^2(\phi+\delta_y) & \cos^2(\phi+\delta_y) \end{pmatrix}^{-1} \qquad (23)$$

The matrix, R', when compared with R (Eq. (12)), is seen to account for the effect of switch leakages (in switch p) only partially in that the factors A_x and A_y and biases B_x and B_y in R are not present in R'. Replacing R of Eq. (11) by R' and substituting into it $T_{A.H}$ and $T_{A.v}$ of Eqs. (22.a) and (22.b), we obtain the current calibration equations given below:

$$\begin{pmatrix} H \\ V \end{pmatrix} = R' \begin{pmatrix} t_h - Q'_{A.H} + \alpha_0(T_K - t_h)N_H \\ t_h - Q'_{A.v} + \alpha_0(T_K - t_h)N_v \end{pmatrix}, \qquad (24)$$

where $Q'_{A.H}$ and $Q'_{A.v}$ are obtained from $Q_{A.H}$ and $Q_{A.v}$ by setting β_{pVA} and β_{pHA} to zero.

4. FURTHER DISCUSSION

A new calibration scheme for the Nimbus-7 SMMR is proposed in this paper. The apparent deviation of the measured radiances from the anticipated symmetry in scan angle is explained by this model through including leakages at the polarization selector switch. This feature, asymmetry in scan angle, is caused by the interference between horizontally and vertically polarized fields which, having traveled different paths and merging at the polarization selector switch, have path differences comparable with the wavelengths'of the microwave radiation.

Another important aspect of the new calibration model is that it offers a possibility of deriving "absolutely calibrated" radiances. Absolute calibration is not important in retrieving geophysical parameters because the retrieval algorithms may be tuned afterwards. However, models for surface and atmosphere can not be validated without absolutely calibrated radiances. The current calibration algorithm cannot separate both polarizations.

It is clear from Eqs. (11) and (21) that the values of transmissivities and leakage factors must be known prior to implementing the new model in the SMMR data processing. Only a few transmissivities had been determined before launch, and none of leakage factors have been measured on the assembled instrument. Attempts to determine some leakage factors using inflight data were unsuccessful because of the difficulties of estimating the surface brightness, H and V of Eq. (11), to the accuracy required for such an analysis. Without means of obtaining acceptable values of α's and β's, it is practically impossible to derive absolutely calibrated brightness temperatures from inflight measurements alone. If ground measurements are made and a good atmospheric model is available, then it is possible to achieve absolute calibration.

The new calibration equations, however, may be used to understand the evolutionary nature of the aging instrument's problems. Instead of attempting to determine α's and β's, we start with a set of preassigned values of α's and β's and compute from measured quantities, N_H and N_V, the surface brightness temperatures in accordance with Eqs. (11) and (21). The new brightness temperatures can be scrutinized whether their long-term behavior is satisfactory or not, or may be used to derive geophysical parameters whose validity can be tested against ground measurements. If comparisons show that the new brightness temperatures produce better results than the current ones, we may adopt the presumed values of α's and β's as representing the instrument's status. In this manner, we may identify and assess the cause(s) of some of the instrument's problems.

ACKNOWLEDGMENTS

The authors would like to thank Drs. Erik Mollo-Christensen and Per Gloersen for helpful discussions.

REFERENCES

Gloersen, P., 1983: Calibration of the Nimbus-7 SMMR: II Polarization mixing corrections, TM 84976, NASA/Goddard Space Flight Center, Greenbelt, Maryland.

Gloersen, P., and F. T. Barath, 1977: A scanning multichannel microwave radiometer for Nimbus-G and SeaSat-A, IEEE J. Oceanic Eng., OE-2, 172-178

Gloersen, P., D. J. Cavalieri, and H. V. Soule, 1980: An alternate algorithm for correction of the scanning multichannel microwave radiometer polarization radiances using Nimbus-7 observed data, TM 80672, NASA/Goddard Space Flight Center, Greenbelt, Maryland.

DISCUSSION

Gautier: Is this calibration being applied yet to SMMR Nimbus-7 data?

Han: No. It is not in the current Nimbus-7 data. But we are trying to implement this.

Gautier: Do you have any idea when this will happen?

Han: I hope within two years. After that we don't have enough funding to do this kind of work. By the way, I forgot one important thing. These kinds of problems can be prevented if the feedhorn rotates with the antenna. The second thing is that if both polarizations have independent radiometers then this problem will be prevented just like with the SSMI case.

Kleespies: Just a comment. Even though the lack of absolute calibration may prohibit absolute derivation of geophysical parameters, we have looked at multispectral imagery created from the SMMR and can see the signal of the geophysical parameters. You can see the precipitation over the ocean, you can see the ice edge. It is there in the signal.

Han: Yes there is signal there, definitely. Original plans for the instrument state that the requirement was relative calibration, not absolute calibration. That was correct. But without absolute calibration it is difficult to develop a new model.

INCIDENT ANGLE EFFECT ON GEOPHYSICAL PARAMETER RETRIEVAL FROM THE NIMBUS-7 SMMR

Daesoo Han
NASA Goddard Space Flight Center
Greenbelt, Maryland 20771, USA

D.S. MacMillan, C.C. Fu, and S.T. Kim
ST Systems Corporation
Lanham, Maryland 20706, USA

ABSTRACT

The attitude of the Nimbus-7 spacecraft has varied over its lifetime. One of the effects of these attitude variations is to change the incident angle at which the Scanning Multichannel Microwave Radiometer (SMMR) instrument views the surface of the earth. The ocean surface microwave emissivity is quite sensitive to incident angle variation near the SMMR incident angle which is about 50 deg. This sensitivity has been estimated theoretically for a smooth ocean surface and no atmosphere. A one-degree increase in the angle of incidence produces a 2.9 deg C increase in retrieved sea surface temperature and a 5.7 m/sec decrease in retrieved sea surface windspeed. A clear example of this effect occurred when the Nimbus-7 spacecraft was pitched downward by about 0.4 deg in January 1984. There was a sudden jump in the bias errors of SMMR retrievals of sea surface temperature and sea surface wind speed. An incident angle correction is applied to the SMMR radiances before using them in the geophysical parameter retrieval algorithms. The corrected retrieval data is compared with data obtained without applying this incident angle correction.

1. INTRODUCTION

The Scanning Multichannel Microwave Radiometer (SMMR) on the Nimbus-7 spacecraft measures microwave radiances from the earth's surface and its surrounding atmosphere at five frequencies (6.6 GHz, 10.7 GHz, 18 GHz, 21 GHz, 37 GHz) in both horizontal and vertical polarizations. A description of the SMMR instrument was given by Gloersen and Barath (1977). Radiances are collected with a 42 degree offset parabolic reflector, which scans 25 degrees to the left and right of the satellite flight direction. If there were no variation of the spacecraft attitude, the SMMR angle of incidence at the earth's surface would be about 50.4 deg. However, there is significant orbital and long-term variation of the spacecraft attitude. In this paper, we examine the effect of this variation on SMMR retrievals. Then we make an incident angle correction to the brightness temperatures. This correction is based on the incident angle dependence of radiances for a smooth ocean surface and a clear atmosphere.

RSRM '87: ADVANCES IN
REMOTE SENSING RETRIEVAL METHODS
A. Deepak, H.E. Fleming, and J.S. Theon (Eds.)   115

In Section 2, we first describe the incident angle behavior of the Nimbus-7 SMMR. Then we discuss the dependence of the sea surface emissivity on frequency, polarization, and the angle of incidence. Although the incident angle variation is small, its impact on the retrieved geophysical parameters computed from the Nimbus Experiment "Team Algorithms" is quite large. We demonstrate how large these effects are for Nimbus-7 SMMR retrievals in Section 3. A simple scheme to compensate for the incident angle variation is applied to SMMR data. In the last section, we discuss the implications of attitude variation on retrievals and the limitations of our correction scheme.

2. INCIDENT ANGLE BEHAVIOR OF THE NIMBUS-7 SMMR

Over its lifetime, there have been variations of the attitude of the Nimbus-7 spacecraft. There are both long-term and orbital components of these variations. The long-term behavior of the resulting incident angle at the center scan positions is shown in Fig. 1. This behavior was computed from monthly averages of the incident angle for ocean latitudes between 60 deg S and 60 deg N. In January 1984 (month 61), the spacecraft was pitched downward by 0.4 degrees. This produced a 0.5 deg decrease in the incident angle. However, one can see that prior to this date, the behavior of the

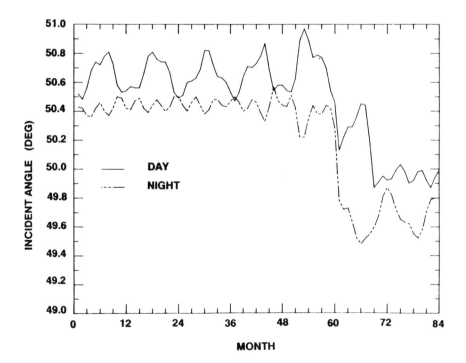

FIGURE 1. Monthly average SMMR incident angle for ocean latitudes between 60 deg S and 60 deg N.

incident angle was already changing. Figure 1 also shows that the
annual cycles of the day and night incident angles also changed
after 1983 (month 60). Within an orbit, the incident angle varies
by about 1 deg. The average night incident angle is less than the
average day incident angle by about 0.2-0.3 deg. This difference
can probably be attributed to the inaccuracy of attitude
determination in the descending segment (night time) of each orbit
when only the horizon sensors are used. When we look at the orbital
behavior of the incident angle computed from the spacecraft attitude
angle, we can see a discontinuous jump when the spacecraft passes
into or out of the sun.

Both in the long-term and within each orbit, we expect to see
the effect of incident angle variation on the measured brightness
temperatures. This is because the measured microwave radiances are
quite sensitive to changes in the incident angle. To simplify our
calculation, we have estimated this dependence for a specular sea
surface of temperature 20 deg C and no atmosphere. The ocean
surface emissivities were computed from the Fresnel reflection
coefficients for horizontal and vertical polarization. An
expression for the sea water dielectric constant was taken from
Chang and Wilheit (1979), and a salinity of 35 °/oo was used.
Figure 2 shows the resulting radiances at three of the five SMMR
frequencies as a function of the angle of incidence. The

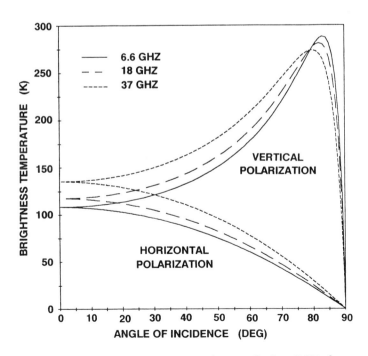

FIGURE 2.   Brightness temperature at three of the SMMR frequencies
             for a smooth ocean surface and no atmosphere.

derivatives of these radiances with respect to the angle of incidence at (the nominal SMMR incident angle) 50.4 degrees are given in Table 1. We see that this sensitivity is about -1.5 K/deg for the horizontal channels and 2.2 K/deg for the vertical channels.

TABLE 1.

Channel	$\dfrac{\delta T_B}{\delta\phi}$	SST		Wind Speed		Water Vapor	
		$\dfrac{\delta SST}{\delta T_B}$	$\dfrac{\delta SST}{\delta\phi}$	$\dfrac{\delta W}{\delta T_B}$	$\dfrac{\delta W}{\delta\phi}$	$\dfrac{\delta WV}{\delta T_B}$	$\dfrac{\delta WV}{\delta\phi}$
	$\dfrac{K}{deg}$	$\dfrac{deg\ C}{K}$	$\dfrac{deg\ C}{deg}$	$\dfrac{m/sec}{K}$	$\dfrac{m/sec}{deg}$	$\dfrac{cm}{K}$	$\dfrac{cm}{deg}$
6.6H	-1.3	-0.37	0.48	-	-	-	-
6.6V	2.1	1.70	3.57	-	-	-	-
10.7H	-1.3	-0.44	0.57	1.81	-2.35	-	-
10.7V	2.1	0.65	1.37	-0.86	-1.81	-	-
18H	-1.4	-0.17	0.24	-	-	-0.05	0.07
18V	2.1	-0.10	-0.21	-	-	-0.02	-0.04
21H	-1.5	0.10	-0.15	-	-	0.06	-0.09
21V	2.2	-	-	-	-	0.05	0.10
37H	-1.6	-	-	0.13	-0.21	-0.03	0.04
37V	2.2	-	-	-0.60	-1.32	0.03	0.07
TOTAL EFFECT			2.87*		-5.69		0.15

* Includes an explicit algorithm dependence of -3ϕ.

3. INCIDENT ANGLE EFFECT ON SMMR RETRIEVALS

Sea surface wind speed, sea surface temperature (SST), and atmospheric water vapor content are three of the geophysical parameters that are derived from SMMR radiances. Table 1 gives the sensitivities of the retrieval algorithms to radiance variations of 1 K. These sensitivities were evaluated for typical average global radiances. When these sensitivities are multiplied by the brightness temperature sensitivities to incident angle, we obtain the retrieval algorithm sensitivities to incident angle change shown in Table 1. A one degree increase in incident angle yields an increase of 2.9 deg C in retrieved SST, a decrease of 5.7 meter/sec in retrieved windspeed, but only a 0.15 cm increase in water vapor content.

Coincident wind speed observations from ships and from SMMR have been compared for the period January 1979 to December 1984 (months 1-72). The average monthly differences of these coincident observations are shown in Fig. 3. It is clear that the decrease in incident angle between 1983 and 1984 produced the observed increase in retrieved wind speeds. A 0.5 deg decrease in incident angle causes an increase in retrieved wind speed of about 2.9 meter/sec.

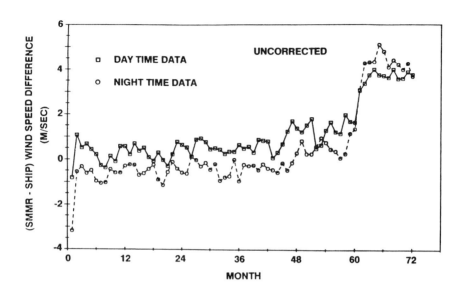

FIGURE 3.   Monthly average difference between SMMR retrieved wind
            speeds and coincident ship measured wind speeds. No
            incident angle correction has been made.

We computed the monthly average difference between retrieved
SST and climatological SST.   The retrieval algorithm used is a
prelaunch version with only an initial tuning.   The SST difference
is plotted in Fig. 4.   We mention several characteristics of this
plot.   There is a long-term downward trend (about a 1 deg C/3 yr)
caused by a corresponding downward trend in the 6.6 GHz V brightness
temperature, which is the principal surface channel for SST
retrieval.   The decrease in the SST difference about October 1983
(month 58) reflects the behavior of the 21 GHz H channel.   The long-
term drift behavior of this channel changed after May 1983 (month
53).   The additional decrease between 1983 and 1984 is caused by the
incident angle decrease.   According to Table 1, a 0.5 deg decrease
in the incident angle produces a decrease of about 1.5 deg C in
retrieved SST.   This decrease occurs despite the presence of an
explicit incident angle dependence of -3 deg C/deg in the algorithm.

We have applied a simple approximate correction to the SMMR
radiances before putting them into the geophysical parameter
retrieval algorithms.   The corrected brightness temperatures are
given by

$$T_B(\text{corrected}) = T_B(\text{uncorrected}) - \alpha(\phi - \phi_o), \quad (1)$$

where ϕ is the incident angle, ϕ_o is the nominal SMMR incident angle
(50.4 degrees), and $\alpha = \delta T_B/\delta\phi$ is taken from Table 1.   This
correction adjusts the radiances to radiances at the nominal

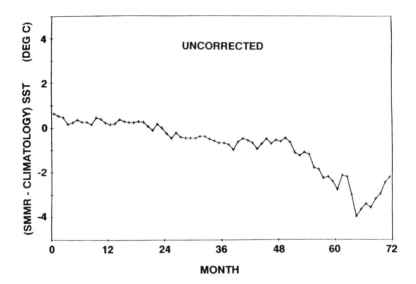

FIGURE 4.   Monthly average difference between SMMR retrieved SST and
            a climatological SST. An incident angle correction has
            not been made.

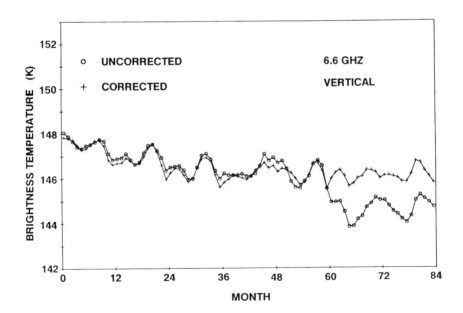

FIGURE 5.   The monthly average SMMR 6.6 GHz vertical brightness
            temperature. Nighttime radiances have been averaged over
            ocean areas from 60 deg S to 60 deg N. The radiances are
            shown before and after incident angle correction.

incident angle. As an example, in Fig. 5, we show the long-term global average behavior of the nighttime 6.6 GHz vertical radiances before and after the correction. We see that the downward trend beginning in 1984 (month 61) is removed. The retrieval comparisons after making this radiance correction are shown in Figs. 6 and 7. We can see a clear improvement in the SST and wind speed retrievals after 1983. We noted above that the nighttime incident angles are systematically low by about 0.2 deg. Therefore, we added 0.2 deg to the nighttime incident angles when we corrected the wind speed retrievals. It can be seen that this reduced the difference between the day and night wind speed bias errors.

4. DISCUSSION AND CONCLUSIONS

We have seen that incident angle variations of less than 1 deg have a large effect on geophysical retrievals. Radiance corrections based on a smooth ocean surface and no atmosphere may be sufficient to correct radiances in an average sense, but individual retrievals require more careful consideration.

To account for incident angle variation, we adjusted the brightness temperatures with the incident angle sensitivities for a smooth ocean surface and no atmosphere:

$$\delta T_B / \delta \phi = T_s \, \delta \epsilon_o / \delta \phi,$$

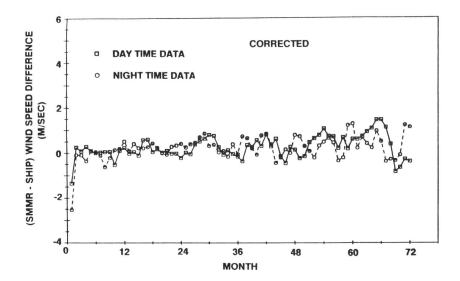

FIGURE 6. Monthly average difference between SMMR retrieved wind speeds and coincident ship measured wind speeds. The retrievals have been corrected for incident angle variation.

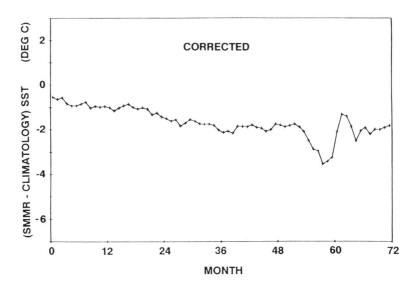

FIGURE 7. Monthly average difference between SMMR retrieved SST and
climatological SST. The incident angle correction has
been made.

where T_s is the SST fixed at 20 deg C and ϵ_o is the smooth surface
emissivity. The situation is more complicated, however, since the
ocean surface is rough and there are atmospheric effects. For
example, if the atmosphere is isothermal then the radiances are

$$T_B = T_s[1-(1-\epsilon)e^{-2\tau}] + T_{sp}(1-\epsilon)e^{-2\tau},$$

where $\tau = \tau_o/\cos\phi$, τ_o is the vertical atmospheric opacity, and T_{sp}
is the 2.7 K cosmic background temperature. Here ϵ is the rough
surface emissivity. The sensitivity with respect to the angle of
incidence is

$$\delta T_B/\delta\phi = (T_s - T_{sp})e^{-2\tau}[\delta\epsilon/\delta\phi + 2\tau(1-\epsilon)\tan\phi]. \qquad (2)$$

This depends upon the incident angle sensitivities of both the
surface emissivity and the atmospheric path length. For fixed
surface conditions (wind speed and SST), this sensitivity depends
only on τ, which is approximately linear in T_B.

The correction of the form in Eq. (1) is sufficient to correct
the global average radiances if α is adjusted for global average
atmospheric and surface conditions. For individual retrievals, the
incident angle sensitivity must reflect the specific local
atmospheric and surface conditions. Radiance modelling indicates
that the vertical channel sensitivities are much less dependent on
atmospheric opacity and surface windspeed than the horizontal

channel sensitivities. This is expected because near a 50 deg incident angle, the horizontal radiances are much more dependent on windspeed than are the vertical channels (Stogryn, 1967). As a function of atmospheric opacity, the incident angle sensitivity also varies much less for the vertical channels than for the horizontal channels. This can be seen for the isothermal atmosphere case of Eq. (2) in which the derivatives of the horizontal and vertical emissivity have opposite signs.

We can see that in order to correct the actual radiances to the radiances at the nominal (50.4 deg) incident angle, we need to compute the incident angle sensitivity. However, this sensitivity depends on the surface conditions and the atmospheric opacity, which are the quantities that we are trying to retrieve. An iterative procedure might be employed in the retrieval whereby the radiances are corrected with the retrieved wind speed and the wind speed is recomputed with the corrected radiances. In addition, one would need a surface emissivity model that adequately describes the surface roughness, foam effects, and incident angle dependence.

ACKNOWLEDGMENTS

The authors would like to thank Drs. Erik Mollo-Christensen and Per Gloersen for helpful discussions and Mr. Jim Beauchamp for computing the wind speed corrections.

REFERENCES

Chang, A. T. C., and T. T. Wilheit, 1979: Remote Sensing of Atmospheric Water Vapor, Liquid Water, and Wind Speed at the Ocean Surface by Passive Microwave Techniques from the Nimbus-5 Satellite, Radio Sci., 14, 793-802.

Gloersen, P., and F.T. Barath, 1977: A Scanning Multichannel Microwave Radiometer for Nimbus-G and Seasat-A, IEEE J. Oceanic Eng., OE-2, 172-178.

Stogryn, A., 1967: The Apparent Temperature of the Sea at Microwave Frequencies, IEEE Trans. on Antennas and Propagation, AP-15, 278-286.

DISCUSSION

Falcone: I have two questions and a comment. The first question is: do you attribute the difference between the day and night variation of radiance to the solar effect on the radiation or to the solar effect on the antenna? And question number two is: did you use Fresnel's equation for smooth surfaces?

Han: Okay. For the first question, day and night difference, there may be some effects from solar radiation but this one is independent of the solar heating cycle. Even though there is a solar heating cycle, it was designed to cancel out. Still, the equation was not perfect, so we see some residual effect there. Second question, we tried to use a different model. We used the Fresnel's equation and then we added some surface roughness and some other effects for atmospheric, water vapor and so on. We found that the vertical channel is almost insensitive.

Falcone: The comment is the same. Problems you have here will pop up in the SSMI if there is a change because it is all algorithm derived.

Han: In the SSM/I case, it doesn't have the lower frequencies 6.6 and 10.7, but it has 19, 22, and higher frequencies and if you try to compute sea-surface temperature or sea-surface wind speed from SSM/I then what you will have is this. Dynamic ranges of sea surface or wind speed are approximately from 0 to 40 meters per second. The signal sensitivity that you can have from 18 or 37 GHz is much less than that. Half of that, probably. That means, if you have an uncertainty of two degrees you will have an uncertainty of 4 meters per second in wind speed. So, yes, in a sense, if you tried to retrieve sea-surface temperature and sea-surface wind speed, then you would have the same problem.

Goroch: I notice you have a seasonal dependence, but I assume that everything has been averaged all over the globe. Did you look at that seasonal dependence?

Han: I'll show you two plots. This one shows seasonal fluctuation. If you plot different months on the time axis, then you will get the same shape. That means when we average (sometimes we average this much data, sometimes we average this much), it is shifting all the time. Because daytime was defined as being when the satellite is in the Sun and both surface and satellite see the Sun, and nighttime is defined as in the shade.

TOWARD A CORRECTION PROCEDURE FOR RADIOSONDE LONG-WAVE RADIATION ERRORS

M.J. Harris

SM Systems and Research Corporation
Landover, Maryland 20785, USA

L.M. McMillin

National Environmental Satellite, Data and Information Service
Washington, D.C. 20233, USA

ABSTRACT

Errors in radiosondes due to long-wave radiative cooling are investigated to determine the corrections to be made before using them as the ground truth for satellite retrievals. The radiative effects are known to be dependent upon the temperature profile experienced by the thermistor. A method is presented for calculating the resulting errors in a known profile. This method requires knowledge of the value of the thermistor's convective heat transfer coefficient H.

In a first approximation H was calculated from the theory of fluid flow across a cylinder. Our calculations of long-wave radiation errors show coolings up to several K (at high levels near 10 mb). Our results for the VIZ thermistor are compared with laboratory data, with measurements made in a parallel-ascent balloon experiment, and with bias measurements made using satellite retrievals as a transfer standard.

1. INTRODUCTION

It is important to correct systematic radiosonde errors as far as possible before using them as ground truth in satellite retrievals. A major systematic error in radiosonde temperature measurements arises from radiative cooling of the thermistor. In addition to convective heat transfer from the surrounding air (as required for temperature measurement), the thermistor also responds to the ambient radiation field. An error is thus introduced which (for long-wave radiation considered alone) can amount to a cooling of several K at the highest levels attained by balloons.

The long-wave flux incident upon the thermistor depends on the temperature profile. For this reason, in standard parallel-ascent radiosonde intercomparisons, since the two instruments experience the same profile the errors measured are understated due to partial cancellation. Nor can such intercomparisons be carried out in such diverse profiles as will be encountered in any radiosonde correction scheme for retrievals. From the point of view of retrievals, a better method of measuring relative biases for any profiles, using

RSRM '87: ADVANCES IN
REMOTE SENSING RETRIEVAL METHODS
A. Deepak, H.E. Fleming, and J.S. Theon (Eds.)   125

satellite retrievals as a transfer standard, has been developed by McMillin et al. (1987, hereafter MGSS).

The long-wave radiative cooling may also be calculated theoretically, and used to estimate the errors in individual radiosondes (rather than relative biases). Here we compare the theoretical results with available independent measurements of radiative errors for the U.S. VIZ thermistor, whose intrinsic properties are known and for which such measurements are available.

2.   THEORY OF RADIATIVE COOLING

The error due to radiative cooling of a radiosonde thermistor at atmospheric level i is given by conservation of energy:

$$H\ \delta T^{rad}(i) = -\varepsilon A\ \sigma T(i)^4 + \varepsilon R + \alpha S \tag{1}$$

where A is the thermistor's surface area, ε and α are its long-wave and short-wave emissivities, H is its convective heat transfer coefficient and R and S are the incident long-wave and short-wave powers. The thermistor properties H, A, ε and α are known for the VIZ instrument, but are not expected to be known in general. The theory may be tested by considering the long-wave effects alone, and we shall in general neglect the short-wave radiation.

For a cylindrical thermistor of the VIZ type we may rewrite Eq. (1) as

$$\delta T^{rad} = -(\varepsilon A/H)\ \sigma T^4 + (\varepsilon A/\pi H)\ F \tag{2}$$

where F is the ambient long-wave flux. Our method is to calculate these fluxes F at each level in a known profile using transmittances calculated for the profile by the method of McMillin and Fleming (1976) and Fleming and McMillin (1977).

The remaining unknown quantity in Eq. (2), the convective heat transfer coefficient H, has been analyzed by Williams and Acheson (1976), who showed that

$$H = (A/\pi)\ (kC/D)\ (\rho v D/\mu)^m \tag{3}$$

where D is the thermistor diameter, μ is the absolute viscosity of air, ρ is its density and k its thermal conductivity, and C and m are approximate constants which depend on the Reynolds number of the flow across the cylinder. The velocity of ascent v was assumed to be a constant 5 m s^{-1}, since much of the available laboratory data employed a ventilation rate equivalent to this value. We calculated k and μ for each level in the profile, using the approximations for these quantities recommended in the U.S. Standard Atmosphere (1962), and substituted the resulting values of H into Eq. (2).

3. COMPARISON WITH EXPERIMENTAL DATA

3.1 LABORATORY DATA

An order-of-magnitude check on the validity of this theoretical treatment can be made by using the known time constant τ of the VIZ thermistor. Williams and Acheson (1976) present laboratory measurements of τ made under standard-atmosphere conditions, and show that

$$\tau = \frac{m\ c_T}{H} \tag{4}$$

where m is the thermistor's mass and c_T its specific heat. We obtained an estimate of H from Eq. (4) by calculating c_T from the thermistor's known bulk composition. The results are compared in Table 1 with the theoretical values H_{calc} obtained from Eq. (3). It is seen that H_{calc} is about a factor 2.5 lower than the value implied by τ, implying that the theory would overestimate the cooling by this factor. However the uncertainty in c_T renders this comparison unreliable.

TABLE 1. COMPARISON OF THEORETICAL VALUE OF H WITH VALUE IMPLIED BY
          TIME CONSTANT τ

Pressure, mb	1000	100	10
H_{calc}	1.00×10^{-2}	3.35×10^{-3}	1.32×10^{-3}
H (from τ)	2.25×10^{-2}	9.22×10^{-3}	3.62×10^{-3}

Detailed laboratory tests on the VIZ thermistor were performed by Ney, Maas & Huch (1961) by heating it in an evacuated bell jar and measuring the temperature change. Rather than presenting their results in terms of H, they simulated the long-wave temperature error δT^{rad} in an atmospheric profile measured by Gergen (1957). They used a simplified version of Eq. (2) in which the emissivity and the geometrical factors were taken to be 1.

Having determined that our calculation of the long-wave flux F for Gergen's profile, as described in the previous section, agreed closely with Gergen's measured flux, we proceeded to calculate δT^{rad} from Eq. (3) and the same simplified treatment of Eq. (2). Our results are compared with those of Ney, Maas & Huch (1961) in Fig. 1. We find that our calculated values become steadily larger than the experimental values with increasing altitude, being about 50% too large at 100 mb and a factor 2 too large at 20 mb. This suggests that our theoretical values of H_{calc} are underestimated by these factors.

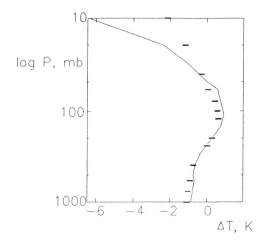

FIGURE 1.   Temperature error δT as a function of pressure,
            calculated for Gergen's (1957) profile from a simplified
            form of Eq. (2), from laboratory measurements on the VIZ
            thermistor by Ney, Maas and Huch (1961: points), and from
            radiative cooling theory as in the text (full line).

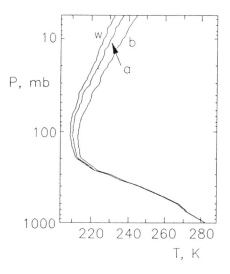

FIGURE 2.   Smoothed daytime temperature profiles measured on 12/8/86
            at Wallops by three VIZ thermistors with different
            coatings (w, white; a, aluminum; b, black) by Schmidlin
            et al. (1986).

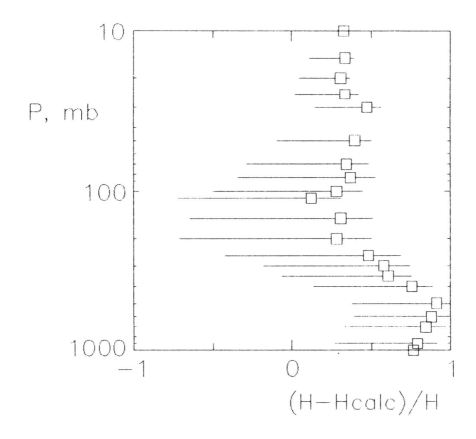

FIGURE 3.   Values of convective heat transfer coefficient H at
selected TOVS levels, from the data of Schmidlin et al.
(1986) (Fig. 2), compared with theoretical value H_{calc}.

3.2   PARALLEL-ASCENT EXPERIMENTS

    Schmidlin et al. (1986) have performed parallel-ascent
experiments specifically designed to measure radiative effects upon
the VIZ thermistor.   Three thermistors are flown together, having
coatings with different long-wave and short-wave emissivities -
white, aluminum and black.   Typical results are shown in Fig. 2.   It
is then possible to set up three equations of the form (1), or, for
coating c:

$$H (T_c - T) = -\varepsilon_c A\sigma T_c^4 + \varepsilon_c R + \alpha_c S \tag{5}$$

where T is the true temperature.   Since we are able to calculate the
long-wave power R for a given profile, as described in section 2, we
are then able to solve three simultaneous equations in the unknowns
T, H and S.

    The values of H thus derived are extremely sensitive to the
mutual differences of the thermistor temperatures T_c.   For this
reason we found it necessary to apply a smoothing to the three
temperature profiles (Fig. 2).   Even after smoothing, at pressures
greater than a few hundred mb, where these differences are very
small, the results for H are unreliable.   We compare them with our
theoretical values from Eq. (3) in Figure 3, where they suggest a
value $(H - H_{calc})/H \sim 0.35$, i.e. that H_{calc} is too small by a
factor of 1.5.

3.3   SATELLITE MEASUREMENT OF RADIOSONDE BIAS

    MGSS have measured the biases between several samples of
nighttime   radiosondes of different types in different mean
temperature profiles.   In Fig. 4 we show the mean profiles
experienced by two such samples, one of VIZ instruments and the
other of the Finnish Vaisala type.

    We calculated the radiative cooling for the mean profile for the
VIZ instruments as described in section 2.   We also calculated the
cooling for the Vaisala instruments in order to compare the theory
with these bias measurements (VIZ minus Vaisala).   We assumed that
Eqs. (2) and (3) hold for the Vaisala radiosonde, except that the
value of H was 1/0.65 times that from Eq. (3), which is the inverse
ratio (in accordance with Eq. (4)) of the time constants for VIZ and
Vaisala thermistors measured by Phillips et al. (1981).   The
emissivity of the aluminum-coated Vaisala thermistor was measured by
Schmidlin et al. (1986) to be 0.22.   It can therefore be seen from
Eq. (2) that the Vaisala instrument has a much smaller long-wave
cooling error than the VIZ ($\varepsilon = 0.86$).   The measured bias is
therefore mainly due to the VIZ error.

    We compare the theoretical and measured biases in Fig. 5, in
which the calculated radiative error for both instruments is
multiplied by 0.5.   It is seen that the calculated results are in
good agreement with the measurements at most atmospheric levels
(within about 0.5 K r.m.s.), with two exceptions.   Both at the

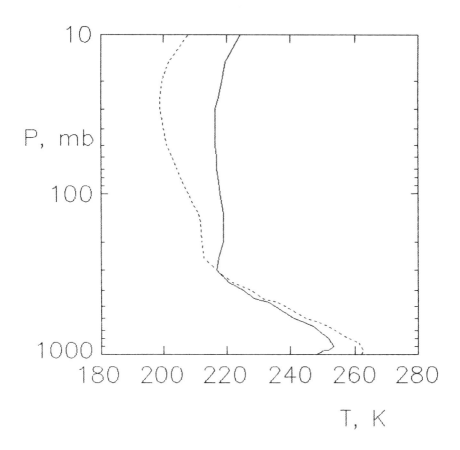

FIGURE 4.   Mean profiles experienced between 1/8/86 and 2/4/86 by
            night-time samples of Vaisala (dash line) and VIZ (full
            line) radiosondes between 60°N and 90°N, after MGSS.

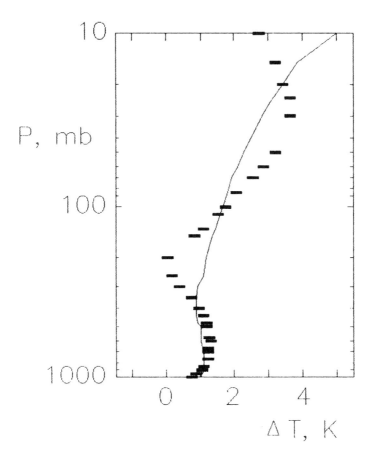

FIGURE 5.   Temperature bias between the Vaisala and VIZ samples in
             Fig. 4, as measured by MGSS (points), and as calculated
             from radiative cooling theory.   The full line represents
             the calculated bias if the value of H from Eq. (3) is
             multiplied by 0.5.

uppermost levels (pressures less than ~30 mb) and near 200 mb our calculations over-state the bias. The first discrepancy is probably an artifact of the MGSS method of measuring biases, in which the bias between two instruments is forced to zero at the top of the atmosphere. In the profile of Fig. 4 the 200 mb level corresponds to the tropopause, a region where satellite retrievals are susceptible to errors. Similar agreement was found with the other MGSS bias samples if H_{calc} is multiplied by 0.5 (McMillin & Harris 1987, corrected for their overestimate of ε for the Vaisala thermistor). The necessary factor 0.5 suggests that the theoretical value H_{calc} is too low by about a factor 2.

4.   CONCLUSIONS

We conclude that the theory of long-wave radiative cooling as described in section 2 agrees with experimental data from diverse sources if the convective heat transfer coefficient H calculated from Eq. (3) is multiplied by a factor close to 2. In particular, the results of MGSS's measurements of radiosonde bias with satellite retrievals as a transfer standard are in agreement with long-wave radiative cooling theory as thus amended. The latter therefore provide a basis for correcting radiosondes before use as ground truth for retrievals, with the theory providing absolute values of errors from these biases.

These errors tend to be somewhat larger than the values ~1 K which are frequently quoted on the basis of parallel-ascent radiosonde intercomparisons (c.f. values ~5 K at the highest levels in Fig. 5). We believe that this can be explained by the profile-dependent nature of the long-wave errors, leading to partial cancellation of errors in the bias between two instruments which share the same profile. In addition the recent WMO intercomparisons were performed in profiles at Wallops and in England for which it appears from Gergen's (1957) and MGSS's measurements that large radiative coolings would not be expected (Personal communication, Nash, J. 1987).

For the purpose of performing corrections to radiosondes it will be necessary to extend the theory by including short-wave radiation in Eq. (2), and to determine the physical properties of some widely-used radiosonde types other than VIZ. A further planned refinement is to make allowance for clouds in the calculation of the long-wave and short-wave fluxes. The MGSS method of measuring biases may be expected to provide a large amount of data against which the theory can be further tested.

ACKNOWLEDGMENTS

We are very grateful to Mr. F. Schmidlin and Mr. J. Luers for providing us with preliminary results from their parallel-ascent experiments.

REFERENCES

Fleming, H. E., & McMillin, L. M., 1977: Atmospheric Transmittance of an Absorbing Gas. 2: A Computationally Fast and Accurate Transmittance Model for Slant Paths at Different Zenith Angles, Appl. Opt., 16, 1366.

Gergen, J. L., 1957: Atmospheric Infrared Radiation Over Minneapolis to 30 Millibars, J. Meteor., 14, 495.

McMillin, L. M., & Fleming, H. E., 1976: Atmospheric Transmittance of an Absorbing Gas: A Computationally Fast and Accurate Transmittance Model for Absorbing Gases with Constant Mixing Ratios in Inhomogeneous Atmospheres, Appl. Opt., 15, 358.

McMillin, L. M., Gelman, M. E., Sanyal, A. & Sylva, M., 1987: A Method to Use Satellite Retrievals as a Transfer Standard to Determine Systematic Radiosonde Errors, Mon. Wea. Rev., in press (MGSS).

McMillin, L. M. & Harris, M. J., 1987: Physical Treatment of Profile-Dependent Radiosonde Biases: Preliminary Results, to appear in Passive Microwave Observing from Environmental Satellites, ed. J. Fischer, (Washington, D.C.: NOAA-NESDIS), p. 46.

Ney, E. P., Maas, R. W. & Huch, W. F., 1961: The Measurement of Atmospheric Temperature, J. Meteor., 18, 60.

Phillips, P. D., Richner, H., Joss, J., & Ohmura, A., 1981: ASOND-78: An Intercomparison of Vaisala, VIZ and Swiss Radiosondes, Pageoph, 119, 259.

Schmidlin, F. J., Luers, J. K. & Huffman, P. D., 1986: Preliminary Estimates of Radiosonde Thermistor Errors, NASA Tech. Paper 2637 (Springfield, Va.: Nat. Tech. Information Service).

U.S. Standard Atmosphere 1962, U.S. Committee on Extension to the Standard Atmosphere (NASA).

Williams, S. L. & Acheson, D. T. 1976: Thermal Time Constants of U.S. Radiosonde Sensors Used in GATE, NOAA Tech. Memorandum EDS CEDDA-7 (Washington, D.C.: U.S. Dept. of Commerce).

DISCUSSION

Revercomb: Is the rate of change of temperature in your equation truly negligable?

Harris: In principle, we ought to take the correction which comes from the time constants, put it in and do the whole profile again. In practice, the effect of that turns out to be very small indeed. So we neglected it.

Hurley: That concludes Session B on Calibration. Thank you for your attention.

MATHEMATICAL INVERSION ALGORITHMS
FOR OPTICALLY-THICK REMOTE SENSING APPLICATIONS

N.J. McCormick
Department of Nuclear Engineering
University of Washington
Seattle, Washington 98195, USA

ABSTRACT

Inversion techniques based on the radiative transfer equation are described for applications in which the scattering phenomena cannot be analyzed with a low-order multiple collision analysis. The applications examined include the characterization of the optical properties of the medium, specifically the angular scattering coefficients of the phase function, or the estimation of the albedo of a surface obscured by a multiple-scattering medium. Both passive and active sources are considered.

Most of the work to date has involved the analytical development of inversion algorithms and their numerical testing using direct solutions of the radiative transfer equation under conditions with and without simulated random experimental fluctuations. In addition, the time-dependent inversion algorithm has been tested with experimental measurements of the backscattered radiance from a very thick homogeneous target.

1. INTRODUCTION

If possible, retrieval of information in a remote sensing application should be done under conditions in which the medium is optically thin. In this way the effects of multiple scattering of the radiation are minimized, both for the measured radiances or irradiances and for the development of the inversion algorithm used to infer the desired information. Most algorithms, such as that for lidar, are based on the fact that the dominant portion of the measured signal is due to singly-scattered radiation, and depend on multiple scattering events amounting to perhaps at most 20% of the signal. Thus these inversion algorithms can be classed as "single-scattering algorithms".

The objective of this paper is to summarize the status of the development and testing of algorithms for possible application when the medium is optically thick and the dominant portion of the measured signal is due to multiply-scattered radiation. Since such algorithms are based on the validity of the radiative transfer equation, which means the radiation is taken to be incoherent electromagnetic energy in the form of photons, this class of algorithms can be termed inverse "radiative transfer algorithms" or "multiple-scattering algorithms."

A basic distinction in inverse radiative transfer problems is whether the medium itself is the object of interest, as in the "medium characterization problem," or whether the multiple-scattering medium is an obscurant in front of a surface or another multiple-scattering medium of interest, as in the "hidden object problem."

Another distinction is whether the radiation re-emitted after scattering is incorporated as part of the source term or is treated explicitly. An "inverse source algorithm" for the measurement of temperature profiles in atmospheres (c.f., Chahine, 1982; King, 1985) is an example in which the scattered radiation is treated as part of the source term, while

RSRM '87: ADVANCES IN
REMOTE SENSING RETRIEVAL METHODS
A. Deepak, H.E. Fleming, and J.S. Theon (Eds.)

medium characterization algorithms for estimating the angular dependence of the scattered radiation must explicitly treat the scattering term (McCormick, 1979, 1986).

Still another distinction in inverse problems is whether the sensing of active or passive radiation is done. Inverse source problems are passive sensing applications, by their very nature, while inverse medium characterization problems can involve either sensing of radiation due to a passive external source, such as the sun's illumination of a cloud or the ocean surface, or radiation due to an active source, such as a laser illumination. Laser external sources are either time-independent (i.e., continuous-working or "cw" lasers) or time-dependent (i.e., pulsed lasers).

In this review only the inverse radiative transfer algorithms for medium characterization or hidden object detection will be considered. First the time-independent algorithms will be summarized (Sec. 2) and then the time-dependent applications will be covered; the latter involve both the medium characterization problem (Sec. 3) and the hidden object problem (Sec. 4). Previous reviews (McCormick, 1981, 1984, 1986) that amplify upon some of the earlier work presented here also may be consulted, as well as the bibliography containing additional references not cited.

2. TIME-INDEPENDENT ALGORITHMS FOR CHARACTERIZING A MEDIUM

The objective of inverse transport algorithms for estimating scattering parameters is to infer the coefficients f_n in the Legendre polynomial expansion of the single-scattering normalized angular cross section,

$$\sigma_s(\underline{\Omega}\prime \bullet \underline{\Omega})/\sigma = (4\pi)^{-1} \sum_{n=0}^{N} (2n+1) f_n P_n (\underline{\Omega}\prime \bullet \underline{\Omega}). \tag{1}$$

Here σ is the extinction or total cross section, as measured in cm^{-1}, and $\underline{\Omega}\prime \bullet \underline{\Omega}$ is the cosine of the angle between the directions of travel before and after a scattering event, and N is the degree of scattering anisotropy. The single-scattering albedo is f_0 and f_n/f_0 are the coefficients of the angular phase function normalized such that

$$(\sigma f_0)^{-1} \int_{\underline{\Omega}} \sigma_s(\underline{\Omega}\prime \bullet \underline{\Omega}) \, d\underline{\Omega} = 1;$$

for example, f_1/f_0 is the mean cosine of the scattering angle.

Early work on this problem (Kagiwada and Kalaba, 1967; Kagiwada et al., 1975) was based on iterative schemes that were generally shown to perform poorly (Fymat and Lenoble, 1979). This early work, as well as the new approach for developing inverse time-independent algorithms, are based on the transfer equation for plane geometry (Chandrasekhar, 1950),

$$(\mu\partial_x + 1) I(x,\mu,\phi) = \sigma^{-1} \int_0^{2\pi} d\phi\prime \int_{-1}^{1} d\mu\prime \sigma_s(\underline{\Omega}\prime \bullet \underline{\Omega}) I(x,\mu\prime,\phi\prime), \tag{2}$$

for $x_- \leq x \leq x_+$, where the radiance $I(x,\mu,\phi)$ is a function of the dimensionless spatial coordinate x the cosine of the polar angle μ is taken with respect to the x-axis, and the azimuthal angle is ϕ. The thickness of the target, $(x_+ - x_-)$, can be optically very large and need not be known.

The inverse algorithms to estimate the coefficients f_n rely on the use of the information available from the azimuthal dependence. This means the incident illumination on the target must not be azimuthally symmetric, as would be the case, for example, with a monodirectional beam normally incident on the surface.

Normally the physical problem can be made symmetric with respect to the reference angle ϕ_r, such as with the boundary conditions

$$I(x_-, \mu, \phi) = \delta(\mu - \mu_0)\,\delta(\phi - \phi_r), \quad 0 \leq \mu \leq 1,$$
$$I(x_+, \mu, \phi) = 0, \quad -1 \leq \mu \leq 0;$$

then we can take

$$I(x, \mu, \phi) = \sum_{m=0}^{N} \cos m(\phi - \phi_r)\, I^m(x, \mu) + I_{>N}(x, \mu, \phi), \tag{3}$$

where the $I_{>N}$−term accounts for any portion of the radiance that is not included in the expansion. The Fourier components of the radiance, $I^m(x, \mu)$, are defined by

$$I^m(x, \mu) = [\pi(1 + \delta_{m0})]^{-1} \int_0^{2\pi} \cos m(\phi - \phi_r) I(x, \mu, \phi)\, d\phi. \tag{4}$$

(For an experiment with general boundary conditions such that there is no reference angle for symmetry, see McCormick, 1986.)

Equations (2) and (3) and the spherical harmonics addition theorem permit Eq. (2) to be written as a set of $2(N+1)$ uncoupled transfer equations in terms of the I^m (Chandrasekhar, 1950). It is these equations that are used to obtain the inverse equations for $m = 0$ to N (McCormick, 1979, 1986)

$$\sum_{n=m}^{N} A_{mn}^i f_n^i = S_m^i, \quad 0 \leq m \leq N, \tag{5}$$

for $i = 0, 1$, where the unknowns in the sets of equations are

$$f_n^i = f_n, \quad i = 0,$$
$$= f_n/(1 - f_n), \quad i = 1. \tag{6}$$

The coefficients that must be evaluated involve differences of the intensities on the surfaces at $x = x_-$ and $x = x_+$,

$$S_m^i = \int_{-1}^{1} \mu^{2i} I^m(x, \mu) I^m(x, -\mu) d\mu \ \big|_{x_-}^{x_+} \tag{7a}$$

$$A_{mn}^i = (-1)^{n-m} \beta_n^m\, I_n^{m,i}(x)\, I_n^{m,i}(x) \ \big|_{x_-}^{x_+} \tag{7b}$$

where

$$I_n^{m,i}(x) = \int_{-1}^{1} \mu^i\, I^m(x, \mu)\, P_n^m(\mu)\, d\mu, \tag{7c}$$

$$\beta_n^m = (2n+1)(n-m)!/2(n+m)!. \tag{7d}$$

There are two combinations of i, for $i = 0, 1$, that give two sets of inverse equations for $m = 0$ to N that are sufficient to estimate the f_n-coefficients; thus there are always at least twice the number of equations as unknowns. It is best to use the $i = 0$ equations as one inverse transport algorithm and the $i = 1$ equations as a second algorithm.

The inverse transport algorithms have been studied numerically and when solved without using a constrained linear inversion procedure, for example, have been shown to be numerically unstable to small simulated numerical errors (Oelund and McCormick, 1985), especially for strongly anisotropic scattering.

3. TIME-DEPENDENT ALGORITHM FOR CHARACTERIZING A MEDIUM

The objective of this algorithm also is to estimate the albedo of single scattering f_0 and as many of the coefficients f_n, $n \geq 1$, as possible. The inverse radiative transfer algorithm is based on the time-dependent generalization of Eq. (2) and uses Fourier azimuthal moments $B^m(\mu, t)$ of the radiance backscattered from the surface of a plane geometry semi-infinite medium,

$$B^m(\mu, t) = [\pi(1 + \delta_{m0})]^{-1} \int_0^{2\pi} d\phi \, I(0, -\mu, \phi, t) \cos m(\phi - \phi_r), \quad 0 \leq \mu \leq 1, \quad (8)$$

Long after a very short incident pulse centered about time $t = 0$, these moments asymptotically decay according to (McCormick, 1982)

$$B^m(\mu, t) \cong C^m(\mu) \, t^{-\frac{3}{2}} \, \exp[-v(1 - f_m)t], \quad (9)$$

where v is the inverse of the mean time between photon collisions with the scattering centers and $C^m(\mu)$ is a constant when the polar angle is fixed. Analytical corrections to the algorithm for a thick but finite medium and for a broadened pulse (Duracz and McCormick, 1986) are of the order of the approximation made in deriving Eq. (9).

The advantage of this time-dependent algorithm compared to the time-independent ones in Eq. (5) is that a single azimuthal moment will yield an estimate of a single f_m-coefficient; thus errors from higher-order Fourier moments do not influence the estimation of the coefficients for smaller values of m. Also, measurements need be made only in directions with a fixed value of μ, rather than for a set of values. Disadvantages of Eq. (9) compared to Eq. (5), however, are that the algorithm itself is an approximation, that it requires measurements asymptotically long after the incident pulse when the signal strength is lower, and the pulse width cannot be so broad as to interfere with the dominant portion of the backscattered radiance.

While early numerical tests proved encouraging (Hunt and McCormick, 1985), a recent theoretical and numerical study (Duracz and McCormick, 1987) has indicated that the algorithm is often sensitive to small experimental errors or imperfections. For this reason only the coefficients f_0, the albedo of single scattering, and $f_1/f_0 = g$, the mean cosine of the scattering angle (i.e., the scattering asymmetry factor), can be expected to be estimated with some precision, and for this reason it was essential that the algorithm be tested with experimental data.

Scattering experiments to test the algorithm have been performed on systems of both nonabsorbing latex spheres and very weakly absorbing polymer dynospheres (Elliott et al., 1988). These tests showed that it is possible even in the presence of experimental noise to estimate the single scatter albedo with a relative error of less than 1%. This is primarily due to the fact that the asymptotic regime is many collision times long, which contributes to the stability of the estimated decay rate and the corresponding value of the single scatter albedo. It must be emphasized, however, that in all cases considered the absorption was very weak or nonexistent; work is underway to test how well the algorithm performs for a more strongly absorbing medium.

It was also shown that the coefficient f_1 is much more difficult to estimate from the backscattered pulse because the asymptotic regime is only a few collision times long. For very small latex spheres, errors of 60% were obtained using even a numerically simulated backscattered radiance, while no estimate could be obtained from the experimental data.

For larger latex spheres that scatter much more anisotropically, and for which f_1 is much larger, an error of less than 1% was achieved in some cases.

Because the estimates of f_1 were much more sensitive to numerical and experimental errors than those for f_0, a fact that is compounded when the coefficients are small, there appears to be little hope that good estimates of higher order f_n-coefficients could be obtained.

4. TIME-DEPENDENT ANALYSIS FOR DETECTING AN OBJECT

The objective is to identify optical depths and/or surface albedos for which an external detector measuring the emerging irradiance could not detect the presence of an object obscured by the atmosphere. The ability to detect the presence of the obscured object depends upon the direction and time characteristics of the incident radiation pulse, the single scattering albedo and phase function of the atmosphere, the albedo of the object and its distance from the surface of the atmosphere, the characteristics of the detector located external to the atmosphere, and the magnitude of the difference in signals when the object is present or absent.

The TIMEX program (Hill and Reed, 1985), which is based on the time-dependent generalization of the radiative transfer equation, was used by Duracz and McCormick (1988a) to analyze the propagation of a pulse of radiation striking uniformly over the surface at $x = 0$ a homogeneous atmosphere of thickness X. For simplicity the incident illumination was assumed to be normally directed at the atmosphere (i.e., azimuthally symmetric) and of infinitesimal duration, while the atmosphere was described by the Rayleigh phase function and a variable single scattering albedo; the reflection of radiation at the object's surface at $x = X$ was assumed to be isotropic (i.e., Lambertian reflection) such that

$$I(X, -\mu, t) = 2A \int_0^1 \mu\, I(X, \mu, t)\, d\mu, \tag{10}$$

where A is the albedo of the object. For the detector, it was assumed that the backscattered irradiance,

$$E(t) = \int_0^1 \mu\, I(0, -\mu, t)\, d\mu, \tag{11}$$

could be measured for a time interval after the pulse varying from $2 \le t \le 20$ mean collision times, i.e.,

$$Q = \int_{2/v}^{20/v} E(t)\, dt.$$

The object detectability is a function of the magnitude of Q with and without the object present. In principle this quantity could be used to infer either the surface albedo of an object when its depth of location is known, or the location when its albedo is known.

Fig. 1 (Duracz and McCormick, 1988a) is a map of the time-integrated backscattered irradiance Q as a function of the object's surface albedo A and its location X. For any value of Q there is an isocline of possible values of $[X, A]$. The upper left-hand portion of the figure shows a region for objects with albedos $A > 0$; the lower left-hand portion shows a region for objects with $A \ge 0$. These two regions indicate the presence of an object. When there is no object, the value of $Q = 0.423$ and the line which corresponds to that value lies between the two regions and goes asymptotically to $X = \infty$. Thus the map provides a global picture of the detectability of an object behind a uniform obscuring atmosphere.

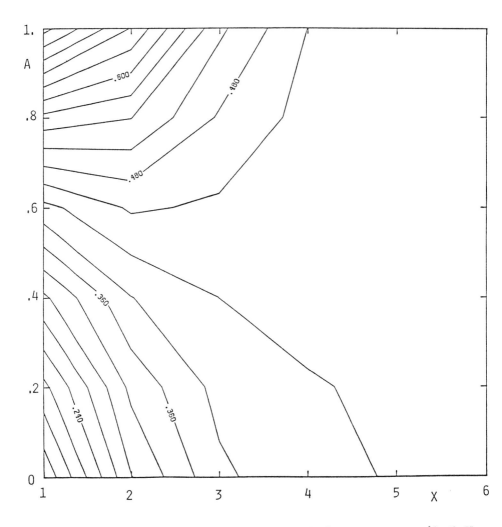

Figure 1. Isoclines of Q, the time-integrated irradiance, in the parameter space [depth X, surface albedo A] for an obscured object (from Duracz and McCormick, 1988a)

The results shown here represent a near-worst-case for attempting to detect an object through a thick obscuring atmosphere in the sense that if the atmospheric scattering is more anisotropic than Rayleigh, or if the surface scattering is not isotropic, detection would be somewhat easier. The effects of different atmospheric scattering functions, different surface albedos, and inhomogeneous atmospheres have recently been investigated (Duracz and McCormick, 1988b).

ACKNOWLEDGMENT

This work was supported by the U.S. Army Research Office.

REFERENCES

Chahine, M.T., 1982: Remote Sensing of Cloud Parameters, J. Atmos. Sci. **39**, 159-170.

Chandrasekhar, S., 1950: Radiative Transfer, Oxford Univ. Press.

Duracz, T. and N.J. McCormick, 1986: Analytical Error Estimates for the Time-Dependent Radiative-Transfer Inverse Problem, J. Opt. Soc. Am. A **3**, 1871-1875.

Duracz, T. and N.J. McCormick, 1987: Numerical Study of the Time-Dependent Radiative Transfer Inverse Problem, J. Opt. Soc. Am. A **4**, 1849-1854.

Duracz, T. and N.J. McCormick, 1988a: Radiative Transfer Calculations for Detecting a Target Behind Obscuring Atmospheres, Proceedings of the 1987 Chemical Research, Development and Engineering Center Conference on Obscuration and Aerosol Research (Aberdeen, MD June 22-26, 1987).

Duracz, T. and N.J. McCormick, 1988b: Radiative Transfer Calculations for Characterizing Obscured Surfaces Using Time-Dependent Back-Scattered Pulses, Appl. Opt. (submitted).

Elliott, R.A., T. Duracz, N.J. McCormick, and D.R. Emmons, 1988: Experimental Test of a Time-Dependent Inverse Radiative Transfer Algorithm for Estimating Scattering Parameters, J. Opt. Soc. Am. A **5**, 366-373.

Fymat, A.L. and J. Lenoble, 1979: Inverse Multiple Scattering Problems–III. Inadequacy of Certain Limb Darkening and Phase Curves for Retrieving Atmospheric Information and Limitations of Approximate Scattering Models, J. Quant. Rad. Transfer **21**, 75-83.

Hill, T.R. and W.H. Reed, 1985: TIMEX: One Dimensional, Time-Dependent Multigroup Explicit Discrete Ordinates Radiation Transport Code System with Anisotropic Scattering, Radiation Shielding Information Center Computer Code CCC-274, Oak Ridge National Laboratory.

Hunt, K.K. and N.J. McCormick, 1985: Numerical Test of an Inverse Method for Estimating Single-Scattering Parameters from Multiple-Scattering Experiments, J. Opt. Soc. Am. A **2**, 1965-1971.

Kagiwada, H.H. and R.E. Kalaba, 1967: Extimation of Local Anisotropic Scattering Properties Using Measurements of Multiply Scattered Radiation, J. Quant. Rad. Transfer **7**, 295-303.

Kagiwada, H.H., R. Kalaba, and S. Ueno, 1975: Multiple Scattering Processes; Inverse and Direct, Addison-Wesley (Reading, MA).

King, J.I.F., 1985: Theory and Application of Differential Inversion to Remote Temperature Sensing. In A. Deepak, H.E. Fleming, and M.T. Chahine (Eds.), Advances in Remote Sensing Retrieval Methods, Deepak Publishing, Hampton, Virginia, 437-446.

McCormick, N.J., 1979: Transport Scattering Coefficients from Reflection and Transmission Measurements, J. Math. Phys. **20**, 1504-1507.

McCormick, N.J., 1981: A Critique of Inverse Solutions to Slab Geometry Transport Problems, Prog. Nucl. Energy **8**, 235-245.

McCormick, N.J., 1982: Remote Characterization of a Thick Slab Target with a Pulsed Laser, J. Opt. Soc. Am. **72**, 756-759.

McCormick, N.J., 1984: Recent Developments in Inverse Scattering Transport Methods, Transport Theory and Statist. Phys. **13**, 15-28.

McCormick, N.J., 1986: Methods for Solving Problems for Radiation Transport – An Update, Transport Theory and Statist. Phys. **15**, 759-772.

Oelund, J.C. and N.J. McCormick, 1985: Sensitivity of Multiple-Scattering Inverse Transport Methods to Measurement Errors, J. Opt. Soc. Am. A **2**, 1972-1978.

BIBLIOGRAPHY

Time-Independent Algorithms for Characterizing a Medium

Larsen, E.W., 1981: Solution of the Inverse Problem in Multigroup Transport Theory, J. Math. Phys. 22, 158-160.

Larsen, E.W., 1975: The Inverse Source Problem in Radiative Transfer, J. Quant. Rad. Transfer 15, 1-5.

Larsen, E.W., 1984: Solution of Multidimensional Inverse Transport Problems, J. Math. Phys. 25, 131-135.

McCormick, N.J., 1987: Inverse Radiative Transfer with a Delta-Eddington Phase Function, Astrophys. Space Sci. 129, 331-334.

McCormick, N.J. and R. Sanchez, 1981: Inverse problem transport calculations for anisotropic-scattering coefficients, J. Math. Phys. 22, 199-208.

McCormick, N.J. and R. Sanchez, 1983: Solutions to an Inverse Problem in Radiative Transfer with Polarization II, J. Quant. Spectrosc. Rad. Transfer 30, 527-535.

Sanchez, R. and N.J. McCormick, 1981: General Solutions to Inverse Transport Problems, J. Math. Phys. 22, 847-855.

Sanchez, R. and N.J. McCormick, 1982: Numerical Evaluation of Optical Single-Scattering Properties Using Multiple-Scattering Transport Methods, J. Quant. Spectrosc. Rad. Transfer 28, 169-184.

Sanchez, R. and N.J. McCormick, 1983: Solutions to Inverse Problems for the Boltzmann-Fokker-Planck Equation, Transport Theory and Statist. Phys. 12, 129-155.

Siewert, C.E., 1978: On a Possible Experiment to Evaluate the Validity of the One-Speed or Constant Cross Section Model of the Neutron-Transport Equation, J. Math. Phys. 19, 1587-1588.

Siewert, C.E., 1979: On the Inverse Problem for a Three-Term Phase Function, J. Quant. Spectrosc. Rad. Transfer 22, 441-446 (1979).

Siewert, C.E., 1983: Solutions to an Inverse Problem in Radiative Transfer with Polarization–I, J. Quant. Spectrosc. Rad. Transfer 30, 523-526.

Siewert, C.E. and W.L. Dunn, 1982: On Inverse Problems for Plane-Parallel Media with Nonuniform Surface Illumination, J. Math. Phys. 23, 1376-1378.

Siewert, C.E. and J.R. Maiorino, 1980: The Inverse Problem for a Finite Rayleigh-Scattering Atmosphere, J. Appl. Math. (ZAMP) 31, 767-770.

ON THE DIFFERENTIAL INVERSION METHOD
FOR TEMPERATURE RETRIEVALS

Kuo-Nan Liou and S.C. Ou
Department of Meteorology, University of Utah
Salt Lake City, Utah 84112, USA

Jean I.F. King
Air Force Geophysics Laboratory
Hanscom Air Force Base
Bedford, Massachusetts 01731, USA

ABSTRACT

The differential inversion method (DIM) is reviewed in the context of the fundamental physics and mathematics governing the
transfer of radiation for plane-parallel atmospheres in local thermodynamic equilibrium. In the Laplace inverse plane the Planck intensity is linearly related to upwelling radiance weighted by the
weighting function. By applying the inverse transform, the local
Planck intensity can be exactly expressed by a linear combination of
the derivatives of upwelling radiances in the logarithmic pressure
coordinate. Using seven HIRS channels, we perform numerical analyses
of the DIM for temperature retrievals. Results based on distinct
U.S. standard and tropical profiles show that the DIM converges to
the true temperature solution with an accuracy of 1-2 K for tropospheric temperatures using a fifth-order polynomial function to fit
seven HIRS radiances. The DIM is free from the need for a priori
data basing and requires no constraints in the retrieval. Finally,
it is pointed out that the key to the success of the DIM for
practical applications appears to depend on whether an appropriate
curve-fitting program can be developed for observed radiances.

1. INTRODUCTION

The sounding of temperature and moisture fields has been
routinely performed from orbiting meteorological satellites in the
last 15 years, with some success. However, the techniques and procedures used for inversion of the temperature profile, as well as atmospheric parameters from observed radiances, are mostly statistical in
nature and are not based on the fundamental physics governing the
transfer of radiation.

The upwelling radiance at the top of the atmosphere can be
expressed by a Fredholm equation of the first kind involving the
Planck intensity and weighting function. This equation is known to
be mathematically ill-conditioned. If a suitable transform can be
made, the integral may be removed and the ill-conditioned nature of
the inverse problem may be resolved: that in the inverse space the
inverse problem is linear and requires no specific constraint for
retrieval. This is the basic concept that constitutes the so-called
differential inversion method proposed by King (1985).

RSRM '87: ADVANCES IN
REMOTE SENSING RETRIEVAL METHODS
A. Deepak, H.E. Fleming, and J.S. Theon (Eds.)

In this paper, we explore the generalization and practicality of the DIM for temperature retrievals. In section 2, we review the fundamentals of the DIM. Applications of this method for practical temperature retrievals are then made using HIRS channels, and are given in section 3. Finally, a summary is presented in section 4.

2. THE PHYSICAL FUNDAMENTALS OF THE DIFFERENTIAL INVERSION METHOD

The upwelling radiance at the top of the atmosphere R_ν for a given channel may be derived from the basic radiative transfer equation for plane-parallel atmospheres in local thermodynamic equilibrium. In the pressure coordinate, we have

$$R_\nu = B_\nu(T_s)\, T_\nu(p_s) + \int_{p_s}^0 B_\nu(p)\, \frac{\partial T_\nu(p)}{\partial p}\, dp \quad , \tag{1}$$

where B_ν is the Planck intensity, T_ν the transmittance, p_s the surface pressure, and T_s the surface temperature. The first and second terms represent, respectively, surface and atmospheric contributions.

We shall consider the pressure integration from p_s to $p \to \infty$ and write

$$\Delta R_\nu = \int_\infty^{p_s} B_\nu(p)\, \frac{\partial T_\nu(p)}{\partial p}\, dp \quad . \tag{2}$$

The layer below the surface may be viewed as an infinite isothermal emitter with a temperature T_s. Since $T_\nu(\infty) = 0$, ΔR_ν is exactly the same as the surface term $B_\nu(T_s)\, T_\nu(p_s)$ in the solution of the radiative transfer equation. In the band center $\Delta R_\nu \to 0$, whereas in the wing of a band, ΔR_ν could be an important source of upwelling radiance.

The weighting function, signifying the weight of the Planck intensity contribution to upwelling radiance, may be defined in the logarithm of pressure in the form

$$W_\nu(p) = -\frac{\partial T_\nu(p)}{\partial \ell np} \quad . \tag{3}$$

For each sounding channel ν, there must be a peak weighting function located at $p = \bar{p}$, which gives the maximum contribution of Planck intensity to the upwelling radiance. Since the spectral transmittance is related to an integration of the exponential function involving the optical depth, we may write $W_\nu(p) \equiv W(p/\bar{p})$. Over a small spectral interval, the Planck intensity does not vary significantly with the wavenumber ν. For simplicity of analysis, we may omit the wavenumber index in the Planck intensity. On the basis of the preceding discussion, the upwelling radiance may be rewritten in a simple mathematical form

$$R(\bar{p}) = \int_0^\infty B(p) \; W(p/\bar{p}) \; dp/p \quad , \tag{4}$$

where $\bar{p}$ replaces the wavenumber index ν. Equation (4) is a well-known Fredholm equation of the first kind. The universal inversion problem is to derive a profile of $B(p)$, given $R(\bar{p})$ and $W(p/\bar{p})$ for finite $\bar{p}$ values. In practice, since only finite $\bar{p}$ values may be chosen, the solution of $B(p)$ from the forward radiative transfer equation is mathematically ill-conditioned. Constraints of one kind or another are required to obtain physically meaningful temperature profiles. However, if the integration can be removed by a proper inverse transformation, such constraints in the inversion problem are no longer required.

We shall approach the inverse problem by a transformation of variables. In view of the term $dp/p = d\ln p$ in Eq. (4), we may introduce the following variables: $\bar{\pi} = -\ln\bar{p}$, $\pi = -\ln p$, and $v = \pi - \bar{\pi} = -\ln(p/\bar{p})$. In the new coordinate $-\ln p$, Eq. (4) may be rewritten in the form

$$R(\bar{\pi}) = \int_{-\infty}^\infty B(\pi) \; W(-v) \; d\pi \quad . \tag{5}$$

This form allows us to perform the bilateral Laplace transform. By virtue of the convolution theory, we find (Widder, 1971)

$$r(s) = b(s) \; w(-s) \quad , \tag{6}$$

where s denotes the transform variable, and

$$r(s) = \int_{-\infty}^\infty e^{-s\bar{\pi}} \; R(\bar{\pi}) \; d\bar{\pi} \quad , \tag{7}$$

$$b(s) = \int_{-\infty}^\infty e^{-s\pi} \; B(\pi) \; d\pi = \int_{-\infty}^\infty e^{-s\bar{\pi}} \; B(\bar{\pi}) \; d\bar{\pi} \quad , \tag{8}$$

$$w(-s) = \int_{\infty}^{-\infty} e^{+sv} \; W(-v) \; dv \quad . \tag{9}$$

In Eq. (8), we may replace π by $\bar{\pi}$. This allows us to evaluate the Planck intensity at discrete local levels when an inverse transform is made. The term $1/w(-s)$ may be expanded into a Maclaurin series (Pearson, 1974) so that

$$\frac{1}{w(-s)} = \sum_{k=0}^\infty \lambda_k \; s^k \quad , \tag{10}$$

where the coefficient is related to the kth derivative of the function $1/w(-s)$ at $s = 0$ in the form

$$\lambda_k = \left[\frac{1}{w(0)}\right]^{(k)} / k! \quad . \tag{11}$$

It follows that the inverse Planck intensity defined in Eq. (6) may be expressed by an infinite power series in the form

$$b(s) = \sum_{k=0}^{\infty} \lambda_k s^k r(s) \quad . \tag{12}$$

In the inverse space, the Planck intensity may be related to the upwelling radiance by the function defined in Eq. (12): that an upwelling radiance defines a local Planck intensity. To get the Planck intensity, we shall perform the inverse Laplace transform using Eq. (8) so that

$$B(\overline{\pi}) = L^{-1}[b(s)] = \sum_{k=0}^{\infty} \lambda_k L^{-1}[s^k r(s)] \quad . \tag{13}$$

In the last expression, the inverse Laplace transform can be performed analytically. When $\pi \to -\infty$, i.e., $\overline{p} \to \infty$, the upwelling radiance does not exist in physical space. This allows us to set $R(\pi \to -\infty) = 0$. Thus, the inverse Laplace transform leads to

$$B(\overline{\pi}) = \sum_{k=0}^{\infty} \lambda_k R^{(k)}(\overline{\pi}) \quad . \tag{14}$$

The Planck intensity at a given level is now expressible in terms of the linear sum of radiance derivatives at that level. In the context of infinite summations, the solution is exact and no assumption is made. This solution requires no constraint and is self-limited to the highest order of recoverable derivatives.

To obtain the inverse coefficient λ_k, we rewrite Eq. (9) in the form

$$w(-s) = \int_0^{\infty} (p/\overline{p})^{-s} W(p/\overline{p}) \, dp/p \quad . \tag{15}$$

In principle, if the weighting functions $W(p/\overline{p})$ are known, $1/w(-s)$ and its derivatives may be evaluated numerically. Subsequently, λ_k can be computed from Eq. (11). However, for the purpose of analysis, it is desirable to develop an analytic form that can approximate the shape of the weighting functions associated with sounding channels. A generalized weighting function was proposed by King (1985), viz.,

$$W_m = m^{m-1} \Gamma^{-1}(m) (p/\overline{p}) \exp\left[-m(p/\overline{p})^{1/m}\right] \quad , \tag{16}$$

where Γ is the Gamma function and m is an index controlling the sharpness of the weighting function. When m = 1 and 0.5, the weighting functions follow, respectively, the Goody-statistical and Elsaessor-regular band models. The weighting function so defined is normalized to unity. Figure 1 depicts the generalized weighting function for m = 0.5, 1, and 2. As m increases, it becomes broader.

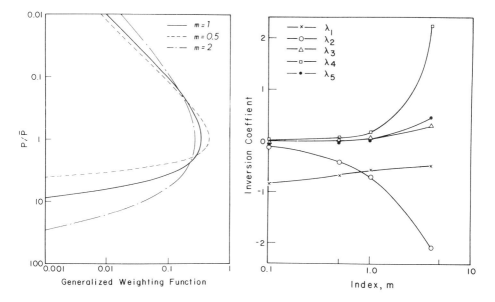

FIGURE 1. The generalized weight-   FIGURE 2. Inversion coeffi-
           ing function as a func-   cients λ_i (i=1-5) as
           tion of $p/\bar{p}$ for the   a function of the
           parameter m of 0.5, 1,   index m computed from
           and 2.   Eqs. (11) and (17).

Using this weighting function, we can show from Eq. (15) that

$$w(-s) = \Gamma[m(1-s)] / \left[\Gamma(m) \, m^{-ms}\right] \ . \tag{17}$$

An evaluation of the inversion coefficient λ_k requires the derivations of Γ-functions, which are related to digamma and polygamma
functions. Tables of these functions are given in Abramowitz and
Stegun (1968). Figure 2 shows the inversion coefficients λ_k (k = 1-
5) as functions of the index m. For any given m, λ_k may be obtained
from this figure.

     The foregoing discussions constitute our interpretation of the
differential inversion method for temperature retrievals originally
proposed by King (1985). In the next section, we shall perform
synthetic analyses utilizing HIRS channels in the 15 μm CO_2 band in
order to physically understand the generalization and determine the
practicality of this method for temperature retrievals.

3.   APPLICATIONS OF DIM FOR TEMPERATURE RETRIEVALS USING HIRS CHANNELS

     In order to investigate the potential applicability of the DIM,
we have used HIRS channels to perform the temperature retrieval. The

TABLE 1. HIRS CHANNEL CHARACTERISTICS

Channel	ν (cm^{-1})	ν_1	ν_2	$\Delta\nu$	Principal Absorbers	Level of W_{max}
1	668	666	670	4	CO_2	30
2	679	674	684	10	CO_2	60
3	690	685	697	12	CO_2	100
4	702	696	712	16	CO_2	250
5	716	708	724	16	CO_2	500
6	732	724	740	16	CO_2/H_2O	750
7	748	740	756	16	CO_2/H_2O	900

characteristics of these channels are listed in Table 1. To compute the transmittances, we have used the absorption coefficient sets derived by Chou and Kouvaris (1986) from line-by-line data (Rothman et al., 1983) based on the k-distribution method (Arking and Grossman, 1971). Over a small spectral interval, the order of absorption coefficients k has no bearing on the transmittance calculation. For a homogeneous path, we may express the spectral transmittance in the k-domain as follows:

$$T_{\bar{\nu}}(u) = \int_{\Delta\nu} e^{-k_\nu u} \, d\nu = \int_{k_{min}}^{k_{max}} e^{-ku} \, f(k) \, dk \quad , \tag{18}$$

where $f(k) = (dk/d\nu)^{-1}$ is the probability density function, which can be obtained by ranking the absorption coefficients in the spectral interval $\Delta\nu$.

The absorption coefficients are available for each 0.002 cm^{-1} covering 540-800 cm^{-1} in the 15 μm CO_2 band. These values are calculated at 19 pressure levels. Temperature adjustments were also included using the following empirical equation:

$$\ln k_\nu = a + b \, \Delta T + c \, \Delta T^2 \quad , \tag{19}$$

where $\Delta T = T - 200$, and the empirical coefficients a, b, and c are pressure and wavenumber dependent.

Using the preceding absorption coefficients, the spectral transmittance for a given level p_i for each HIRS channel may be computed in inhomogeneous atmospheres in the form

$$T_{\bar{\nu}}(p_i) = \sum_{n=1}^{N} \exp\left[-\sum_{j=1}^{i} k_{nj} \, q_j \, \Delta p_j/g\right] \Delta\nu_n \quad , \tag{20}$$

where q denotes the mixing ratio, g the gravitational acceleration,

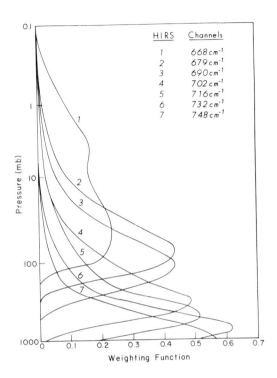

FIGURE 3.   The weighting function of seven HIRS channels in the loga-
            rithmic pressure coordinate. The peaks of these weighting
            functions are listed in Table 1.

$\Delta\nu_i$ = 0.002, and N = $\Delta\nu/\Delta\nu_i$. Figure 3 shows the computed weighting
functions for seven HIRS channels. The peaks of these weighting
functions are exactly the same as those listed in Table 1.

It is possible to compute λ_k directly from Eqs. (11) and (15)
once numerical values for the weighting function are known. However,
the computations could become very tedious. With the use of the
generalized weighting function defined in Eq. (17), λ_k can be evaluated by known mathematical functions. We have developed a numerical
method to fit HIRS weighting functions to the generalized weighting
function to obtain the index m. We define the sum of the weighted
square error at a given level i in the form

$$E = \sum_{i=1}^{M} \left[W_m(p_i/\bar{p}) - W(p_i/\bar{p}) \right]^2 \varepsilon_i \quad , \tag{21}$$

where M denotes the total atmospheric levels used in transmittance
calculations, $\bar{p}$ the pressure level at which the weighting function W
has a maximum value, and ε_i is a function defined by

$$\varepsilon_i = \begin{cases} \exp\left(-p_i/\bar{p}\right) & , \ p_i/\bar{p} > 1 \\ \\ \exp\left(-\bar{p}/p_i\right) & , \ p_i/\bar{p} < 1 \end{cases} \tag{22}$$

With the factor ε_i, E is particularly dominated by error near the maximum weighting function. We then proceed to minimize E, viz.,

$$\frac{\partial E}{\partial m} = \sum_{i=1}^{N} 2 \left[W_m(p_i/\bar{p}) - W(p_i/\bar{p}) \right] \varepsilon_i \left\{ (1-1/m) \right.$$

$$\left. - (p_i/\bar{p})^{1/m} \left[1 - \ell n(p_i/\bar{p})/m \right] + \ell nm - \psi(m) \right\} = 0 \ , \tag{23}$$

where $\psi(m)$ is the digamma function. The bisection method is subsequently used to solve for m (Hamming, 1973). Table 2 lists the values of the parameter m for each HIRS channel, RMS error, and error near the peak of the weighting function. Also listed are values for the peak weighting function. The RMS errors are sufficiently small to have a significant effect on the upwelling radiance calculation. For channel 1, a large m is found since it has a broad weighting function. For other channels, m ranges between 0.2 and 0.7. It is apparent that neither the random model nor the regular model can fit HIRS transmittances well. Once the parameter m is known, the inversion coefficient λ_k may then be evaluated for each channel (see Fig. 2).

It is necessary to develop a curve-fitting method to fit the seven upwelling radiance values since high-order derivatives are required to calculate the Planck intensity. We have used a fifth-order polynomial function as a first approach to fit seven HIRS radiances, i.e.,

$$R(\bar{\pi}) = \sum_{n=0}^{5} a_n \bar{\pi}^n \ , \tag{24}$$

TABLE 2.   VALUES OF PARAMETER m, RMS ERROR, AND ERROR NEAR
          THE PEAK WEIGHTING FUNCTION FOR THE HIRS CHANNELS

Channel	m	RMS error	W_{max}*	error near peak
1	2.8370	0.0141	0.234	-0.0036
2	0.6410	0.0137	0.433	0.0073
3	0.6668	0.0107	0.436	0.0017
4	0.4570	0.0237	0.487	-0.0129
5	0.4273	0.0137	0.517	-0.0048
6	0.2305	0.0159	0.614	0.0076
7	0.3160	0.0118	0.566	-0.0004

*W_{max} denotes the maximum value of the weighting function.

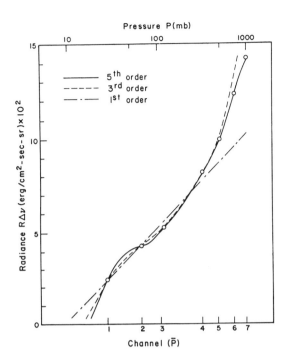

FIGURE 4.   The computed upwelling radiances (circles) for seven HIRS
channels in the logarithmic pressure coordinate, and curve
fittings to the computed values using first-, third-, and
fifth-order polynomials. The U.S. standard temperature is
used in the computation.

where $\bar{\pi} = -\ln\bar{p}$. Derivatives up to the fifth-order may then be
carried out. Newton's interpolation formula (Hamming, 1973) was used
to obtain the coefficients a_n. Figure 4 shows the fitting of seven
HIRS spectral radiances using first-order (straight line), third-
order, and fifth-order polynomials. The U.S. standard temperature
profile was used in the radiance calculation. The fifth-order
fitting shows fluctuations for channels 1-3. This suggests that a
smooth curve-fitting method could be advantageous in the inversion
exercises because high-order derivatives of the curve are required.

At this point, since the inversion coefficients λ_k are known, as
are the derivatives, the Planck intensity at the peak of the weighting function $\bar{p}$ can be computed from Eq. (14). From the Planck intensity for each channel, the temperature can be evaluated. Figure 5
shows retrieval results for the first-, third-, and fifth-order polynomial fittings to synthetic HIRS radiances using the U.S. standard
temperature profile. It is quite encouraging to find that the
retrieval results converge to the true solution as high-order radiance derivatives are incorporated in the calculation. For the fifth-
order approximation, errors in the retrieved temperatures in the
lower atmosphere corresponding to channels 4-7 are within about 2 K.

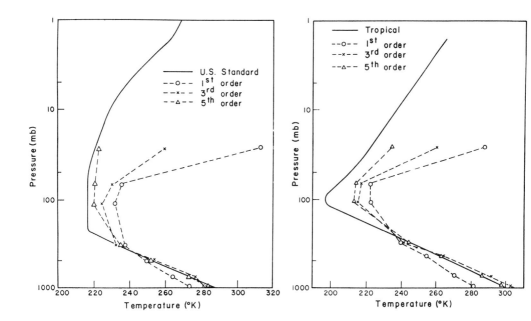

FIGURE 5.   Results of DIM temperature   FIGURE 6.   Same as Fig. 5,
            retrievals using first-,   except for the
            third-, and fifth-order   tropical temper-
            expansions for the U.S.   ature profile.
            standard temperature
            profile.

Retrieved temperatures associated with channels 1-3 show larger deviations from true values. These deviations are due to the unsatisfactory fitting described previously. We have also used the tropical
temperature profile to perform the retrieval. As shown in Fig. 6,
the shape of this profile differs significantly from that of the U.S.
standard profile. Again, the retrieved solution converges to the
true profile when higher-order radiance derivatives are included in
the inversion calculation. The retrieved temperatures are within
about 1 K for channels 4-7. However, large errors are found in the
upper levels corresponding to channels 1-3. The large retrieval
errors are in part caused by the large curvature occurring at about
100 mb, and by the unsatisfactory fitting of radiances for these
channels.

     Finally, we perform retrieval analyses by adding random errors
in the radiance values. Maximum random errors of 2 and 5% were used.
The resulting temperature errors are shown in Fig. 7. Except for
channel 1, the addition of random errors does not produce significant
errors in temperature retrievals. This is particularly evident for
channels 4-7, where peaks of the weighting function are in the lower
atmosphere. The inability to perform retrievals for channel 1 when
random errors are added is due to the nature of the broad weighting

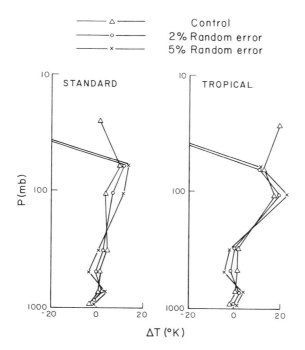

FIGURE 7.   Analyses of errors for DIM temperature retrievals with and
             without random errors. Maximum random errors of 2 and 5%
             are used in the calculation.

function. In this case, the parameter m ≅ 2.8, and λ_k have large
negative and positive values as shown in Fig. 2. As a result, small
errors in radiance values are amplified and propagate onto the
retrieved Planck intensity, leading to unstable results for this
channel.

4. SUMMARY

The DIM for temperature retrievals has been reviewed in terms of
the theory of radiative transfer. It is shown that in the Laplace
transform space the Planck intensity is directly related to the upwelling radiance weighted by the weighting function. After expanding
the transformed weighting function in a Maclaurin series and performing the inverse transform, the Planck intensity in real space can be
exactly expressed by a linear combination of the derivatives of
radiances.

Using seven HIRS channels in the 15 μm CO_2 band, numerical
analyses for temperature retrievals have been carried out. These
involve the curve-fitting of seven radiance values. We demonstrate
that a fifth-order polynomial fitting to radiance values is adequate
to yield correct retrieval results for temperature and that the DIM
converges to the true temperature solution. In retrieval exercises,

two distinct U.S. standard and tropical temperature profiles were used. The retrieved temperatures in the troposphere corresponding to channels 4-7 are accurate within 1-2 K. In general, an addition of random errors with a maximum value of 5% to radiance values does not introduce instability in the inversion exercises. One exception is channel 1, corresponding to the center of the 15 μm CO_2 band, which has a broad weighting function. This appears to suggest that the DIM is particularly useful and practical for sharp weighting functions. Our preliminary conclusion is that the DIM is a powerful retrieval method with no constraints required. Future research will involve an improved fitting program for upwelling radiances, an objective error analysis, and tests of the method using observed radiances and temperature profiles.

ACKNOWLEDGMENTS

The research was supported by the Air Force Geophysics Laboratory under Contract F19628-87-K-0042. Sharon Bennett typed and edited the manuscript.

REFERENCES

Abramowitz, M. and I.E. Stegun, 1968: Handbook of Mathematical Functions. Dover, New York, 1045 pp.

Arking, A. and K. Grossman, 1971: The influence of the line change and band structure on temperatures in planetary atmospheres. J. Atmos. Sci., 29, 937-949.

Chou, M.D. and L. Kouvaris, 1986: Monochromatic calculations of atmospheric radiative transfer due to molecular line absorption. J. Geophys. Res., 91, 4047-4055.

Hamming, R.W., 1973: Numerical Methods for Scientists and Engineers. 2nd Edition, McGraw-Hill, New York, 721 pp.

King, J.I.F., 1985: Theory and application of differential inversion to remote temperature sensing. In Advances in Remote Sensing Retrieval Methods, A. Deepak, H.E. Fleming and M.T. Chahine (Eds.), Deepak Publishing Co., New York, 437-444.

Pearson, C.E., 1974: Handbook of Applied Mathematics. Van Nostrand Reinhold Co., New York, 87-89.

Rothman, L.S., R.R. Gamache, A. Barbe, A. Goldman, J.R. Gillis, L.R. Brown, R.A. Toth, J.M. Flaud and C. Camy-Peyret, 1983: AFGL atmospheric line parameters compilation: 1982 edition. Appl. Opt., 22, 2247-2256.

Widder, D.V., 1971: An Introduction to Transform Theory. Academic Press, New York, 253 pp.

DISCUSSION

Smith: I think it is an exciting approach especially if you are dealing with a near continuum of radiance observations as future instruments are going to be doing. I have a comment though. First of all, you denoted the radiance as a function of P or pressure. Since its structure is directly related to the temperature structure as a function of pressure, the higher order the radiance information, the better off you are as far as temperature profile information content. However, in looking at the results that you've shown for a very limited number of measurements, it didn't appear that you are conserving energy. I was wondering what was causing that.

Liou: What do you mean by conserving energy?

Smith: Well, if you integrate the radiative transfer equation over your temperature profile it didn't appear that you would end up with the observed radiance being equal to that. You would calculate from your solution profiles. They are fitting very well in the troposphere and then systematically off in the stratosphere. If you integrate the difference between your solution and the truth using the weighting functions for those channels it does not appear that you would end up with zero.

Liou: Well, there is some deviation of temperatures by about a degree here. And perhaps the deviation here can make up for the loss there. Well, I don't understand the meaning of conservation of energy. You performed the radiance calculation using the profile plus the weighting function. Do you mean that the radiance should not deviate that much from this profile because that is where most of the energy comes from?

Rodgers: I wondered if you would like to try on your retrieval method the set of diagnostics I presented yesterday.

Liou: I must apologize, I just got in last night so I didn't attend your talk. Some sort of statistical analysis? I'll get with you later.

Yee: The energy conservation may not exactly apply to this particular problem. Maybe we didn't pay much attention to what you mentioned about localization of this particular method. In other words this may not rely on the entire profile. That means the interference of one particular layer of the retrieval is not inferenced by the other layers very heavily. Because of this differentiation, I have some questions though. For the fifth order of differentiation, you may have to rely on many layers away but I would suspect that their weight would decrease very fast from the first order. So essentially you still rely on a very local point of the radiance. And also that particular layer of temperature. So this problem is very different.

Liou: Right. But Bill mentioned that when you perform integrations over the radiative transfer equation, the retrieved profile used in the radiance calculation, if the retrieval is good, should approximate the overall observed profile. That is the energy conservation principle, if I understand it correctly.

Gille: Is the answer to the point that Bill brought up, that although it appears that your fit suggests low temperatures at the top when you integrate over that broad weighting function you are incorporating only a small part of the low temperature?

A SMART ALGORITHM FOR NONLINEAR INTERPOLATION AND NOISE DISCRIMINATION

Steven J. Leon[1]
Southeastern Massachusetts University
North Dartmouth, Massachusetts 02747, USA

Jean I.F. King
Air Force Geophysics Laboratory
Hanscom Air Force Base
Bedford, Massachusetts 01731, USA

ABSTRACT

Transfer theory relates the upwelling intensity to the integral transform of the Planck intensity. Radiance values can be obtained by remote sensing from satellites. In order to recover the Planck intensity from the data one must solve an integral equation of the first kind. Such equations are notoriously ill-posed and consequently difficult to solve in a numerically stable manner. If one assumes an exponential model for the Planck intensity, then it follows that the radiance data should be represented as a rational function. An algorithm has been developed for interpolating the radiance data by a special form of rational function. The algorithm is stable for both the interpolation and the inversion. Thus small errors in the data will not lead to significant perturbations in the computed interpolating function and the corresponding inverse function. On the other hand, significant errors in the data will cause one of the components of the interpolating function to have a positive pole. The algorithm is intelligent in that it has been designed to detect this situation and then locate and correct those points which are in error. Thus, the algorithm can be used to detect and compensate for a faulty sensing channel.

1. INTRODUCTION

In this paper we are concerned with modelling radiance data which has been obtained by remote sensing from satellites. We introduce an interpolation algorithm for modelling the data by a special form of rational function. The algorithm can be used to detect and correct faulty sensing channels. In theory the poles of the interpolating function should either be complex or lie on the

---

[1]Research sponsored by the Air Force Office of Scientific Research/AFSC, United States Air Force, under Contract F49620-85-C-0013. The United States Government is authorized to reproduce and distribute reprints for governmental purposes notwithstanding any copyright notation hereon.

RSRM '87: ADVANCES IN
REMOTE SENSING RETRIEVAL METHODS
A. Deepak, H.E. Fleming, and J.S. Theon (Eds.)

negative real axis. However, a significant error in one of the data points will cause the computed function to have a positive pole. The location of this pole will indicate which data point is in error. The algorithm can then be used to correct that error and to interpolate the corrected data.

This model can also be used to solve the inversion problem. To determine the Planck intensity from the radiance data it is necessary to solve an integral equation of the first kind. The inverse of the corrected interpolating function can be represented directly in terms of the computed coefficients and poles. The solution of the inversion problem will provide valuable information about the structure of the atmosphere.

2. THE MATHEMATICAL MODEL

The rationale for our model comes from transfer theory. According to this theory the upwelling intensity is the integral transform of the Planck intensity. Our model is based on the assumption that the Planck intensity can be represented as a function of the form

$$B(t) = w_1 t + w_2 + \sum_{j=3}^{n+1} w_j \exp(-c_{j-2} t) \tag{1}$$

We assume also that $B(t)$ is the inverse Laplace transform of $a(s)/s$, where $a(s)$ denotes the upwelling intensity in the nadir secant direction s. It follows that

$$a(s)/s = \mathcal{L}(B(t)) = w_1/s^2 + w_2/s + \sum_{j=3}^{n+1} \frac{w_j}{s + c_{j-2}}$$

If the upwelling intensity is expressed as a function of the nadir cosine direction $u = 1/s$, then

$$a(u) = w_1 u + w_2 + \sum_{j=3}^{n+1} \frac{w_j}{1 + u c_{j-2}} \tag{2}$$

Thus to model the upwelling intensities we must find a function of the form (2) that fits the data. This leads to the following interpolation problem.

Interpolation Problem: Given $2n$ upwelling intensities a_i, $i = 1, ..., 2n$ sensed in arbitrary nadir cosine directions, determine the unique set of coefficients $w_1, ..., w_{n+1}, c_1, ..., c_{n-1}$ such that

$$a(u_i) = a_i, \quad i = 1, ..., 2n$$

The solution to the interpolation problem will also provide the solution to the inversion problem. Once the interpolation coefficients have been determined, the Planck intensity can be constructed using equation (1).

3. THE INTERPOLATION ALGORITHM

For simplicity we will state the algorithm in the special case that $n = 3$, however, it is not difficult to generalize this procedure to arbitrary n.

Algorithm 1.

1. For $i = 1, ..., 6$

$$\beta_{i1} = a_i / \prod_{\substack{j=1 \\ j \neq i}}^{6} (u_i - u_j).$$

2. For $j = 2, 3, 4$ set

$$\beta_{ij} = \beta_{i1} u_i^{j-1}, \quad i = 1, ..., 6.$$

3. For $j = 1, 2, 3, 4$ set

$$M_j = \sum_{i=1}^{6} \beta_{ij}.$$

4. Solve the system

$$M_1 c + M_2 b = - M_3$$

$$M_2 c + M_3 b = - M_4$$

for c and b.

5. Find the roots r_1 and r_2 to the quadratic equation

$$x^2 + bx + c = 0.$$

6. For $i = 1, ..., 6$ set

$$a_{i1} = u_1, \quad a_{i2} = 1, \quad a_{i3} = - r_1/(u_i - r_1)$$

$$a_{i4} = - r_2/(u_i - r_2).$$

7. Solve the linear system

$$A\mathbf{w} = \mathbf{a}$$

for $\mathbf{w}$.

8. Set $c_1 = -1/r_1$ and $c_2 = -1/r_2$.

The interpolating function is then of the form given in equation (2). If the roots r_1 and r_2 turn out to be complex, then equations (1) and (2) can be reformulated so that they involve only real coefficients and the proper modifications can be made to the algorithm so that the new coefficients can be computed using real arithmetic.

4. STABILITY OF THE INTERPOLATION

Versions of this algorithm were tested on simulated data. The test models involved no more than six components. Data was simulated by constructing rational functions and adding low level noise to the function values. In each case the algorithm successfully interpolated the data to the working machine precision.

Further testing was done to determine the effects of small perturbations on the simulated data. The tests showed that the interpolation was stable. The perturbations in the interpolating functions were of the same order as the perturbations in the data. The perturbations in the roots of the characteristic equation, however, were generally much greater than the perturbations in the interpolating functions. The roots of the characteristic equation, computed in step 5 of the algorithm, are the poles of the interpolating function. The weights $w_1, ..., w_{n+1}$ are computed in steps 6 and 7 and hence depend on the roots computed in step 5. Consequently the perturbations in the weights are even greater than the perturbations in the poles. Actually the perturbations in the weights compensate for the perturbations in the characteristic roots. Thus the computation of the interpolating function is stable even though the computation of its coefficients is not.

When the simulated data was perturbed by amounts significantly above low noise levels, the effects were quite noticeable in the resulting interpolating function. In these cases the perturbation in the characteristic roots were large enough so that the computed function would end up with at least one positive pole. Furthermore, a significant perturbation in a_i would lead to a positive pole r_j close to u_i.

5. TEST RESULTS FOR ACTUAL RADIANCE DATA

The interpolation algorithm was tested on eighteen sets of radiance data taken from McClatchey (1979). In each case the interpolating function fitted the data very accurately and in each case one of the poles turned out to be positive. The weight w_i corresponding to the positive pole was always small compared to the other weights. Thus if that w_i were set to 0, the resulting function would still come close to interpolating the data. The 18 data sets were all determined from sensing channels using nadir cosine directions u_i of 0.017, 0.066, 0.170, 0.400, 0.750, 0.920. For each set of radiance values the algorithm was applied and the weight corresponding to the positive pole was set to zero. We will refer to the resulting function f as the zero-weight approximating function. In 17 of the 18 data sets the radiance value a_4 was slightly higher than $f(u_4)$ and in all 18 of the data sets the value of a_5 was slightly smaller than $f(u_5)$. In thirteen of the eighteen sets the maximum deviation occurred at u_4 and the remaining five data sets had maximum deviations at u_5. This provides a strong indication that something is wrong with data obtained from the fourth sensing channel and perhaps the fifth channel as well. Figures 1 and 2 show the graphs of $a(u)$ and $f(u)$ for a typical data set.

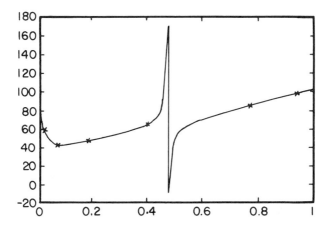

FIGURE 1. Rational Interpolation of Radiance Data

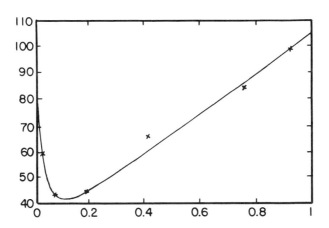

FIGURE 2. Zero-Weight Approximating Functions

6. DATA CORRECTION

Algorithm 1 was expanded to include a data correction component. The new algorithm tests if there is a positive pole with a small weight. If so, that weight is set to 0. The algorithm then checks each data point to determine which point gives the maximum deviation from the zero-weight approximating function.

The a_i value at that point is then replaced by the value of $f(u_i)$. For the data set in Figures 1 and 2 the maximum deviation occurs at $u_4 = 0.400$. Here the radiance value a_4 was 64.3 and $f(u_4)$ turns out to be 58.2971. We can replace a_4 by this new value and rerun the algorithm. The new interpolating function will be referred to as the corrected interpolating function. The graph of this function is shown in Figure 3. The maximum deviation for the corrected data is 2.8×10^{-14}. This was typical for the majority of the data sets. For twelve of the eighteen data sets the maximum deviation was on the order of 10^{-14} after one correction. Four of the data sets required two or three corrections to achieve this accuracy. The corrections failed to bring about improvements in only two cases. In one of these cases the corrections resulted in an additional positive pole very close to 0. The algorithm then sets both hyperbolic weights equal to zero. The resulting linear function did not give a good fit to the nonlinear data. We were, however, able to get a good fit in this case by making a small adjustment to the algorithm.

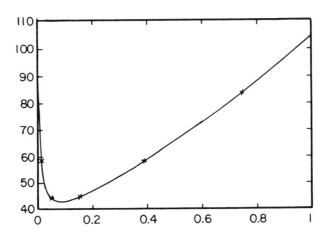

FIGURE 3. Corrected Interpolating Function

7. THE INVERSION PROBLEM

The inversion problem involves solving an integral equation of the first kind. This type of equation is ill-posed in that the solution does not depend continuously on the data. Consequently in the presence of noise it is difficult to obtain a numerically stable solution. The standard approach to solving this type of problem is to discretize the equation and then solve it as a linear system. To overcome the stability problem additional constraints are added. Rather than following this approach we make use of the assumption that the inverse $B(t)$ can be modelled by equation (1). The coefficients in (1) can be determined from the coefficients of our rational interpolating function (corrected if necessary).

To check the stability of our method, tests were made on simulated data to determine the effect that small perturbations in the data would have on the computed inverse. These tests indicated that the inversion is stable for small values of n, but becomes increasingly unstable as n increases. For the McClatchey data, the value of n was 3 (i.e., there were 6 points in each data set). The inversion should be stable in this case. The graph of the computed Planck intensity for one of these data sets is given in Figure 4.

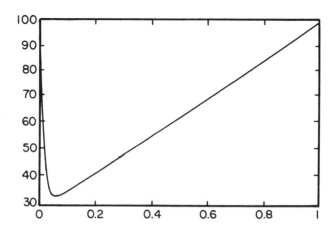

FIGURE 4. Graph of the Planck Intensity

REFERENCES

McClatchey, Robert A., 1979: Satellite Temperature Sounding of the Atmosphere: Ground Truth Analysis, AFGL Technical Report 76-0279, Hanscom AFB, Massachusetts 01731.

BIBLIOGRAPHY

Chandrasekhar, S., 1950: *Radiative Transfer*, Oxford University Press, London.
Delves, L. M. and Walsh, J., 1974: *Numerical Solution of Integral Equations*, Clarendon Press, Oxford.
King, Jean I. F., 1985: Theory and Application of Differential Inversion to Remote Temperature Sensing. In Deepak, A., et al., (eds.) *Advances in Remote Sensing Retrieval Methods*, A. Deepak Publishing, Hampton, Virginia.

DISCUSSION

Fougere: I was curious why you don't use spline functions in the interpolation. You never get wild jumps back and forth.

Leon: The rationale for our method comes from the H-theory inversion of Chandrasekkar. If we only wanted to fit the radiance data, then it would make sense to use cubic splines. Our goal is to be able to reconstruct the slant intensity from the radiance function. To do this in accordance with the Chandrasekkar model, we must represent the radiance data by a rational function.

If this model is correct, then, in theory, the rational function should not have any positive poles. There should be no wild jumps. The presence of such jumps indicate faulty data. The locations of the jumps can be used to determine which data points are in error. This is an important added benefit of our algorithm that can be used to detect fully sensing channels. This is why we refer to it as a smart algorithm.

A possible generalization along the same lines as spline interpolation would be to use local rational fits to the radiance data. Such a generalized spline would have rational components rather than cubic polynomial components.

OPEN DISCUSSION

SESSION C.1 - JEAN KING

King: I would like to work from the general to the particular. First, I don't know if this is the place but there has to be some place to thank Adarsh for the auspices under which we've held this meeting. It has just been a superb forum for exchange.

If you try to draw and take the general view towards the retrieval problem I think you can discern two different approaches. One is the approach from the operational point of view which is characterized by instrumentation, computer intensive, experimental, applied mathematical. And there is the other approach from the standpoint of transfer theory—more theoretical focusing on inversion methods and I think, to a large degree, they have labored in their separate venues. I don't think this is necessarily bad. In fact, I think it is good, because it is good that the experimentalists pushed the theoreticians and vice versa that the theoreticians working somewhat in isolation came up with theoretical requirements which in the traditional leap-frogging way have advanced our knowledge of nature. But I do think that at some point the two camps should interchange ideas and at least arrive at some common goals. The goal setting I think is important even if the goals are unattainable. Because it is often that the pursuit of unattainable goals has by-products that advance the art and solve the problems unrelated to the central problem that is at issue. So, with these thoughts in mind, I got together with Bill Smith and we've come up with a kind of straw man and I call it a retrieval credo. I think credo is better than manifesto. There are four points and there are some that should be added. We would appreciate comments, written or verbal, to add to the list or change the ones we have. I think there is general agreement on the first one: free from work towards retrieval methods, free from the need of a priori data basing. I don't know how anyone could take issue with that. You can. Okay. The second is to work towards minimizing discretization, if there is such a word, and differencing. The point of that, I don't want these to be motherhood statements, is currently the way one (in many methods) solves the inversion problem. You take the radiance interval that is a sum profile and you dissect it, and then you reconstruct it putting in transmittance parameters, and so forth, and the reconstruction then yields another profile, which is the profile desired, namely the Planck temperature distribution. There is nothing that says that is the only way that it has to be done. If one draws back a bit from the problem, for instance, an inverse transform which represents the inference for a monochromatic transmittance there is no dissection involved in that. But the point, if you look at what one is basically doing, is it is a mapping transform. You are viewing the same phenomenon, the phenomenon that you are viewing of course is a photon density. And the photon density, you are viewing it in interspace by conduction, by sticking thermometers in that space. So what you want to do, in the forward problem, is transform from metric involving pressure and so forth to viewing that same phenomenon externally. It is photon density again that you are observing but now you are observing it in a radiation space, you are counting photons essentially. So you can view either the forward or inverse operation as taking a profile that is viewed in one kind of space of a thermal phenomenon and viewing it in another space and backwards for the inverse problem. So, in that sense, I have never been happy with the fact that the problem is characterized as a forward and an inverse problem. Because really, they are just viewing the same phenomenon from different perspectives. Passing to the third and fourth, which were contributions from Bill. To utilize all the information that is inherent in radiance rich high resolution data (in other words, to devise algorithms that make use of all the data that is implicit and improvements that can arise from increasing by factors of 10 the amount of data you have to work with) I think we should work toward methods yielding simultaneous inferencing of all optically active meteorological parameters. There is another one I think that should be up here, and that is I think we should work towards inversion methods which are not pure regression. That is, work towards retrieval methods that have some physics in it, that do not just involve building correlation and matrices between two observed curves. I don't know quite the right words. The idea is to work toward retrieval methods with a maximum amount of physics and the least amount of curve matching in them. So perhaps this would be the time to discuss these.

Wahba: I absolutely agree with points 2, 3, and 4 and disagree with number 1 both from a highly theoretical point of view and also from a practical point of view. Every method of solution for ill-posed problems is going to have some prior information in it, either explicitly or implicitly. The very choice of the method entails assumptions concerning the nature of the solution. In an ill-posed problem like satellite tomography, there is generally going to be far fewer degrees of freedom for signal than data points. If you are going to get an entire curve or function from this data, your solution must inevitably be constrained or regularized so that the solution is determined uniquely. If the solution is constrained to be within some family of curves, how you choose these curves constitutes the prior information. If you use regularization or variational methods, the choice of penalty function carries the prior information. If the prior is not explicit, it is going to be implicit in the numerical method. For example, different iterative schemes for solving operator equations have different "smoothing" properties, if they are run only to fit the data approximately, which is what you would do if the data are not exact. I think that ultimately the proper way to use satellite radiance data in numerical weather prediction, is to use it to update the state variables of the NWP model. If this can be done well, then possibly modest, but positive information is added to the model initial conditions. (Note: Most of the tape covering this comment was unreadable and this comment has been reconstructed from memory.)

King: It is odd, I thought that first one would be the least controversial. I still think that it is valid. Would you agree that we should minimize the need for a priori data basing and also the main problem with being too restrictive in your data base set is that when you end up with your inferred profile, it is very difficult to know whether it is an artifact of what you put in or whether it is new information that you are garnering from the radiance data?

Rodgers: We have to discretize in some form or other simply because we are making a finite number of measurements. Also, because we are only making a finite number of measurements and because our weighing functions have some width there is always a null space and the component of the solution in the null space has to come from a priori in some form or other. The question is not minimizing the need for a priori but understanding the effect that a priori has on the profile.

King: How do you discriminate then between the a priori input and the new information that arises from the radiance observations?

Rodgers: This is what I tried to do in my error analysis paper-to show just what function of the true profile the retrieved profile is.

Johnson: I disagree with point 1, but I could not state it so clearly as Grace and Clive did. But I also have some concerns about a couple of the others. I think the whole group here is promulgating, sort of, not so much a narrowness but a way of doing the things that they have done for a long time. I believe, for example, one thing that would be really useful for this meeting would be to have a larger representation of active remote sensing people. That brings me to my major comment. I think, as Clive says, there are certain things in the system that you are observing that are orthogonal to the way you are observing and to get away from that you shouldn't put your extra effort into tweaking; that is, don't spend twice your time to get another percent or two of information out of that. It would be far better to spend your time observing it in a different way, in a different spectral region, different resolution, different vantage point, etc. And the best way to get a handle on learning more information about the system is to OBSERVE more about the system.

Liou: I'm wondering whether the error analysis is applicable to all the inversion problems. We have been working on the error analysis for twenty years and are still working on it. In what manner is the error analysis related to the forward transfer problem? Earlier it was pointed out that it is important to understand the fundamental physics.

Neuendorffer: I think it is important that you do retrievals with a minimum of a priori data and then with a maximum of a priori data and see what sort of differences you get. If they come out with the same result, great, if they come out with radically different results I think it is important to realize that you have a difference and why you have a difference.

Fleming: I think that misses a very crucial point. If you have a sounding that is close to some climatological average that you put in as a priori data, it will certainly beat something that uses no a priori data. I think the important thing in meteorology is the exception or the deviation from the normal pattern. To do well in that case, I think, is very critical in terms of any inferencing method.

Goroch: I think it may be worthwhile to look at this from a more practical point of view such as an operational weather center. When you are retrieving a satellite sounding, you do not have a priori data available but you have independent data available. You really should use the independent observations as much as possible.

Crosby: I am going to be silly here for a minute. But if one thinks of it mathematically, you can always superimpose a solution which contains extreme oscillations. It is just silly to suggest that you can do a solution without some sort of constraints.

Westwater: I believe, in agreement with Professor Liou, that one way of perhaps getting the true physics into the situation is not from the radiances alone but by coupling the quantities inferred from the radiance with dynamic models which do have the physics of the atmosphere built into the models. So it is really a coupling of the remote sensing observations with the dynamic equations of motion that will really apply the constraints that will allow the inverse problem to be completely solved.

McCormick: Perhaps a compromise on your credo might be to change the wording in the first statement for careful use of a priori data basing. Because I think without use of a priori information almost all practical applications will be lacking.

Thompson: Up to this point we have been pure practitioners of indirect sensing in that we haven't really talked about why we do this sort of business until we got into this discussion of numerical weather forecasting. I remind you that the numerical weather forecasting problem is also an ill-posed problem. It has always impressed me that we optimize those two problems entirely independently, more or less. That is, we come to an indirect sensing conference and we talk about how we can get the error of a profile down to a degree and we will work another 10 years to do that sort of optimization hoping that if we can do that then somehow we have helped optimize the numerical weather forecasting problem. And, I suppose in some sense we do. But, do we ever sit down and consider whether we could develop some sort of retrieval algorithm that would optimize the accuracy of the 14-day weather forecast? That is, do both the problems at once. I certainly don't know how to do it, but the thought of doing that sort of thing raises some new issues about what we do as atmospheric scientists. As far as I know I have never met anybody who wanted a single retrieval of half a degree accuracy. Instead, you want a lot of them in such a way that they produce a better weather forecast or a better sense of the climate or some other product, but a single temperature profile has very little value.

McMillin: This conference has all been highly theoretical and the last couple of comments have gotten back to what I consider to be the real issue. The person who is paying your bill is interested in whether it rains or doesn't when you tell them it is going to rain tomorrow. Although goals on specific methods are nice, the method that goes into operation should be the one that produces the best forecast. If it happens to be one that is unappealing then maybe more work is needed on the appealing methods but they shouldn't be forced into operation.

Rodgers: Just to remind you, weather forecasting isn't the only problem we've got. I'm in stratospheric chemistry. One degree is probably a bit outside the bounds of what we really need for that particular problem but you really do want to look at individual profiles and try to see what is going on in a profile by profile basis.

Knoll: I have a very different interest than most of you here. I am interested in the hardware development and evolution of the measurement science and developing specifically the instrumentation in an infrared region that will add to the retrieval of both the meteorological and composition information. I would almost add a fifth point to this credo, which is error analysis. In my experience this is very important in the sense of defining priorities for the development of the measurement science. In other words, is it more important to have additional spectral channels, or higher radiometric accuracy. These are very important issues because you can develop the hardware in many different directions. So the question then is: what is the order of importance in the evolution of the measurement science as inferred from an understanding of the retrieval curves and retrieval process?

APPLICATION OF THE MAXIMUM ENTROPY METHOD
TO THE GENERALIZED INVERSE PROBLEM OF SPECTRAL ESTIMATION

N.L. Bonavito
NASA Goddard Space Flight Center
Information Analysis Facility
Greenbelt, Maryland 20771, USA

ABSTRACT

Traditional methods of obtaining an estimate of the power spectral density have followed the methods of Blackman and Tukey (1959). These have been modified to take advantage of the Fast Fourier Transform. However, these techniques which employ a smoothing of the autocorrelation function by a time domain window or a smoothing of the squared magnitude of the Fourier Transform do not design windows based on the true spectrum. Two immediate consequences of this are sidelobe leakage in the transfer function of the smoothing window, and a limit on resolution.

A spectral estimation approach first devised by Burg (1967), based on the Maximum Entropy formalism avoids assumptions about the data or its autocorrelation function outside the sample interval. This discipline has its roots in the work of Boltzman (1877), and the basis for it and its application has been thoroughly discussed by Jaynes (1957). For power spectra, the Maximum Entropy estimator retains all the estimated lags without smoothing and for the one-dimensional case, it uses Wiener optimum filter theory to design a prediction filter which will whiten the input time series. From the whitened output power and the response of the prediction filter, it is possible to compute the input power spectrum, thus leading to a very high resolution on short segments of data.

In this presentation, the application of the Burg algorithm to computer generated data and to a time series of oceanic laboratory data is discussed.

Results show that the Maximum Entropy approach exceeds that of the classical estimator and is much superior for short data records than the Fast Fourier Transform.

1. INTRODUCTION

The spirit of the Maximum Entropy formalism can be traced back to the <u>Ars Conjectandi</u> of Jacques Bernoulli in 1713. The essence of his work is the recognition that a probability assignment is a means of describing a certain state of knowledge. If available evidence gives no reason to consider one proposition more or less likely than

Copyright © 1989 A. DEEPAK Publishing
ISBN 0-937194-13-1

another, then the only way to properly describe that state of knowledge is to assign to them equal probabilities. It was for the explicitly stated purpose of finding probabilities when the number of equally possible cases is infinite, that Bernoulli introduced the connection between a theoretical probability and an observable frequency. Given the multiplicity of a proposition A, or the number of ways M, in which it could be true, the total number of equally possible outcomes N, and the number of independent observations n, Bernoulli calculated the probability for observing the proposition to be true m times. Approximately fifty years later, Thomas Bayes, a clergyman, turned this argument around and gave an inverse probability formula, namely, that probability given N, n, and m, that M has various values.

In the latter part of the eighteenth century, Laplace redefined Bayes' work in greater clarity (1812) and then proceeded to apply it with great success to problems in many different areas. Denoting various propositions by A, B, C, etc., and AC and BC the propositions that both A and C, and B and C are true respectively, Laplace's statement of "Bayes' Theorem" can be written.

$$p(A/BC) = p(A/C)\ p(B/AC)/p(B/C) \qquad (1)$$

This equation is a mathematical representation of the process of learning. Here $p(A/C)$ is the prior probability of A, knowing only C. $p(A/BC)$ is the posterior probability, updated as a result of acquiring new information B. The prior information C represents the totality of what we knew about our hypothesis or proposition A before getting the data B. A famous example that Laplace did solve began with the proposition that the unknown mass of planet Saturn should be within a specified interval. Mutual perturbations of Jupiter and Saturn represented data obtained from observatories, while prior information consisted of the common sense observation that Saturn's mass could not be so small that it would lose its rings, nor so large that it would disrupt the solar system. Using equation (1), he estimated the planet's mass to be 1/3512 of the solar mass, with a probability of .99991. Another 150 years' accumulation of data raised the estimate 0.63 percent!

Formally, the principle of the Maximum Entropy can be said to have originated in the work of Boltzman (1877) and Gibbs (1875-1878).

In thermodynamics, the First Law for closed systems determines a class of possible macroscopic states that are permitted. Out of all possibilities permitted by the First Law, the macrostate chosen by Nature (The Second Law) is the one that can be realized in the greatest number of ways, that is, the one that has the Maximum Entropy. This is the principle given by Gibbs (1875-1878) governing heterogeneous equilibrium, and, for one hundred years, physical chemistry has been based on it. Subsequent ergodic attempts to justify Gibb's statistical mechanics by the equality of phase space averages failed to explain the equality of canonical ensemble averages and experimental values.

The major contribution to solving the dilemma was the introduction of the information measure or "entropy",

$$H = -\sum_i p_i \ln p_i \qquad (2)$$

by Shannon (1948), where the p_i represent a probability density function. Any field of inquiry which uses the entropy function either as a criterion for the choice of probability distributions to determine the degree of uncertainty about a proposition, or as a measure of the rate of information acquisitions, is using information theory. Jaynes, (1957) proposed that the Shannon measure, or entropy, be used to define the values for probabilities. Prior to Jaynes' work, there was no clear reason for the use of the entropy principle in statistical mechanics other than the after-the-fact conclusion that since "statistical mechanics works," it must be right. Shortly thereafter, it was demonstrated that all of the laws of classical thermodynamics could be defined from Shannon's entropy using Jaynes' Principle of Maximum Entropy (1961). Briefly, this says that the minimally prejudiced probability distribution is that which maximizes the entropy subject to constraints supplied by the given information. Mathematically this can be stated as:

$$\text{Maximize } H = -\sum_i p_i \ln p_i \qquad (3)$$

subject to the constraints:

$$\sum_i p_i = 1$$
$$\sum_i p_i g_r(X_i) = \overline{g}_r \qquad (4)$$

Here $g_r(X_i)$ is a function of the variable X_i. The form of $g_r(X_i)$ is known, and $\overline{g}_r$ is the mean value for each of the functions $g_1(X_i)$, $g_2(X_i)$, ...$g_r(X_i)$. Maximizing the entropy function subject to the given constraints can be done by Lagrange's method of undetermined multipliers. The resulting probability distribution is found to be:

$$p_i = \exp[-\lambda_0 - \lambda_1 g_1(X_i) - \lambda_2 g_2(X_i) - ...] \qquad (5)$$

where the λ's are the undetermined multipliers. There are as many multipliers as there are equations of constraint. More concisely, Jaynes' Principle of Maximum Entropy can be stated as follows:

"If data D are given concerning the outcome of a random experiment, then predictions about the outcome of the experiment should be based upon that distribution, P = (P1, P2, ..., PN), which maximizes the Shannon Information Measure, subject to the constraints imposed by the data D."

This statement then can be considered the starting point for all work in the Maximum Entropy formalism.

2. BURG - ONE DIMENSIONAL SPECTRAL ESTIMATION

In any method of spectral analysis, there are always three major considerations:

1. Resolution: How close in frequency can two spectral components be spaced and be identified?

2. Dynamic Range: How small can a spectral peak be, relative to the largest and still be observed in the spectra?

3. Variance: How accurate is the estimate of the spectra to the true spectra?

Consider a finite waveform of duration T which is sampled at N points at intervals Δt such that T = NΔt. If we wish to create the Fourier Transform in order to investigate the spectral content of the waveform, we must make assumptions about the rest of the time from minus to plus infinity for which we have no information at all. This issue always arises when analysis is applied to finite samples. Many approaches have been proposed over the years which range from assuming zero values for the waveform where the waveform is not sampled, and using "windowing" techniques which attenuate the ends of the sampled waveform so as to minimize the abrupt discontinuities which introduce spurious high frequencies or sidelobes into the apparent spectrum, to various predictive extensions of the sampled waveform to "infinity" based on a knowledge of the physical system involved. A popular example for the latter method is the Discrete Fourier Transform (DFT) Pair where the waveform is simply repeated on both sides of the sample until the periodic wavetrain of period T extends from minus infinity to plus infinity. This method must by used carefully, and requires more knowledge of the system involved than is generally available in spectral analysis.

We have a discrete time series $[Y_0, Y_1, ..., Y_N]$ that has a Fourier Transform

$$Y(f) = (N + 1)^{-1} \sum_{K}^{N} Y_k e^{iwk} \qquad (6)$$

where w = 2πf, and power spectrum

$$S(f) = |Y(f)|^2 = \sum_{K=-N}^{N} R_k e^{iwk} \tag{7}$$

where $R_{-k} = R_k{}^*$ and

$$R_k = \sum_{j=0}^{N-K} Y_j^* Y_{k+j}, \quad 0 \le k \le N, \tag{8}$$

is called the autocovariance or autocorrelation. From the Wiener-Khinchine Theorem, the Fourier Inverse of equation (7) is

$$\frac{1}{2\pi} \int_{-\pi}^{\pi} S(f) e^{-iwk} dw = R_k, \quad -N \le k \le N, \tag{9}$$

where $w_N = \pi$ is the Nyquist frequency above which S(f) repeats itself periodically. The difficulty here is not that all of the R_k are known, but our data D, although exact, comprise only a subset $[R'_{-m}....R'_m]$, where m<N. The problem then, is how to estimate S(f) from this incomplete information.

Burg (1967) made his classical application of Jaynes' Principle to the field of signal processing by observing that under the proper conditions, it is possible to extract the maximum spectral information content of a sampled waveform. The Burg method states: "Given a limited set of autocovariance coefficients, the algorithm is based upon choosing that spectrum which corresponds to the most random or the most unpredictable time series whose autocovariance coincides with the given set of values."

Burg solved the problem in which time series information consists of measured values of the autocorrelation function. He chose the spectrum which corresponds to the most random time series whose autocorrelation coincides with the given set of values. Thus, given the exact first (p + 1) autocorrelation lags $[R_0, R_1,...R_p]$ of a zero mean, stationary Gaussian random process, the one dimensional Maximum Entropy spectrum was shown to be the result of maximization of

$$H = \int_{w_n}^{w_n} \log S(f) df \tag{10}$$

subject to the constraints of equation (9). The solution is

$$S(f) = \frac{\sigma_p^2 \Delta t}{\left| 1 + \sum_{K=1}^{p} a_{pk} \exp(-2\pi i f k \Delta t) \right|^2} \tag{11}$$

or

$$S(f) = \Delta t \sum_{M=-\infty}^{\infty} R_n exp(-2\pi i f n \Delta t) \tag{12}$$

where the frequency f is constrained to be in the Nyquist interval, $-1/2 \Delta t \leq f \leq 1/2 \Delta t$, σ_p^2 is the output power of a (p + 1) long prediction error filter, and the error filter coefficients are (1, a_{p1}, a_{p2},..., a_{pp}).

$$R_n = \begin{cases} \psi_n, & |n| \leq p \\ -\sum_{K=1}^{p} Apk\ R_{n-k}, & |n| > p \end{cases} \tag{13}$$

Here ψ_n is the Blackman-Tukey spectral estimate up to lag p. For $|n| > p$, Burg's result "extrapolates" R_n and hence S(f), to "infinity". This has the effect of eliminating sidelobe leakage arising from windowing in the classical periodogram approach and leads to higher resolution capability. There are several important points concerning this result:

1. The solution yields the spectrum which is of the same analytic form only, as that resulting from an autoregressive (AR) model of order p, driven by white noise.

2. The AR coefficients are the Lagrange multipliers of the Maximum Entropy optimization problem.

3. The Maximum Entropy method determines the order of the resulting AR process.

4. Given autocorrelation lags at some non-uniformly spaced lags, the Maximum Entropy formulation exists, while the AR solution does not.

5. If estimates of the autocorrelation lags are given, the Maximum Entropy power spectral density does not have the analytic form of an AR model. It is non-linear.

6. For cases where the time series data are not the autocorrelation sequence of the process, one approach suggested by Jaynes is to make approximations or estimates of the autocorrelation lags R_n , and then use them in Burg's solution as if they were exact.

One of the strong features of the Maximum Entropy spectral estimation method is the ability to yield higher resolution than Fourier methods. Recall that the Fourier resolution is reciprocal of the observation window ($\Delta f \Delta t \geq 1$). Since the Maximum Entropy "extrapolates" the correlations, it "extrapolates" the window, thereby reducing Δf and improving the resolution. Figure 1a illustrates a Fast Fourier Transform (FFT) spectrum calculated from 25 samples of a signal containing two sinusoids

$$r(n) = e^{2\pi i f_1 nt} + e^{2\pi i f_2 nt + \varphi} + \tilde{v}(n) \qquad (14)$$

where f_1 = -13.3 Hz., f_2 = -15.3 Hz., (Δf = 2Hz.), $\tilde{v}(n)$ = complex white noise, φ = random phase = 0.0 degrees, SNR = 21dB, and the sampling frequency is 100 Hz. Since the observation time is 25/100 = 1/4 sec. = Δt, the resolution Δf = $1/\Delta t$ = 1/1/4 4Hz. > 2Hz. As a result, the FFT spectrum cannot resolve the two sinusoids. On the other hand, Fig. 1b. shows the calculated Maximum Entropy spectrum using the exact same data, in which the two sinusoids are easily resolved. In this case, the filter order is fifteen.

In a recent investigation (Bonavito et al., 1987), the Burg algorithm was applied to water wave height data obtained from an experiment conducted at the NASA Wind/Wave and Current Interaction Facility at Wallops Island, Va. Results of these calculations were compared against those obtained for the same data using a Fast Fourier Transform (FFT) technique. Although the latter method included 64 x 1024 data points, the same power spectral density estimate was obtained with the Maximum Entropy estimator using only 128 points, indicating that the performance of the latter method is much superior for short data records than that of the FFT.

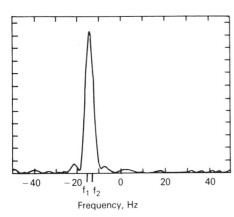
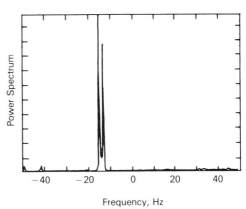

Figure 1a. Fourier spectrum. f_1 = -13.3 Hz, f_2 = -15.3 Hz, sampling frequency = 100 Hz, number of samples = 25, observation time = ¼ sec. No resolution since $\Delta f = 1/\Delta t = 1/¼ = 4$ Hz > 2 Hz.

Figure 1b. Super-resolution with MEM using same data. Here f_1 and f_2 are resolved using a filter order of 15.

3. DISCUSSION

For signal processing, and for one dimensional power spectral estimation, both the Maximum Entropy and FFT give approximately the same results for large uniformly spaced data records. However, for short or uneven data records, or both, the Maximum Entropy result far exceeds that of the FFT.

Analogous to the one dimensional case, the two dimensional Fourier analysis cannot provide sufficient resolution capability for small blocks of data, since the resolution is still dependent upon the window size. The Maximum Entropy measure on the other hand is very suitable for a small number of pixels. However, it does not extend into two dimensions in a straight-forward manner, and consequently remains a highly non-linear method. Attempts are being made to find an analytic solution to this problem. A numerical algorithm which iterates between the correlation and coefficient domains, each time imposing constraints on parameters in the two domains, is presently being tested. The testing has been limited to date, but successful, in cases involving one, two, and three sinusoid computer generated data, of different frequency and noise power and for several sizes of the known autocorrelation array.

REFERENCES

Blackman, R. B. and J. W. Tukey, 1959: The Measurement of Power Spectra. Dover, New York.

Boltzman, L., 1877: Wiener Berichte, $\underline{76}$, p. 373.

Bonavito, N. L. et al., 1987: Submitted to the IEEE, Journal of Acoustics, Speech and Signal Processing.

Burg, J. P., 1967: Maximum Entropy Spectral Analysis. Proceedings of 37th Meeting, Society of Exploration Geophysicists, reprinted in Modern Spectrum Analysis, D. G. Childers, Ed., New York, Wiley, 1978.

Gibbs, J. W., 1875-1878: Heterogeneous Equilibrium. Conn. Academy of Science. Reprinted Longmans, Green, and Co., New York, 1928, Dover, New York, 1961.

Jaynes, E. T., 1957: Information Theory and Statistical Mechanics. Phys. Rev. $\underline{106}$, p. 620 and $\underline{108}$, p. 171.

Jaynes, E. T., 1961: Statistical Theory of Communication. Review of Y. W. Lee, American Journal of Physics. $\underline{29}$, p. 276.

Laplace, P. S., 1812: Theorie Analytique des Probabilities. 2 vols. (Available from Editions Culture et Civilisation, 115 Ave. Gabriel Lebron, 1160 Brussels, Belgium.)

Shannon, C. E., 1948: A Mathematical Theory of Communication. Bell System Technical Journal. $\underline{27}$, p. 379-423, and p. 623-656.

DISCUSSION

NO QUESTIONS

A NEW SPARSE DATA TOMOGRAPHIC IMAGE
RECONSTRUCTION ALGORITHM

C.K. Zoltani, R.T. Smith[1], and G.J. Klem
Ballistic Research Laboratory
Aberdeen Proving Ground, Maryland 21005, USA

ABSTRACT

A new tomographic image reconstruction algorithm for sparse data sets, using finite elements and the maximum entropy formalism has been developed. In a finite dimensional subspace of $\mathcal{L}^2(D)$, approximate solutions of the maximum entropy optimization problem were obtained. With the incorporation of a–priori information consisting of known maximum and minimum densities of the material being scanned, even in the case of twenty views instead of the usual one hundred eighty views, images of acceptable quality were obtained. Unlike conventional maximum entropy reconstructions, where there is a serious degradation of the image when the object is placed near the periphery of the scanned region, this algorithm does not exhibit such limitations. A number of reconstructed images of phantoms are used to illustrate the technique.

1. INTRODUCTION

Some remotely sensed data situations are characterized by sparsity of data, either in the number of views which are available or the number of detectors per view. In either case, the reconstructed image can be severely degraded leading to the question of the best choice of image reconstruction or processing algorithm.

To answer this question a number of approaches have been suggested. Post processing of the image to remove artifacts, i.e. reconstruct the image from the available data and then impose some constraints to eliminate non–physical objects has been tried, see Justice (1986). Alternatively the use of some measure of the statistics of the image to substitute for missing data has claimed some adherents. A corollary of this approach is to extrapolate from the known to the unknown regions. Other strategies are discussed by Hanson, (1987), Potter et al. (1987), Little et al. (1987), as well as in the compendium of papers which appeared in Stark (1987).

Here, from the point of view of computed tomography, we present an algorithm which is an extension of the maximum entropy methodology as discussed, for example, by Minerbo (1977). The method reported here, however, uses finite elements with an optimization in a setting which allowed the object to be at the boundary of the scanned region without destroying the reconstructed image as happens with conventional maximum entropy approaches.

---

[1] Millersville University, Millersville, PA 17551–0302

RSRM '87: ADVANCES IN
REMOTE SENSING RETRIEVAL METHODS
A. Deepak, H.E. Fleming, and J.S. Theon (Eds.)   179

2. THE ALGORITHM

2.1 BACKGROUND

Computed tomography enables the determination of density cross–sections of slices of an object in the plane of a radiation source from the recorded absorption levels of the transmitted radiation. The problem can be formulated mathematically as an inverse problem and a unique solution, i.e. a reconstructed image is guaranteed when data from an infinite number of views is available. With a finite number of views, mathematically the problem becomes ill–posed and small changes, i.e. discrepancies, are amplified leading to inaccurate image representation. These shortcomings can be overcome as discussed in the procedure presented in this paper.

2.2 IMPLEMENTATION

Central to the idea of tomographic reconstruction, given a radiation source, a detection system and an object placed in the path of the radiation, is the fact that data is obtained as a set of integrals. Thus, let $f(x,y)$ be the x–ray attenuation at a point (x,y) in the plane. Measured data, G_{jm}, is available in the form

$$G_{jm} = S_{jm}(f) = \int_{S_{jm}}^{S_{jm+1}} \int_{-\infty}^{+\infty} f(s \cos \theta_j - t \sin \theta_j, \ s \sin \theta_j + t \cos \theta_j)$$

$$dtds$$

$$m = 1, \ldots m(j); \ j = 1, \ldots J, \tag{1}$$

where J is the number of projections and $m(j)$ the number of detectors for the j–th view and θ is the scan angle.

The object of interest lies in a finite region $\mathcal{D}$ contained in $\mathbb{R}^2$. The entropy of the image can be defined as

$$\eta(f) = -\int \int_{\mathcal{D}} |f(x,y)| \ln [|f(x,y)| A] \, dxdy \tag{2}$$

where A is the area of $\mathcal{D}$ and $f \in \mathcal{L}^2 (\mathcal{D})$, the set of square integrable functions in $\mathcal{D}$, i.e.

$$\int \int_{\mathcal{D}} f^2 \, dA < \infty \text{ holds.} \tag{3}$$

Deviating from previous approaches, instead of maximizing the entropy with the measured values as constraints, we minimize minus the entropy plus a penalty term subject to some known, a priori bounds. That is, f can be determined as the solution of the constrained optimization problem

$$\inf_{f \in \Sigma} E(f) = \inf_{f \in \Sigma} -\eta(f) + \gamma \sum_{j,m} (G_{jm} - S_{jm}(f))^2 \tag{4}$$

where the constraints set Σ is defined as

$$\Sigma = \{ f \in \mathcal{L}^2(\mathcal{D}): a \leq f \leq b, \text{ and } f = 0 \text{ in } \mathbb{R}^2 \setminus \mathcal{D}\} \tag{5}$$

where we usually require $a > 0$ and $b < \infty$.

From the theory of penalty functions, we know that if we take $\Sigma = \mathcal{L}^2(\mathcal{D})$, then as $\gamma \to \infty$, the solution of the unconstrained minimization problem, see Bazaraa et al. (1979), converges to the solution of the equality constrained problem solved by Minerbo (1977). By taking a sufficiently large value of the penalty parameter γ, the residual error, $\sum (G_{jm} - S_{jm}(f))^2$ can be made sufficiently small to achieve the desired fidelity in the reconstructed image. In practice, of course, the values of G_{jm} are degraded by noise and we wish only to obtain a total residual error within some tolerance determined by the known accuracy of the measurements. Also, we have included a priori information in the problem formulation, by choosing a and b so that the attenuation lies within some known physical limits.

The solution of Eq. (4) can be approximated by solving the problem in a finite dimensional subspace of $\mathcal{L}^2(\mathcal{D})$, S^h. Let $\{\phi_1(x),\ldots,\phi_n(x)\}$ be a basis for S^h and for any $c = (c_1, c_2, \ldots, c_n)$ in $\mathbb{R}^n$ let

$$f_c(x) = \sum_{i=1}^{n} c_i \phi_i(x). \tag{6}$$

As shown in Smith and Zoltani (1987), for a certain class of approximating subspaces (including the one used in the present work), the solution of Eq. (4) out of S^h converges to the solution of the infinite dimensional optimization problem, as h tends to zero. Precise estimates of the deviation of the reconstructed image from that of the true phantom are out of the question at this time, as an analytic relationship between the maximum entropy solution and the true image is unknown.

The finite dimensional constrained optimization problem then, is to determine f_c which minimizes

$$E(f_c) = -\eta(\sum_i c_i\phi_i) + \gamma \sum_{j,m} (G_{jm} - S_{jm}(\sum_i c_i\phi_i))^2 \tag{7}$$

subject to

$$a \leq \sum_i c_i\phi_i(x) \leq b, \text{ for all } x \in \mathcal{D} \text{ and } a > 0.$$

To solve this nonlinear, convex programming problem, a finite element method was devised, see Smith et al. loc. cit. The optimization problem reduces to one with linear inequality constraints, i.e. determine c which minimizes

$$E(f_c) = -\eta(f_c) + \gamma \sum_{j,m}(G_{jm} - S_{jm}(f_c))^2 \tag{8}$$

subject to $a \leq c_k \leq b, \ k = 1,\ldots M.$

The problem then is to calculate the vector $c = (c_1,\ldots c_M)^T$ for given values of γ, a and b.

$$F(c) = -\eta(\sum_{k=1}^{M} c_k\phi_k(x,y)) + \gamma\sum_{j,m} [G_{jm} - \sum_{k=1}^{M} c_k S_{jm}(\phi_k)]^2 \tag{9}$$

and

$$G(c) = \nabla F(c) = \sum_{k=1}^{M} e_k\{h^2 + \int\int_{E^k} \phi_k(x,y) \ln[\sum_{k=1}^{M} c_k\hat{\phi_k}(x,y)] \ dxdy$$

$$-2\gamma\sum_{j,m} [G_{jm} - \sum_{k=1}^{M} c_k S_{jm}(\hat{\phi_k})]S_{jm}(\phi_k) \tag{10}$$

where e_k is the k-th unit vector, i.e. $e_k = (0,0,\ldots1,\ldots0)^T$ and E^k is the support of the k-th basis function, $\phi_k(x,y)$. The solution can be made arbitrarily close to the optimal solution by choosing a large enough value for γ. In this problem the penalty parameter was increased gradually and it was found that $\gamma = 75$ yielded the best result.

3. IMPLEMENTATION OF THE TECHNIQUE

3.1 THE SCANNED OBJECT

A computer code was used to generate x–ray absorption data for mathematical phantoms to be used as input to the reconstruction algorithms. Several objects, assuming a parallel mode of scanning, were studied. A source of monochromatic parallel beam x–rays and a straight line of detectors perpendicular to the x–ray transmission was used. For each of the examples, the phantoms were set within a square target grid of 30 unit cells on each side. The x–ray source was 69 units from the center of the target grid. A line of 25 detectors was placed 16 units from the center of the target grid, opposite the x–ray source. Data was produced from five projection angles equally spaced around the target. A square reconstruction grid of 30 unit cells on each side spans the width of the 25 detector bins.

The first phantom consisted of three identical solid cylinders with unit density on a 0.1 density background, see Fig.1. Note that one of the cylinders is near the boundary of the scanning area. For the second sample problem, two ellipsoids, crossing at right angles were placed at arbitrary positions within the target area. Each had a density of 0.5 while the remainder of the target grid was assigned a background density of 0.1. The volume of intersection, of course, had a density which was the sum of the two ellipsoids occupying that volume.

3.2 RESULTS

The sample phantoms were reconstructed using our new algorithm and then compared with the results of the currently preferred sparse data technique MENT, Minerbo (1977). Figure 2 on the left shows the reconstructed density plot of the first example using the new technique, while on the right the MENT results for the same problem are illustrated. The object becomes unrecognizable as soon as it is placed at or near the periphery of the scanned region. These figures are both three–dimensional grid representations of the reconstructed density (with hidden lines removed). The 31x31 grid used for the reconstruction, the actual finite element grid is shown in the middle of the 51x51 grid, with points outside the 31x31 grid being given density zero.

Figure 3 gives the reconstructed density plot of the second example. The ellipsoids of density of 0.5 stand out vividly against an undulating background of density 0.1. The five projection angles can be picked out as the "ridges" or "humps" which stretch through the object space. It is seen that the algorithm gives good resolution of sharp interior corners even when only five views are available. The code was run on a Cray 2 of the Ballistic Research Laboratory.

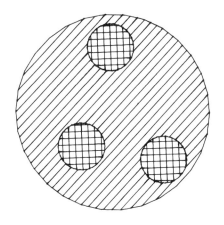

FIGURE 1.   Scanning geometry

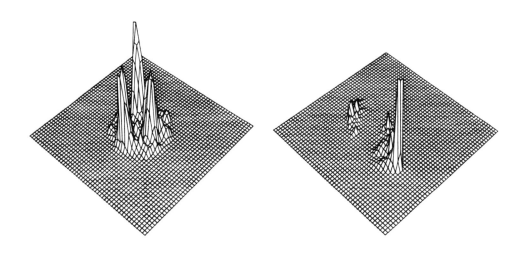

FIGURE 2.   Problem One, 5 views: a. 3D density plot using FEME Code
                                   b. MENT Results from the same data

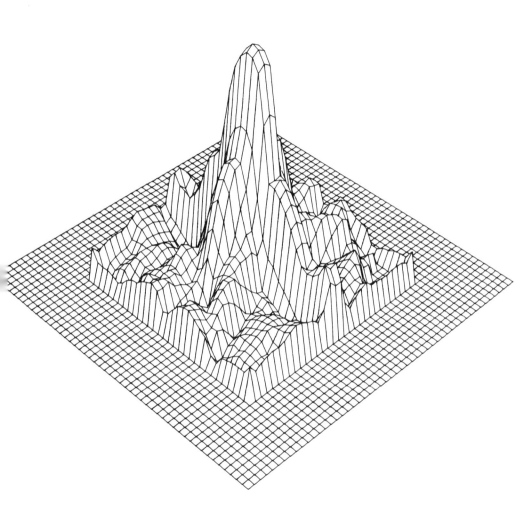

FIGURE 3. Second phantom, 5 views: 3D density plot of two
ellipsoids with their principal axes at right angles.

REFERENCES

Bazaraa M.S. and C.M. Shetty, 1979: Nonlinear Programming, John Wiley, New York.

Hanson, K. M., 1987: Bayesian and related methods in image reconstruction from incomplete data. In Stark, H. (Ed.) Image Recovery: Theory and Application. Academic Press, New York.

Justice, J.H. (Ed.) 1986: Maximum Entropy and Bayesian Methods in Applied Statistics, Cambridge University Press.

Little, R.J.A. and D.B. Rubin, 1987: Stat. Analysis with Missing Data, Wiley, New York.

A.K. Louis, 1981: Approximation of the Radon transform from samples in limited range. Lecture Notes in Medical Informatics, 8, 127–139.

G. Minerbo, 1979: A maximum entropy algorithm for reconstructing a source from projection data. Comp. Graph. Image Proc. 10, 48–68.

Potter, L.C. and K.S. Arun, 1987: Extrapolation of multi–dimensional band–limited sequence using energy concentration information. Proc.1987 IEEE Int. Conference on Acoustics, Speech and Signal Processing, Dallas, TX, 1240–43.

Smith R.T. and C.K. Zoltani, 1987: An application of the finite element method to maximum entropy tomographic image reconstruction. J. Sci. Computation 2, 283–290.

Stark, H. (Ed.) 1987: Image Recovery: Theory and Application. Academic Press, New York.

DISCUSSION

Wahba: Do you have a non-linear problem? What kind of program did you use to solve that? Did you use off-the-shelf software for that?

Zoltani: We used the MINOS code. The choice was arbitrary since several other viable alternatives exist.

OPEN DISCUSSION

SESSION C.2. - GRACE WAHBA

I was chairman of the maximum entropy techniques session so I'll just make a few short remarks. We got a very nice overview of what maximum entropy methods consist of in some cases and some nice examples by Dr. Fougere. I'll just make a few remarks on the third paper because it is somewhat close to my own interests. Dr. Zoltani did a regularization method for reconstructing a picture where the penalty function was entropy. The kinds of penalty functionals that I am used to working with are quadratic functionals rather than entropy. I guess the main remark I want to make is that the penalty functional that you choose in carrying hidden information about what you think about the "truth" that you are trying to recover. So, I think the correct statement about penalty functionals is, that the best choice of penalty functional really depends on what you are trying to recover. It depends very much on the application. Most commonly-used quadratic penalty functionals involve some kind of smoothness assumption about the solution. The entropy functional doesn't really impose smoothness information and it seems to have had its greatest success in recovering pictures, like the face of Lincoln, which we don't want to be smooth.

Fleming: Just a brief comment about trying to line up a speaker for this topic. One person I contacted told me to beware of the true believers in maximum entropy because they want to use maximum entropy for everything.

Rodgers: It seems to me that maximum entropy is really only useful for one thing and that is for estimating a probability density function, because it is a function of probability. It may be useful for estimating the parameters in a quadratic penalty function.

McCormick: I think it is worthwhile to point out that the quadratic functional is actually a special case of a form of the entropy penalty function in the sense that if you have an a priori and a posteriori and the probabilities are close together then you get a quadratic form.

THE SIMULTANEOUS RETRIEVAL OF ATMOSPHERIC TEMPERATURE AND WATER VAPOR PROFILES—APPLICATION TO MEASUREMENTS WITH THE HIGH SPECTRAL RESOLUTION INTERFEROMETER SOUNDER (HIS)

W.L. Smith, H.M. Woolf[1], H.B. Howell[1], H.-L. Huang,
and H.E. Revercomb
Cooperative Institute for Meteorological Satellite Studies
[1]NOAA/NESDIS Systems Design and Applications Branch
Madison, Wisconsin 53706, USA

ABSTRACT

The High resolution Interferometer Sounder (HIS) is the first of a new generation of passive remote sensors for achieving high vertical resolution sounding information. An aircraft version of HIS is a Michelson Interferometer with a spectral resolving power ($\lambda/\Delta\lambda$) of approximately 2000 covering a spectral range from 3.7-16.7µm. A spacecraft version of the instrument is under development for operational geostationary applications beginning around 1995. In this paper, a technique is described for the simultaneous retrieval of atmospheric profiles from the 3000 spectral radiance observations provided by the HIS. Results achieved from spectral radiances observed from the NASA U2 and ER2 aircraft are compared with radiosondes to demonstrate the improved sounding performance achieved with the HIS.

1. INTRODUCTION

The retrieval procedure presented here was developed for application to the airborne version of the High resolution Interferometer Sounder (HIS) (Smith et al., 1986) flown on the NASA U2/ER2 aircraft. The HIS spectra are achieved at a resolution of about 0.5 cm^{-1} from 600-1100 cm^{-1} (9.1-16.7µm), and 1.0 cm^{-1} resolution from 1100-2700 cm^{-1} (3.7-9.1µm). The ground resolution and spacing of the HIS observations is 2 km from the nominal aircraft altitude of 65,000 feet. The noise equivalent temperature and calibration accuracy are both about 0.1-0.2°C over most of the spectrum (Revercomb et al., 1987).

Figure 1 (Band 1) shows a clear sky spectrum of radiance observed during a flight on 19 June 1987 over northern Alabama. Also shown are radiance spectra calculated from a nearly coincident radiosonde observation for the actual "moist" atmospheric condition (Band 2) and for a hypothetical "dry" atmospheric condition (Band 3). The theoretical calculation is achieved using the line-by-line transmittance model "FASCODE" (Clough et al., 1986). For the calculation of radiance, the total transmittance is obtained considering all major optically active constitutents (H_2O, CO_2, N_2O, CH_4, O_3, N_2, NO, CO) simultaneously. The spectrum shown in Figure 1 (Band 3) illustrates the influences of

RSRM '87: ADVANCES IN
REMOTE SENSING RETRIEVAL METHODS
A. Deepak, H.E. Fleming, and J.S. Theon (Eds.)

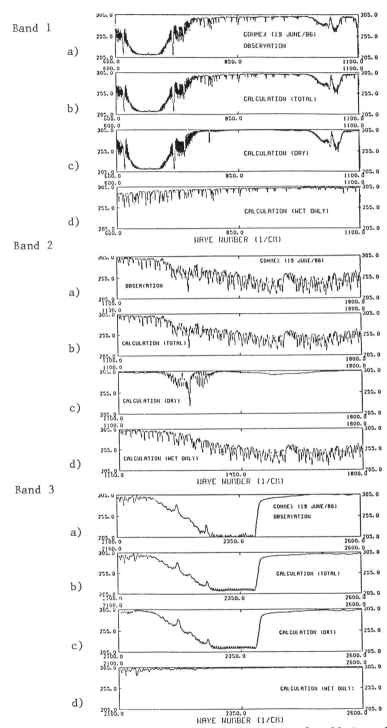

FIGURE 1.   Observed and calculated spectra for 19 June 1987.

all the gases with the exception of water vapor whereas the spectrum
shown illustrates the radiance contribution by water vapor alone. As
can be seen, much of the spectrum is due to spectrally overlapping
radiance contributions by water vapor and the more uniformly mixed
"dry" atmospheric constituents. Thus, it is desirable to utilize a
temperature and moisture profile retrieval algorithm which simultaneously accounts for the spectrally overlapping radiance contributions
by water vapor and the "dry" atmospheric constituents. Such an
algorithm is developed in this paper.

2. SOUNDING RETRIEVAL PROCEDURE

The perturbation form of the radiative transfer equation has been
shown by Smith et al. (1984) to be

$$\delta T_B = \delta T_s f_s \hat{\tau}_s - \int_o^P {}^s \delta T f \frac{d\hat{\tau}}{dp} dp + \int_o^P {}^s \delta \tau f \frac{dT}{dp} dp \tag{1}$$

where δT_B is the deviation of the observed brightness temperature, T_B,
for a given wavelength, from that corresponding to a reference or
guess atmospheric temperature and water vapor profile condition, δT_s
is the deviation of the actual surface skin temperature from the
reference condition, δT is the deviation of the true atmospheric
temperature, T, from its reference, $\hat{T}$, $\delta \tau$ is the deviation of the true
spectral transmittance profile from that corresponding to the refer-
ence water vapor profile condition, $\hat{\tau}$, and f is the ratio
$(\partial B/\partial T)/(\partial B/\partial T_B)$ where B is Planck radiance.

With the HIS, we have 3000 spectral channels sensitive to the
surface temperature and the temperature and water vapor profile. The
sensitivity of the radiances to temperature and water vapor varies
greatly across the spectrum due to the degree of overlap between water
vapor absorption lines and lines due to uniformly mixed gases such as
CO_2 and N_2O. Also, the Planck function dependence upon wavelength and
temperature contributes to the spectrally varying sensitivity.
Because of the mutual dependence of radiance on temperature and water
vapor, it is desirable to solve for the temperature and water vapor
profile simultaneously from the complete set of spectral observations.
A simultaneous solution is achieved by noting that for the very high
HIS spectral resolution, the atmospheric transmittance can be represented as the product of those for water vapor and dry air. Thus,

$$\hat{\tau} = \tau_d \hat{\tau}_w \tag{2}$$

In Eq. (2), τ_d is the transmittance of the "dry" atmosphere which is a
function of the uniformly mixed gases whose concentrations are known
and $\hat{\tau}_w$ is the spectral transmittance due to water vapor alone. (Ozone
is included in τ_d.) Using Eq. (2) in Eq. (1) yields

$$\delta T_B = f_s \delta T_s \hat{\tau}_s - \int_o^{P_s} f\delta T \hat{\tau}_w \frac{d\tau_d}{dp} dp$$

$$- \int_o^{P_s} f\delta T \tau_d \frac{d\hat{\tau}_w}{dp} dp + \int_o^{P_s} f\tau_d \delta\tau_w \frac{dT}{dp} dp \tag{3}$$

In order to solve Eq. (3) for the detailed structure of the water vapor profile, a methodology based upon an idea first expressed by Rosenkranz et al. (1982) is developed. Let

$$\delta\tau_w \cong \frac{d\hat{\tau}_w}{d\hat{U}} \delta U \tag{4}$$

where U is the precipitable water vapor profile. Furthermore, we note that

$$\delta U = \frac{d\hat{U}}{dT} (T - T_w) \tag{5}$$

where T_w is the temperature profile as a function of the true precipitable water vapor profile U whereas T is the temperature profile as a function of the reference precipitable water vapor profile $\hat{U}$. Substituting Eq. (5) and Eq. (4) into Eq. (3) yields

$$\delta T_B = f_s \delta T_s \hat{\tau}_s - \int_o^{P_s} f\delta T \hat{\tau} \frac{d\tau_d}{dp} dp - \int_o^{\hat{U}_s} f\delta T_w \tau_d \frac{d\hat{\tau}_w}{d\hat{U}} d\hat{U} \tag{6}$$

where $\delta T = T - \hat{T}$ and $\delta T_w = T_w - \hat{T}$. Equation (6) describes the perturbation of the observed brightness temperature spectrum from a reference condition in terms of three variables: the deviation of the true surface skin temperature from the reference, the deviation of the atmospheric temperature profile from the reference profile as a function of pressure and the deviation of the atmospheric temperature profile from the reference profile as a function of atmospheric precipitable water vapor content. Note that if the reference precipitable water vapor concentration as a function of pressure is correct then $\hat{\tau}_w = \tau_w$, $\delta T_w = \delta T$ so that the two integrals combine to yield the normal form of the radiative transfer equation ($f\delta T_s \tau_s - \int_o^{P_s} f\delta T$ $(d\tau/dp)dp$). Otherwise, $\delta T_w \neq \delta T$ so that the T and $\hat{T}$ profiles resulting from the solution of Eq. (6) will differ from each other depending upon the error in the presumed water vapor condition.

The numerical solution of Eq. (6) is achieved by using a basis function representation of δT and δT_w; namely,

$$\begin{matrix} \delta T(p) \\ \text{and} \\ \delta T_w(p) \end{matrix} = \sum_{i=1}^{N} C_i \phi_i(p) \qquad (7)$$

where $\phi_i(p) = (p/p_i*) \, \text{Exp}(-p/p_i*)$ where p is atmospheric pressure and p_i* is the pressure of the "standard" atmospheric pressure levels (1000, 850, 700, 500, 400, 300, 250, 200, 150, 100, 70, 50 mb). The functions $\phi_i(p)$ can be shown to be equivalent to Planck radiance weighting functions $(d\tau/d\ln p)$ of a uniformly mixed absorbing constituent peaking at the pressure p_i*. Substituting Eq. (7) into Eq. (6) gives the system of equations

$$\delta T_{Bj} = \sum_{i=0}^{M} C_i A_{ij} \qquad j = 1, 2, , , K \qquad (8)$$

where K is the number of spectral channels and

$$C_0 = \delta T_s \, , \, A_0 = f_{sj} \hat{\tau}_{sj}$$

$$A_{ij} = \int_0^{P_s} \phi_i(p) f_j \hat{\tau}_{wj}(p) d\tau_{dj}(p)$$

$$i = 1, 2, , , M/2$$

$$A_{ij} = \int_0^{P_s} \phi_i(p) f_j \hat{\tau}_{dj}(p) d\hat{\tau}_{wj}(p)$$

$$i = M/2 + 1, M/2 + 2, , , M$$

where M/2 is the number of "standard" atmospheric pressure levels. In the case of the aircraft HIS experiment, K is approximately 2500 and M is 24. Thus, Eq. (8) represents a system of 2500 equations with 25 unknowns.

The solution of Eq. (8) is achieved using the conditioned least squares inverse solution

$$\vec{C} = (A^T A + \gamma I)^{-1} A^T \vec{t}_b = A^* \vec{t}_b \qquad (9)$$

where $\vec{t}_b$ is the vector of brightness temperature observations (δT_B), A^* is the solution matrix, I is the identity matrix, and γ is a Lagrangian multiplier $(=10^{-3})$ used to condition the matrix $A^T A$ for inversion. The superscript T denotes matrix transposition.

When utilizing brightness temperature observations it is proper to account for the errors in δT_B due to measurement errors and to uncertainties in the atmospheric transmittance observations used to calculate T_B from $T(p)$ and $U(p)$. This error can be defined as

$$\varepsilon_j = \sqrt{\rho_j^2 + \eta_j^2} \tag{10}$$

where $\rho = r/(\partial B/\partial T_B)$ is the expected random brightness temperature measurement error (r_j being the radiance error) specified from the observations when viewing a constant calibration source, and η_j is the random difference between the observed and calculated brightness temperature for known atmospheric conditions. In order to account for the spectral dependence of ε_j, both sides of Eq. (8) are scaled by ε_j (i.e., $\delta T_{Bj} = \delta T_{Bj}/\varepsilon_j$ and $A_{ij}=A_{ij}/\varepsilon_j$). Thus, the spectral regions whose observation error and atmospheric transmittance uncertainties are smallest carry the greatest weight in the solution for the atmospheric profiles.

Having determined the coefficient vector $\vec{C}$, the atmospheric profiles $T(p)$ and $T_w(p)$ are obtained from Eq. (7). The water vapor concentration profile $U(p)$ is then obtained using Eq. (5) in the form

$$U(p) = \hat{U}(p) + \frac{d\hat{U}(p)}{dT(p)} [T(p) - T_w(p)] \tag{11}$$

or

$$U(p) = \hat{U}(p) [1+ \frac{T(p) - T_w(p)}{d\hat{T}/d\ln\hat{U}}] \tag{12}$$

The mixing ratio profile is then calculated using the relation

$$q(p) = \hat{q}(p) \frac{dU(p)/dp}{d\hat{U}(p)/dp} = \hat{q}(p) \frac{dU(p)}{d\hat{U}(p)} \tag{13}$$

where all the vertical derivations are computed as centered finite differences.

3. ATMOSPHERIC TRANSMITTANCES AND CALCULATED RADIANCE SPECTRA

The atmospheric transmittance functions for water vapor, $\tau_w(p)$, ar $\tau_d(p)$ are calculated using the line-by-line transmittance model "FASCODE" (Clough et al., 1986). For the calculation of the radiance corresponding to the reference sounding condition, the total transmittance is calculated considering all the major optically active constituents simultaneously. In order to represent the HIS spectra, the transmittances are calculated at high spectral resolution ($\sim$0.06 cm^{-1}) and then transformed to an interferogram incorporating the spectral response and finite field of view properties of the airborne

HIS instrument. The interferogram is

$$\tau_x(p) = \int_0^\infty \tau_\nu(p)\phi_\nu \ \frac{\sin 2\pi\nu x\alpha^2/4}{2\pi\nu x\alpha^2/4} \ \cos(2\pi\nu x)d\nu \qquad (14)$$

where x is delay, $\tau_\nu(p)$ is the spectral transmittance as a function of wavenumber, ν, α is the half angle field of view of the HIS and ϕ_ν is the spectral response function due primarily to the interference filter used to limit the incoming radiation. Then $\tau_\nu(p)$ for the HIS resolution is achieved using a cosine transform of $\tau_x(p)$ using precisely the same delay cutoff and interferogram apodization function used to process the HIS interferogram radiance observations.

$$\tau_\nu(p) = 2 \int_0^{X_m} \tau_x(p)\Psi_x \ \cos \ (2\pi\nu x)dx \qquad (15)$$

where X_m=1.4 for band 1 (600-1100 cm^{-1}), X_m=0.9 for band 2 (1100-1900 cm^{-1}), and band 3 (2000- 2700 cm^{-1}). In the processing of the HIS spectra, the apodization is

$$\Psi_x = (1 - x^2/X_m^2)^2 \qquad (16)$$

Returning to Fig. 1, one can see that a continuum of atmospheric levels are descriminated by the HIS observed spectrum, with the temperature and water vapor sensitivity varying greatly across the spectrum. Although one could limit the spectral regions used for the retrieval process to those which possess distinctly different radiance contributions from water vapor and the "dry" atmospheric gases used for temperature profiling, that is not the strategy employed here. Instead, all wavelengths are utilized simultaneously for the determination of atmospheric temperature and the absorbing constituents.

Differences between observed and radiosonde calculated radiance generally exceed those due to the single sample observation radiance errors of 0.2 mw/m^2-cm^{-1}-sr for band 1, 0.1 mw/m^2-cm^{-1}-sr for band 2, and 0.01 mw/m^2-cm^{-1}-sr for band 3. These differences are due to errors in simulating the spectral characteristics of the HIS observations in the radiance computations, errors in the radiosonde observations, errors in the FASCODE transmittance computations, and errors due to quadrature in the radiative transfer computations. However, these discrepancies are mainly systematic and, therefore, can be accounted for in the retrieval algorithm by correcting the calculated spectra for this systematic discrepancy with observations using the approximation

$$\hat{T}_B^c(\nu) = \hat{T}_B(\nu) + \beta(\nu) \qquad (17)$$

where $\hat{T}_B^c(\nu)$ is the corrected radiative transfer computation of the brightness temperature spectrum and $\beta(\nu)$ is the systematic difference between observed and calculated brightness temperature spectra for known atmospheric conditions. The systematic difference between observed and calculated radiance, $\beta(\nu)$, can be determined using a relatively small number of radiance/radiosonde coincident observations. The random difference, $\eta(\nu)$ of Eq. (10), needs to be evaluated on the basis of a large number of intercomparisons between observed and calculated spectra.

4. EXAMPLE RESULTS

The algorithm developed here is being applied to HIS observations achieved during the 1986 COoperative Huntsville Meteorological EXperiment (COHMEX) from the NASA U2 and ER2 aircraft. As an example, results for June 15, 1986 (a U2 case) and June 19, 1986 (an ER2 case) are shown. The retrievals for both days were retrieved using a single retrieval coefficient matrix (A* of Eq. (9)) specified from atmospheric transmittance computations for a weather service forecast temperature and moisture profile for Huntsville, Alabama for 12 GMT June 15, 1986. The variability of the atmospheric transmittances due to variables other than water vapor is small such that a single solution matrix, A*, suffices. The 12 GMT June 15, 1986 forecast profile for Huntsville, Alabama was also used as the "guess" profile for the temperature and moisture solutions. Figure 2 shows a comparison of this first guess profile with an average of the special radiosonde observations for June 15 and June 19, 1986 during the time period of the HIS observations.

Cloud images from the GOES satellite at the central time of the HIS observations reveal that the COHMEX observation area on June 15 possessed small element fair weather cumulus clouds whereas on June 19 the region was largely cloud-free during the time of the HIS observations.

Figure 3 shows two example comparisons of HIS soundings and nearly space coincident radiosonde observations. The observation times differed by as much as three hours. The comparisons are illustrated on a SKEW-T/LOG-P diagram in order to accentuate differences. Despite the lack of time coincidence between the two types of soundings, the agreement is striking. However, the vertical resolution and accuracy may not yet completely meet the theoretical expectations of the HIS sounding approach (Smith et al., 1979, 1981; Spänkuch et al., 1986). Retrieval errors, although small, still exist due to uncertainties in the atmospheric transmittance functions used to calculate the "guess" radiance spectrum and the solution coefficient matrix. In fact, it was just discovered during the preparation of this manuscript that the use of an order of magnitude higher spectral resolution in performing the Direct Fourier Transform (Eq. (14)) of the atmospheric transmittances greatly reduces the discrepancy between observed and calculated radiance. This improvement of the atmospheric transmittance calculations should yield a significant improvement in the vertical resolution of the HIS soundings beyond that displayed in Fig. 3.

Finally, Figs. 4, 5, and 6 show HIS derived temperatures and dewpoint temperatures for several atmospheric pressure levels observed

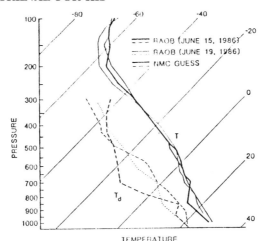

FIGURE 2.   Comparison of NMC "guess" with average of special COHMEX
radiosondes for 15 and 19 June, 1986.

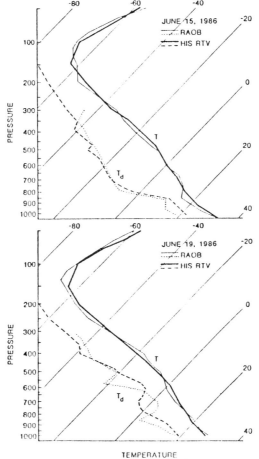

FIGURE 3.   Comparison of HIS temperature (T) and dewpoint (Td)
retrieval (RTV) with radiosondes (RAOB).

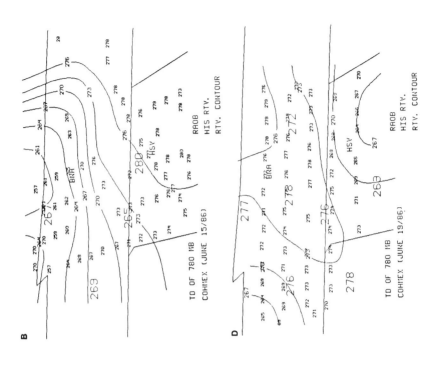

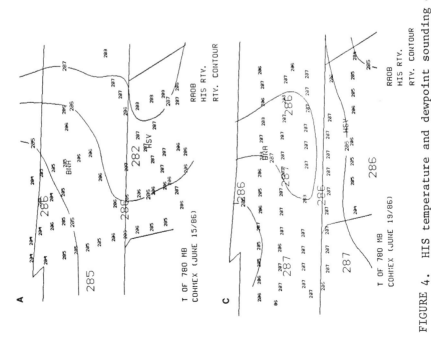

FIGURE 4. HIS temperature and dewpoint sounding data along the aircraft flight track with contour analyses superimposed for 15 and 19 June, 1986. Special COHMEX radiosonde values plotted as large numbers for intercomparison. Figures (a) and (c) are for temperature and (b) and (d) are for

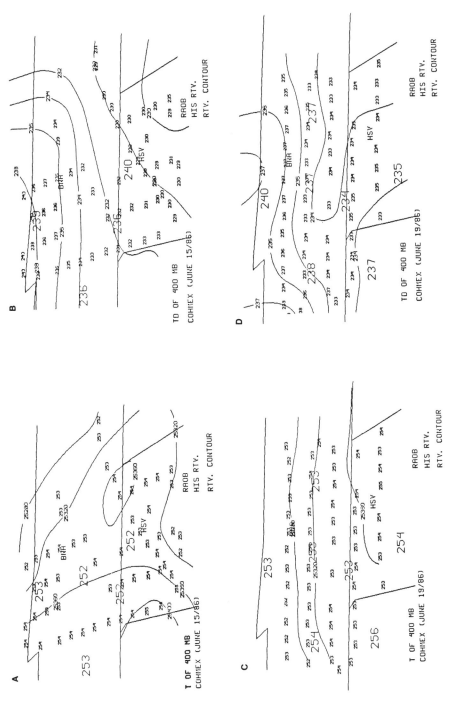

FIGURE 5. Same as Fig. 4, except for the 400 mb level.

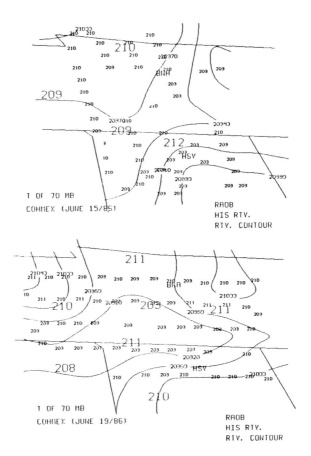

FIGURE 6.   Same as Fig. 4, except for the 70 mb.

along the aircraft flight tracks on June 15 and June 19, 1986.
Available radiosonde observations are plotted for intercomparison. As
can be seen, the HIS retrieval data are spatially consistent to within
1°C. It is particularly noteworthy that the HIS is able to resolve
small scale spatial variations of atmospheric temperature and dewpoint
which are in qualitative agreement with the few and much more coarsely
spaced special radiosonde data. Also note the repeatability of the
HIS soundings over the same geographical locations at intersections of
the aircraft flight tracks.

5.   CONCLUSION AND FURTHER CONSIDERATIONS

The profile retrieval alorithm outlined above appears to suffice
for the inference of atmospheric soundings from the high spectral
resolution HIS interferometer. The errors which remain are believed
to be due entirely to errors associated with atmospheric transmittance
and the forward radiative transfer computation. In fact, during the
writing of this manuscript, it was discovered that significant errors

in the atmospheric transmittances used for profile retrievals result
when insufficient spectral quadrature is used to simulate the observation spectral properties. These numerical errors are currently being
eliminated, and as a result, even better vertical resolution and
higher accuracy of future HIS retrievals are expected.

We believe that the HIS sounding capabilities represent a
revolutionary advance in passive remote sensing. The airborne
instrument has been successfully demonstrated, and has provided unique
observations of mesoscale temperature and moisture structures during
field programs held during 1986. The HIS can serve as an aircraft
facility vertical sounder during the next decade.

Possibly most important is the advance which will be realized in
global weather analysis and forecasting once the HIS is implemented on
satellite platforms. The experimental results being achieved from the
aircraft instrument will hopefully inspire space agencies to proceed
as rapidly as practically possible toward a spacecraft application of
the HIS technology.

Finally, the topic covered in this paper was limited to
temperature and moisture sounding. The same principle of simultaneous
retrieval is being expanded to include ozone profiling. Also, the HIS
spectra contain total abundance and limited vertical profile
information on CH_4, N_2O, CO_2, and NO_x and the simultaneous retrieval
of these important gaseous constituents is being pursued. The high
spectral resolution of HIS will enable cloud effects to be handled
more precisely. An algorithm for the retrieval of cloud altitude,
amount, and emissivity simultaneously with the atmospheric temperature
and absorbing constituent profiles is under development and results
from this complete solution will be reported at the next International
Workshop on Remote Sensing Retrieval Methods.

ACKNOWLEDGMENTS

The authors express the sincere appreciation to James Fischer
(NOAA/NESDIS) and numerous staff members of NASA Headquarters and the
NASA Ames aircraft flight facility for their support of HIS instrument
development and successful aircraft demonstration. We are grateful to
numerous personnel at the University of Wisconsin, University of
Denver, Bomem, Inc., and Santa Barbara Research Center for their
participation in the construction and testing of the HIS. The expert
preparation of this manuscript by Laura Beckett is greatly appreciated. This research was funded by NASA Grant NAS8-36169 and NOAA
Grant NA87AA-H-SP085.

REFERENCES

Clough, S. A., F. X. Kneizys, E. P. Shettle, G. P. Anderson, 1986:
    6th Conf. on Atmospheric Radiation, AMS, Boston, MA, p. 141.

Revercomb, H. E., D. D. LaPorte, W. L. Smith, H. Buijs, D. G. Murcray, F. J. Murcray, and L. A. Sromovsky, 1987: High-altitude aircraft measurements of upwelling radiance: prelude to FTIR from geosynchronous satellite, 6th Intl. Conf. on Fourier Transform Spectroscopy, Vienna, Austria, 24–28 August, in press.

Rosenkranz, P. W., M. J. Komichak and D. H. Staelin, 1982: A method for estimation of atmospheric water vapor profiles by microwave radiometry. J. Appl. Meteor., 21, 1364–1370.

Smith, W. L., H. B. Howell and H. M. Woolf, 1979: The use of interferometric radiance measurements for sounding the atmosphere. J. Atmos. Sci., Vol. 36, No. 4, 566–575.

Smith, W. L., V. E. Suomi, W. P. Menzel, H. M. Woolf, L. A. Sromovsky, H. E. Revercomb, C. M. Hayden, D. N. Erickson, and F. R. Mosher, 1981: First sounding results from VAS-D. Bull. Amer. Meteor. Soc., Vol. 62, No. 2, pp. 232–236.

Smith, W. L., H. E. Revercomb, H. B. Howell, and H. M. Woolf, 1983: HIS - a satellite instrument to observe temperature and moisture profiles with high vertical resolution. Fifth Conf. on Atmospheric Radiation, Baltimore, MD, October 31–November 4.

Smith, W. L. and H. M. Woolf, 1984: Improved vertical soundings from an amalgamation of polar and geostationary radiance observations Conf. on Satellite/Remote Sensing and Applications, Clearwater Beach, FL, June 25–29.

Smith, W. L., H. E. Revercomb, H. B. Howell, H. M. Woolf, and D. D. LaPorte, 1986: The High resolution Interferometer Sounder (HIS) CIMSS View, Vol. II, No. 3, Fall.

Spänkuch, D., J. Güldner and W. Döhler, 1986: Investigations on temperature soundings using partial interferograms. Beitr. Phys Atmosph., Vol. 60, No. 1, 103–122.

DISCUSSION

NO QUESTIONS

A UNIFIED RETRIEVAL METHODOLOGY FOR
THE DMSP METEOROLOGICAL SENSORS[1]

R.G. Isaacs
Atmospheric and Environmental Research, Inc.
Cambridge, Massachusetts 02138, USA

ABSTRACT

An overview is presented for a unified retrieval methodology applicable to the data sets of the Defense Meteorological Satellite Program (DMSP) meteorological sensor payload. Desired quantities include temperature and water vapor profiles, surface temperature and emissivity, and cloud properties. The hybrid retrieval approach employs both statistical and physical retrieval concepts. The approach exploits existing DMSP operational experience with statistical methods to provide a first guess capability. The first guess is upgraded, if necessary, using a physically based, simultaneous retrieval. Required cloud properties are obtained by image processing high spatial resolution visible and infrared data from a colocated imager.

1. INTRODUCTION

The sensor payload of the Defense Meteorological Satellite Program (DMSP) spacecraft of the 1990's will consist of a visible/infrared imager (the operational linescan system or OLS), a microwave temperature sounder (SSM/T-1), a millimeter wave water vapor sounder (SSM/T-2), and a microwave imager (SSM/I). The current OLS imager provides high spatial resolution, global cloud imagery. Notably, all the other sensors are millimeter/microwave instruments. Of these microwave mission sensors, the SSM/T-1 and SSM/I are currently operational (Falcone and Isaacs, 1987). The SSM/T-2 is scheduled for launch in the early 1990s. The attributes of these sensors are summarized in Table 1. Meteorological data requirements tasked to this sensor complement include the acquisition of cloud information, temperature and water vapor profiles, precipitation, and surface properties. One important application of these data products is global numerical weather prediction (Isaacs et al., 1986a).

Current operational analysis procedures for the OLS, SSM/T-1, and SSM/I treat each sensor data stream independent of the others. OLS imagery is processed into global cloud property fields using an automated nephanalysis algorithm (Fye, 1978). The

---

[1]This work sponsored by the Air Force Systems Command, Air Force Geophysics Laboratory under contracts F19628-84-C-0134, F19628-85-C-102, and F19628-86-C-0141.

TABLE 1.   DMSP METEOROLOGICAL SENSORS

Instrument	Frequency or Wavelength	Polari- zation (H or V)	FOV (km)	Response	NEΔT (K)
SSM/T	50.5 GHz	H	200	surface	0.6
	53.2	H	200	T at 2 km	0.4
	54.35	H	200	T at 6 km	0.4
	54.9	H	200	T at 10 km	0.4
	58.825	V	200	T at 16 km	0.4
	59.4	V	200	T at 22 km	0.4
	58.4	V	200	T at 30 km	0.5
SSM/I	19.35	H and V	50	surface	0.6
	22.235	V	50	water vapor	0.6
	37.0	H and V	25	clouds, rain	0.8
	85.5	H and V	12.5	clouds, snow	1.1
SSM/T-2	90.0	V	100	surface, water vapor	0.6
	150.0	V	60	surface, water vapor	0.6
	183.31±1	V	50	water vapor	0.8
	183.31±3	V	50	water vapor	0.6
	183.31±7	V	50	water vapor	0.6
OLS	0.4-1.1μm		0.6	surface/clouds	-
	10.5-12.5μm		2.4	surface/clouds	-

retrieval scheme for temperature sounding by the SSM/T-1 is based on regression of SSM/T-1 brightness temperature data against desired mandatory level temperatures (Rigone and Stogryn, 1977) and an analogous statistical approach will be used in the determination of meteorological parameters such as cloud liquid water content and precipitation (among others) from the SSM/I data (Lo, 1983). These approaches share a common heritage in the "D" matrix technique described by Gaut et al. (1975). Data from the SSM/T-1 and SSM/T-2 will be integrated together in a statistical retrieval of water vapor profiles (cf. Isaacs, 1987).

For a variety of reasons, the operational approach described above could be improved. The main criticisms are: (1) the lack of a multispectral perspective, (2) reliance on statistical retrieval approaches, which produce retrieval fields with reduced variance properties, fail to treat inherent problem nonlinearities, and provide little opportunity to monitor retrieval quality, and (3) neglect of some physical aspects of the retrieval problem, such as the effect of cloud on millimeter wave brightness temperatures. To address these issues, an alternative retrieval scheme has been developed. The approach outlined below is by no means statistically optimal. However, it does attempt to address the difficulties cited above and, in particular, integrates available data sources in a unified, multispectral retrieval constrained by radiative transfer principles.

2.   RETRIEVAL APPROACH

The retrieval approach for the DMSP illustrated in Fig. 1 employs physical considerations and allows for the incorporation of all data sources. Recognizing potential operational constraints, an attempt has been made to build on the attributes of the existing DMSP retrieval capability and experience. The microwave sensor data, $T_b{}^o$, is employed with the "D" matrix

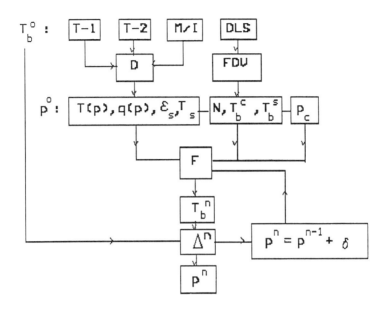

FIGURE 1. Schematic of unified DMSP retrieval scheme.

statistical retrieval to provide first guesses for the desired parameters, P^o: temperature, $T(p)$, and water vapor profiles, $q(p)$, surface emissivity, ϵ_s, and temperature, T_s. Simulated brightness temperatures, $T_b{}^n$, are then evaluated to examine the consistency of the first guesses with the observations. The forward problem calculation (denoted by F in Fig. 1) is accomplished using the RADTRAN simulation code (Falcone et al., 1982) as modified by Isaacs et al. (1985). When the residuals, Δ^n, (i.e. differences between simulated and observed brightness temperatures) are small, the process terminates. However, when residuals are larger than a preset tolerance (usually determined by the sensor noise equivalent brightness temperatures (NEΔT) and scene noise), the procedure goes on to adjust the first guess profiles. This adjustment is accomplished by using the residuals in a simultaneous physical retrieval.

Cloud or precipitation in the field-of-view (FOV) of the microwave sensors can be problematic. Precipitation will generally preclude soundings of temperature and moisture and the determination of surface properties. Quality control flags for precipitation (as well as precipitation amounts) can be obtained from the SSM/I statistical retrieval (Lo, 1983; Jin and Isaacs, 1987). Isaacs and Deblonde (1987) have discussed the potential impact of cloud on statistical millimeter wave water vapor retrievals and evaluated the sensitivity of these channels to cloud presence.   Cloud fields from the DMSP OLS (using an appropriately spatially averaged subset of visible and infrared imagery) aid in cloud/no cloud discrimination.   To determine first guess cloud properties within the FOV necessary to accomplish the physical retrieval step, the high spatial resolution OLS imager data is utilized.   Image processing of this data within the relatively larger microwave footprint provides first guess cloud coverage, N, and equivalent brightness temperatures (EBTs) for cloud top and surface, $T_b^{c,s}$.   Cloud top brightness temperature along with the first guess temperature profile yields a first guess cloud top pressure, p_c.   Over the oceans, cloud properties derived from the OLS imager data are supplemented by information on cloud integrated liquid water content (ILWC) available from the SSM/I.   Cloud ILWC provides a parameterization of cloud optical thickness and emissivity.   These cloud properties are required to treat the effect of cloud on the SSM/T-2 sensor data and therefore are input to the forward problem.

3. STATISTICAL (FIRST GUESS) RESULTS

The integrated scheme described above has been tested in simulation assuming cloud free conditions.   These results are described in the next two sections.   First guesses for temperature profile, integrated water vapor profile, surface temperature, and emissivity are based on a statistical retrieval employing SSM/T-1, SSM/T-2, and SSM/I data.   Data was simulated for these sensors using the RADTRAN microwave radiative transfer code (Falcone et al., 1982) and ensembles of midlatitude and tropical atmospheres from a radiosonde data set consisting of a total of 400 soundings.   Sensor noise was introduced by sampling Gaussian distributions with zero means and standard deviations equal to the NEΔT for each channel given in Table 1.   Scene noise was ignored as were the spatial averaging effects of differing sensor FOVs.   Emissivity values were calculated from an ocean surface reflectance model assuming calm seas.   The resultant values were varied assuming a Gaussian distribution with the calculated value as mean and an assumed standard deviation.   While surface emissivity per se is not a required meteorological parameter,   it provides information on surface winds and sea ice over the ocean, and vegetation, soil moisture, and snow cover over the land (Isaacs et al, 1986b).   The surface temperature was set equal to the 1000 mb atmospheric temperature.

The retrieval results for various instrument combinations are shown in Figs. 2 and 3, for midlatitude and tropical cases, respectively.   Shown are fraction of unexplained variance (FUV) values for each instrument combination.   Perfect retrievals have an FUV of 0.0 while those which perform no better than climatology have FUVs of 1.0. (FUVs greater than 1.0 mean that climatology is better than the retrieval.)   Fig. 2 shows that the combination of T-1 and T-2 data provide a more accurate integrated water vapor profile than the use of T-2 data alone.   Especially note the potential improvement in temperature profile retrieval near the

surface obtained when combining SSM/T-1 data with that from the
SSM/T-2 or SSM/I. The importance of the SSM/I data combined with
the SSM/T data for surface parameter retrievals is also reflected
in the surface temperature and emissivity results. These results

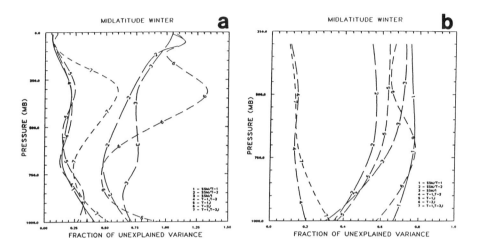

FIGURE 2.   First guess retrieval FUVs for various sensor
             combinations (midlatitude statistics).

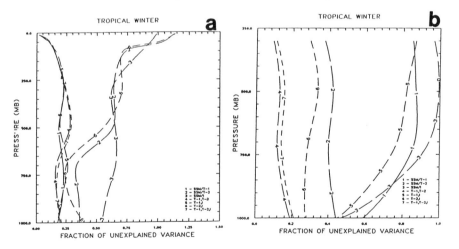

FIGURE 3.   First guess retrieval FUVs for various sensor
             combinations (tropical statistics).

are shown in Table 2.   These midlatitude results can be compared
to those for the tropics (Fig. 3). Due to the smaller variance of
tropical temperature profiles near the surface compared to
midlatitude counterparts, there is little advantage to the
combined data set. However, in the tropics the combination of

SSM/T and SSM/I data still provide the most accurate retrievals of surface emissivity.

TABLE 2.   STANDARD DEVIATION OF FIRST GUESS RETRIEVAL ERRORS FOR SURFACE EMISSIVITY AND TEMPERATURE

PARAM.	SSM/T1	SSM/T2	SSM/I	1&2	1&I	2&I	1,2&I	P S.D.	P MEAN
				MIDLATITUDE STATISTICS					
T_s	4.4223	5.6669	5.7308	4.1072	3.5782	6.6991	5.0054	6.8623	277.7990
ϵ_s	0.0201	0.0219	0.0188	0.0171	0.0128	0.0235	0.0185	0.1232	0.8041
				TROPICAL STATISTICS					
T_s	2.2754	3.2836	3.9874	2.4830	2.2967	3.4474	2.4270	5.4435	296.2430
ϵ_s	0.0173	0.0252	0.0116	0.0137	0.0092	0.0119	0.0108	0.1232	0.8041

4.   PHYSICAL RETRIEVAL RESULTS

4.1   THEORY

Statistical first guess results, P^0, are assessed for consistency with the brightness temperature observations, $T_b{}^0$, by calculating a set of nth guess synthetic sensor data, $T_b{}^n$. The RADTRAN algorithm is employed as the forward problem generator, F. Adjustments to the first guesses are accomplished using the brightness temperature residuals Δ^n and physical retrieval concepts based on both Susskind et al. (1984) and the simultaneous retrieval method of Smith et al. (1985). Monitoring residuals also provides the means to quality control each retrieval.

The radiative transfer equation for microwave frequencies is:

$$T_{b\nu} = [\epsilon_s T_s + (1 - \epsilon_s) \int_0^{P_s} T(p) d\tau'_\nu] \; \tau_\nu(p_s) + \int_{P_s}^0 T(p) d\tau_\nu \quad (1)$$

where

$$\tau_\nu(p) = \exp [- \int_0^p k (\nu, p') dp'/\mu] \quad (2)$$

and

$$\tau_\nu'(p) = \exp [- \int_p^{P_s} k(\nu, p') dp'/\mu]. \quad (3)$$

Here, μ is the cosine of the path zenith angle, τ_ν and τ'_ν are the upward and downward transmission functions, respectively, and p_s is the surface pressure. Specular surface reflection is assumed.

Differentiating equation (1) with respect to the desired variables U, T, T_s and ϵ_s and dropping the frequency indices, one obtains:

$$\Delta^n = \frac{\partial T_b}{\partial \epsilon_s} \delta \epsilon_s + \frac{\partial T_b}{\partial T_s} \delta T_s + \frac{\partial T_b}{\partial T} \delta T + \frac{\partial T_b}{\partial U} \delta U \tag{4}$$

where:

$$\frac{\partial T_b}{\partial \epsilon_s} = [T_s - \int_0^{p_s} T(p) d\tau'] \tau_{p_s} \tag{5}$$

$$\frac{\partial T_b}{\partial T_s} = \epsilon_s \tau_{p_s} \tag{6}$$

$$\delta T \frac{\partial T_b}{\partial T} = \tau_{p_s}(1 - \epsilon_s) \left\{ \int_0^{p_s} \delta T d\tau' \right\} + \int_{p_s}^0 \delta T d\tau . \tag{7}$$

$$\delta U \frac{\partial T_b}{\partial U} = (\epsilon_s - 1) [T_s - \int_0^{p_s} T(p) d\tau' + \tau_s T(0)] \frac{\partial \tau_s}{\partial U} \delta U$$

$$+ \int_0^{p_s} [\frac{\partial \tau}{\partial U} + \tau_s(\epsilon_s - 1) \frac{\partial \tau'}{\partial U}] \delta U dT \tag{8}$$

Parts of δU $(\partial T_b/\partial U)$ have been obtained using integration by parts. Note that $(\partial \tau(p_s)/\partial T)$ and the variation with respect to surface pressure have been ignored.

The quantities δU and δT are expanded in series of the eigenvectors of the covariance matrices (EOFs) of U and T, with N_u and N_t terms, respectively.

$$\delta U(p) = \sum_{j=1}^{N_u} A_j \phi_j(p); \qquad \delta T(p) = \sum_{j=N_u+1}^{N_u + N_t} A_j \phi_j(p) \tag{9-12}$$

$$\delta T_s = A_{N_u + N_t + 1}; \qquad \delta \epsilon_s = A_{N_u + N_t + 2}$$

Upon substitution of (9 - 12) into (4) a linear equation in the coefficient A_j is obtained:

$$\Delta^n = \sum_{j=1}^{N_u + N_t + 2} A_j \Phi_j \tag{13}$$

where the Φ_j's are functions of the terms in equations (5-8). The desired difference terms, δ, in the relaxation equation:

$$p^n = p^{n-1} + \delta \tag{14}$$

are available by solving (13) for the A_j's.

The ridge stabilized, least squares solution is given by:

$$A = (\Phi^T \Phi + \sigma H)^{-1} \Phi^T \Delta^n \tag{15}$$

where σ is the ridge parameter and the diagonal elements of the H matrix are the inverse of the fractional variance due to each EOF.

4.2 PHYSICAL RETRIEVAL RESULTS

Results of the physical retrieval for the tropical tempera-ture and integrated water vapor profiles are shown in Figs. 4a,b, respectively. The statistical first guesses described in the previous section are employed and the necessary EOFs of tempera-ture and water vapor are evaluated from the radiosonde data set used in the generation of the first guess retrieval statistics. Physical adjustment of the temperature profile results in a slight increase of RMS error in the vicinity of the tropopause and a small improvement in accuracy near the surface. The tropopause problem is attributable both to EOF truncation representation errors and the generally weak contribution functions in this region. There is a much more noticeable improvement on the accuracy of water vapor results. Retrieval of low level moisture is considerably improved by the physical adjustment applied to the statistical first guess results. This is understandable since the physical retrieval process directly treats the inherently nonlinear dependence of channel brightness temperatures on water vapor which is not well represented by the first guess linear regression.

5. PROCESSING OF IMAGERY DATA FOR CLOUD PROPERTY RETRIEVAL

In cloudy areas, cloud coverage and cloud top height first guesses are necessary for the physical retrieval step (Equation 1 is modified.) Colocated with the microwave sensors aboard the DMSP spacecraft, the Operational Linescan System (OLS) provides both visible and infrared imagery at high spatial resolution. With much higher spatial resolution, the visible and infrared data from the OLS imagery can be used to characterize the uniformity of the much larger microwave footprints. In those areas where the contributions from the atmosphere to microwave brightness temperature are small (i.e., nonprecipitating situations), visible or infrared data is able to provide guidelines on the uniformity of the surface observed within a field of view. When clouds obscure portions of the microwave field of view, the imager data provides the complementary capability of cloud property determination.

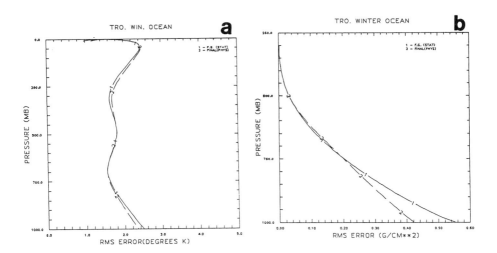

FIGURE 4. Physical retrieval results for the tropics.

Classically, techniques to infer FOV non-uniformity have been referred to as texture analysis methods, A number of approaches can be used for texture analysis including: (a) examination of the spatial power spectum of an image through Fourier decomposition, (b) edge enhancement, and (c) spatial coherence. We have chosen the spatial coherence approach (Coakley and Bretherton, 1982) for the determination of both cloud and surface properties from OLS data. The statistics evaluated are the local mean (I) and local standard deviation (LSD) of radiance (or gray shade) values. The LSD is calculated for n x n sets of pixels.

A plot of LSD_k vs. $\bar{I}_k$ gives the cloud coverage fraction within the microwave footprint and the EBT of the surface and effective cloud top. Figs. 5 a,b illustrate LSD_k vs. $\bar{I}_k$ frequency plots for a partially cloudy microwave FOV over the ocean evaluated using GOES visible and infrared imagery. The spike in the visible result (Fig. 5a) denotes the surface reflectance value due to clear visible pixels. The higher reflectivity, signatures with nonzero standard deviation result from partially cloudy pixels. In the infrared data (Fig. 5b), the highest and lowest EBTs correspond to the surface and cloud top emission, respectively. The cloud top EBT is used with the first guess temperature profile to obtain the first guess effective cloud top pressure, p_c.

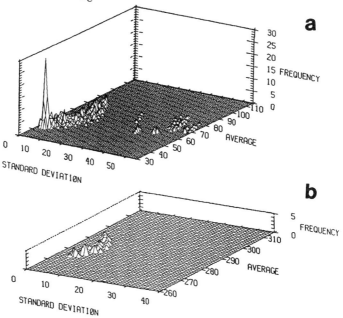

FIGURE 5.   Spatial coherence results for partially cloudy microwave field-of-view using: (a) visible and (b) infrared GOES imager data.

6. CONCLUSIONS

This paper prescribes a unified retrieval approach tailored to the DMSP meteorological sensor suite. All of the available data is utilized in a multispectral sense. The existing operational statistical retrieval capability is exploited to provide parameter first guesses. First guess derived brightness temperature simulations are then compared with the original data and adjustments, based on sensor channel residuals, are made to the first guesses as required. A physically based simultaneous retrieval provides the appropriate parameter adjustments. Results of a retrieval simulation calculation illustrate the improvements possible to the statistical first guesses by utilizing all sensor data multispectrally and to the first guess parameters by employing the physical adjustment step. Recognizing the potential effect of clouds on the millimeter wave data, a method is outlined to introduce required first guess cloud information by image processing visible/infrared imager data. This procedure provides the opportunity to characterize the uniformity (both atmospheric and surface) of the relatively large microwave FOV. This provides potential insights into both cloud and surface type classification with implications for determining their emissivities.

While the retrieval approach outlined has been formulated as a stand alone system, its most important application may be to provide data for use in a numerical weather prediction model. In this context it is noted that modifications to the procedure outlined are desirable. Operationally, for example, it may be advantageous to obtain some first guess elements such as the temperature and moisture profiles from model predicted fields. Furthermore, in the context of a prediction model application, forecast error covariances may be used as constraints on the adjustment process (J. Eyre, personal communication). These changes are not inconsistent with the retrieval described and should in fact improve its effectiveness. Furthermore, the method provides estimates of each retrieval's accuracy.

ACKNOWLEDGEMENTS

The author wishes to thank V. J. Falcone and Dr. K. R. Hardy of the Air Force Geophysics Laboratory for their support and encouragement of this effort. I am indebted to Drs. R. N. Hoffman of AER and J. R. Eyre of UKMO for stimulating discussions on retrieval approaches and applications to NWP.

REFERENCES

Coakley, J. A., and F. P. Bretherton, 1982: Cloud cover from high-resolution scanner data: Detecting and allowing for partially-filled fields of view. J. Geophys. Res., 87(C7), 4917-4932.

Falcone, V. J., L. W. Abreu, and E. P. Shettle, 1982: Atmospheric attenuation in the 30-300 GHz region using RADTRAN and MWTRAN. Proc. Soc. Photo Opt. Instrum. Eng., 337, 62-66.

Falcone, V. J. and R. G. Isaacs, 1987: The DMSP microwave suite (1987): Proceedings, NOAA Conference on Passive Microwave Observing from Environmental Satellites. Williamsburg, VA, 1-4 June.

Fye, F. K., 1978: AFGWC automated cloud analysis model. AFGWC-TM-78-002, Air Force Global Weather Central, 97 pp.

Gaut, N. E., M. G. Fowler, R. G. Isaacs, D. T. Chang and E. C. Reifenstein, 1975: Studies of microwave remote sensing of atmospheric parameters. Air Force Cambridge Research Laboratories. AFGL-TR-75-0007, (NTIS #ADA008042).

Isaacs, R. G., G. Deblonde, and R. D. Worsham, 1985: Millimeter wave moisture sounder feasibility study: Effect of cloud and precipitation on moisture retrievals. AFGL-TR-85-0040, 60 pp. (NTIS # ADA162231).

Isaacs, R. G., R. N. Hoffman, and L. D. Kaplan, 1986a: Remote sensing of meteorological parameters for global numerical weather prediction. Rev. Geophys., 24,4,701-743.

Isaacs, R. G., Y.-Q. Jin, G. Deblonde, R. D. Worsham, and L. D. Kaplan, 1986b: Remote sensing of hydrological variables from the DMSP microwave mission sensors, Proceedings, Second

Conference on Satellite Meteorology/Remote Sensing and Applications. American Meteorological Society, pp. 243-248.

Isaacs, R. G., 1987: Review of 183 Ghz moisture profile retrieval studies. AFGL-TR-87-0127 (NTIS #A182417).

Isaacs, R. G. and G. Deblonde, 1987: Millimeter wave moisture sounding: The effect of cloud, Radio Science, 22, 3, 367-377.

Jin, Y.-Q., and R. G. Isaacs, 1987: Simulation and statistical retrieval for inhomogeneous, nonisothermal atmospheric precipitation, J. Quant. Spectrosc. and Radiat. Transfer, 37, 5, 461-468.

Lo, R. C., 1983: A comprehensive description of the mission sensor microwave imager (SSM/I) environmental parameter extraction algorithm. NRL Memorandum report 5199, Naval Research Laboratory, 52pp. (NTIS # ADA134052).

Rigone, J. L., and A. P. Stogryn, 1977: Data processing for the DMSP microwave radiometer system. In Proc. Eleventh International Symp. Remote Sensing of the Environment. Univ. of Michigan, Ann Arbor, MI, pp. 1599-1608.

Smith, W. L., H. M. Woolf, and A. J. Schreiner, 1985: Simultaneous retrieval of surface and atmospheric parameters, a physical and analytically-direct approach. Advances in Remote Sensing Retrieval Methods, A. Deepak, H. E. Fleming, and M. T. Chahine (eds.) Deepak Publishing. pp. 221-230.

Susskind, J., J. Rosenfield, D. Reuter, and M. T. Chahine, 1984: Remote sensing of weather and climate parameters from HIRS2/MSU on TIROS-N. J. Geophys. Res., 89(D3), 4677-4697.

COMBINED GROUND- AND SATELLITE-BASED RADIOMETRIC REMOTE SENSING

E.R. Westwater, M.J. Falls, and J. Schroeder
NOAA/ERL/Wave Propagation Laboratory
Boulder, Colorado 80303, USA

D. Birkenheuer[1] *and J.S. Snook*[2]
NOAA/ERL/PROFS
Boulder, Colorado 80303, USA

M.T. Decker
CIRES, University of Colorado/NOAA
Boulder, Colorado 80303, USA

ABSTRACT

The Wave Propagation Laboratory is currently operating a ground-based 5-channel microwave radiometer at Stapleton International Airport, Denver, Colorado. Combined soundings from the NOAA TOVS and the ground-based radiometer are shown to determine temperature profiles with rms errors less than 2.0 K from the surface to 300 mb. Operational VAS data acquisition and real-time sounding processing are now being performed by PROFS. Observations are presented from the ground-based radiometer, VAS soundings within 100 km of Denver, and ground truth provided by radiosondes. Combined thermal retrievals from the VAS and ground-based system are also presented. The Wave Propagation Laboratory also operates a limited network of four ground-based dual-frequency radiometers that measure precipitable water vapor and cloud liquid. An example illustrates how data from this system can be used to constrain the precipitable water analysis obtained from VAS.

1. INTRODUCTION

The Wave Propagation Laboratory operates a five-channel zenith-viewing microwave radiometer that measures lower altitude temperature profiles, precipitable water vapor, and cloud liquid (Hogg et al., 1983). The instrument is located at Stapleton International Airport (Denver, Colorado), operates continuously in near all-weather con-

---

[1]On contract with the Cooperative Institute for Research in the Atmosphere and the CSU Research Foundation, Ft. Collins, Colorado.

[2]On contract with T. S. Infosystems, Inc., Lanham, MD 20706.

RSRM '87: ADVANCES IN
REMOTE SENSING RETRIEVAL METHODS
A. Deepak, H.E. Fleming, and J.S. Theon (Eds.)   215

ditions, and provides data every 2 minutes. Twice-a-day radiosondes
provide ground truth for radiometric soundings, and statistical
accuracy evaluations are available. These accuracies enter directly
in our algorithms for combining data from various sources, including
polar-orbiting and geostationary satellites.

The idea of combining ground- and satellite-based thermal
retrievals was discussed by Westwater and Grody (1980). Their simulations suggested that the accuracy of combined retrievals in the lower
troposphere could substantially exceed that of satellite retrievals
alone. Conversely, the accuracy of combined retrievals in the upper
troposphere could be substantially better than those of the ground-
based system alone. Estimates of combined-retrieval accuracy showed
maximum rms errors of less than 2.0 K from the surface to 300 mb.
Later, ground-based thermal soundings were combined with (1) NESDIS
(National Environmental Satellite, Data, and Information Service)
operational retrievals using the TOVS (TIROS-N Operational Vertical
Sounder) instruments on the NOAA 6/7 satellites (Westwater et al.,
1984), and (2) data from the Microwave Sounding Unit (MSU) on the same
satellites (Westwater et al., 1985).

A primary focus of NOAA's Program for Regional Observing and
Forecasting (PROFS) is to develop products from alternative data
sources and to implement and evaluate their analysis and forecast utility, specifically for the improvement of short-range local forecasts.
Soundings of temperature and moisture are derived from radiance data
collected by the Visible and Infrared Spin-Scan Radiometer (VISSR)
Atmospheric Sounder (VAS). These data are now available in real time
at PROFS through a totally automated acquisition and processing system
(Snook, 1987). Research is being conducted to identify effective ways
of combining VAS and ground-based radiometer data.

2. THE ACCURACY OF THE GROUND-BASED RADIOMETRIC PROFILER

The location of the WPL microwave radiometer (Radiometric Profiler) is only about 20 m from the National Weather Service (NWS)
radiosonde launch facilities. Consequently, ground truth is routinely
available at 12-h intervals. The solid curve in Fig. 1 shows the temperature retrieval accuracy of the Profiler for the period October
1985 to June 1986. Note that the rms accuracies are less than 2.0 K
only below 500 mb. Our retrieval method uses a priori data to
generate retrieval coefficients and is independent of current radiosonde data or forecasts. To assess the effect of instrument calibration errors and errors in modeling atmospheric transmission, we also
show in Fig. 1 the retrieval accuracy that results when brightness
temperatures (T_b) calculated from radiosondes are substituted for
measured T_b in the retrieval algorithm. An observable difference between the two curves (about 1 K rms) is evident from the surface to
300 mb.

The ground-based system also measures precipitable water vapor
(PWV) with an accuracy that approaches that of the radiosonde (Hogg et

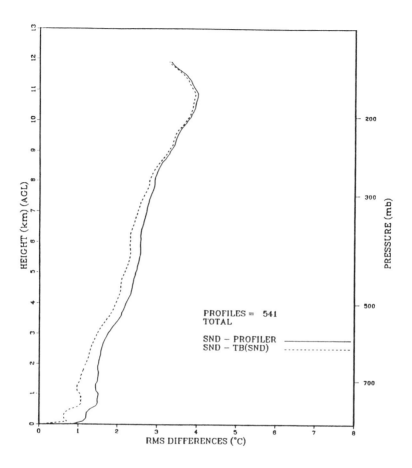

Figure 1.   Rms temperature retrieval errors for the ground-based
            Radiometric Profiler, at Denver, Colorado, August 1985 to
            June 1986.   Solid line shows rms errors when measured data
            are used; dashed line shows rms errors when brightness tem-
            peratures calculated from radiosondes are used.

al., 1983).   We show in Fig. 2 a scatter plot of PWV measured by
radiometers and by NWS radiosondes.   The radiometric determinations
are completely independent of the radiosonde, the calibrations being
determined by the so-called tipping curve method (Hogg et al., 1983).
Radiosonde measurements of humidity become questionable when the relative humidity falls below 20 percent, resulting in underestimated PWV.
In January 1987, the 52.85 GHz channel on the ground-based system
failed.   Since that time, only soundings from a five-channel instrument have been available.

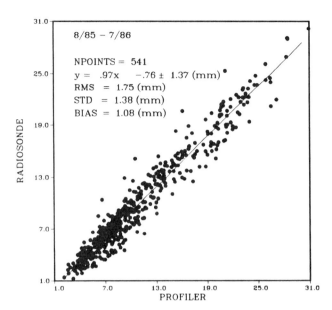

Figure 2.   Scatter plot of PWV measured by Radiometric Profiler vs.
            PWV determined by NWS radiosondes, at Denver, Colorado,
            August 1985 to July 1986.

3.   COMBINED GROUND-BASED AND POLAR ORBITER SOUNDINGS

   The material summarized in this section was published by
Westwater et al. (1984; 1985).   In these articles, TOVS temperature
soundings from NOAA 6/7 polar-orbiting satellites were both compared
and combined with ground-based thermal retrievals.   To make meaningful
comparisons with radiosondes, a temporal window of ±3 h around
radiosonde release time was chosen for the satellite data, and the
region of geographical coverage was restricted to roughly ±1 degree in
latitude and longitude around Denver, Colorado.

3.1   COMBINATION WITH NESDIS OPERATIONAL RETRIEVALS

   Radiometric Profiler and NESDIS operational retrievals are
completely independent, being obtained by different instruments and by
different methods.   Although TOVS-measured radiances were available,
it was more expedient to combine thermal retrievals.   One particularly
simple method of combining two independent observations is by inverse
covariance weighting of the individual observations (Rodgers, 1976).
We define the individual thermal retrieval vectors to be $\underset{\sim}{T}_P{}'$ and $\underset{\sim}{T}_S{}'$
(primed quantities refer to departures from the average) and the
covariance matrices describing their accuracy to be S_P and S_S.   Then
the combined retrieval $\underset{\sim}{T}_C{}'$ is given by

$$\underset{\sim}{T}_C{}' = (S_P^{-1} + S_S^{-1})^{-1} (S_P^{-1} \underset{\sim}{T}_P{}' + S_S^{-1} \underset{\sim}{T}_S{}') \qquad (1)$$

where S_x^{-1} is the matrix inverse of S_x. To implement this method, we corrected for biases (the TOVS retrievals were biased over Denver) and then determined the relevant covariance matrices at 26 fixed pressure levels that ranged from 780 to 10 mb. The results, based on 460 cases during 1981/1982, are shown in Fig. 3. The combined retrievals are everywhere better than the satellite or the Profiler retrievals alone, and are less than 2.0 K from the surface to about 250 mb.

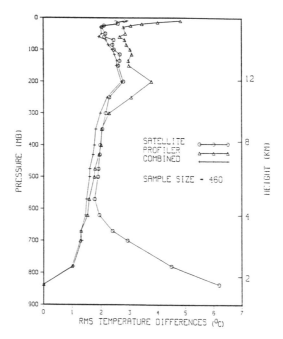

Figure 3.   Rms temperature differences between the radiosonde and (1) the NOAA 6/7 satellites (NESDIS operational retrievals using the TOVS instrument), (2) the Radiometric Profiler, and (3) the TOVS-Profiler combination. (After Westwater et al., 1984).

3.2   COMBINATION WITH MSU

To determine the accuracies that a purely microwave sounding system could achieve, we also combined the Radiometric Profiler data with data from only the microwave component of TOVS, the MSU (Westwater et al., 1985). Relative weighting functions for this system are shown in Fig. 4. When combining the microwave data, we used a different retrieval method than the one discussed in Section 3.1. If we let $\underset{\sim}{T}'$ be the estimate of the profile vector $\underset{\sim}{T}'$,

$$\hat{\underset{\sim}{T}}' = S_{Td}\, S_d^{-1}\, \underset{\sim}{d}' \; . \tag{2}$$

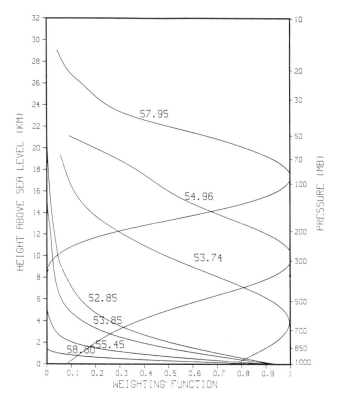

Figure 4. Temperature weighting functions, in relative units, for the
Radiometric Profiler and the MSU. The upper three curves
are for the MSU; the lower four curves are for the
Profiler. The numbers labeling the curves are the channel
frequencies (GHz). (After Westwater et al., 1985).

In Eq. (2), $\underset{\sim}{d}$ is a vector whose components contain radiance measured
both by ground-based and satellite instruments, as well as surface
meteorological data. The well-known process leading to Eq. (2) is
known as minimum rms estimation. The algorithm is physical, in the
sense that it requires the ability to calculate radiance from profiles, but it is also statistical, since it requires an a priori data
base. The covariance matrices S_{Td} and S_d are estimated from climatology and from known noise characteristics of the respective instruments. The temperature retrieval accuracies resulting from applying
Eq. (2) to a year of combined MSU and Radiometric Profiler measurements are shown in Fig. 5. The rms retrieval errors of the combined
system are less than 2.0 K from the surface to about 350 mb. A rather
surprising result of our analysis was that the combined system could
determine geopotential heights up to 300 mb with an accuracy roughly
equivalent to that of present-generation radiosondes.

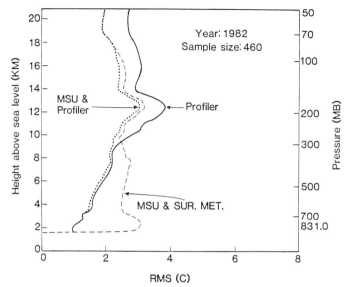

Figure 5.   Rms temperature errors for the MSU, Radiometric Profiler,
             and combination of MSU and Profiler. (After **Westwater** et
             al., 1985).

4. OPERATIONAL VAS SOUNDINGS AT PROFS

4.1 PROCEDURE TO GENERATE VAS SOUNDINGS

    This section summarizes results by Snook (1988). The VAS
sounding generation at PROFS uses a direct, physical retrieval
algorithm developed at the University of Wisconsin-Madison (Smith,
1983). The approach uses radiance measurements from various infrared
wavelengths collected on board the GOES-7 platform to modify a first
guess provided by the PROFS Mesoscale Analysis and Prediction Program.
Temperature and mixing ratio profiles are derived for 40 levels
ranging from the surface to 0.1 mb. The surface first guess is produced hourly from Surface Aviation Observations (SAO).

    VAS radiance data may be contaminated by clouds, so a scheme to
eliminate their effects is necessary. The technique currently used by
PROFS subtracts the 11.2 μm window-band brightness temperature from
the colocated surface temperature (Snook, 1987). If the difference is
greater than 10 K, the pixel is flagged as cloudy.

    Soundings are generated for an 80 km by 80 km field of view com-
posed of an 11 × 11 matrix of pixels to reduce the effects of signal
noise on the radiance measurement. The sounding generation system
uses transmission functions to relate measured and calculated radiance
values for each infrared wavelength. Since GOES-7 transmittance functions were not immediately available for the summer of 1987, GOES-6
functions were used for the data reported here. This substitution
undoubtedly introduced some systematic bias to the VAS soundings.

4.2 THE ACCURACY OF VAS SOUNDINGS OVER DENVER, COLORADO

The accuracy of VAS soundings was evaluated through a direct comparison with the nearest radiosonde sounding (RAOB) in space and time. Discrepancies can occur with this type of comparison since the RAOB represents a point observation, whereas the VAS observation represents a volume. Terrain effects are also important contributors to RAOB-VAS differences. Figure 6 compares the 1200 UTC Denver RAOB and the 1245 UTC VAS temperature soundings for 27 days during June, July, and

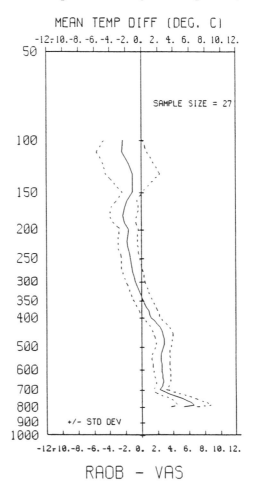

Figure 6.   Mean vertical temperature (°C) comparison for 1200 UTC Denver RAOB and 1245 UTC VAS sounding using 27 clear-to-partly-cloudy cases from June, July, and August 1987. The VAS sounding is located approximately 30 km south of the Denver RAOB. The solid line depicts the mean temperature difference of RAOB minus VAS. The surrounding dashed lines indicate the positive and negative standard deviations. (After Snook, 1988).

August 1987. Although the biases are substantial, the standard
deviations are less than 2.0 K below 150 mb. Thus, reasonable temperature soundings can be achieved after removal of the consistent
biases. A similar analysis was applied to dew point temperature differences but is not given here. However, in terms of precipitable
water vapor, the bias was 6.16 mm with a standard deviation of 2.85.
In the combinations and comparisons with the Radiometric Profiler data
in the next section, all VAS biases are taken into account.

5. COMBINED VAS AND RADIOMETRIC PROFILER RETRIEVALS

During the summer of 1987, a Mesoscale observation experiment was
conducted in the vicinity of the Radiometric Profiler site in Denver.
Consequently, a substantial number of extra radiosonde soundings were
available for analysis. Temperature retrievals at 700, 500, and 300
mb are shown in Fig. 7 for a 26-h period in July during which there
were (1) five RAOB ascents; (2) ten VAS soundings with at least 95
percent clear conditions; and (3) an almost continuous time series of
Profiler data. (As a note of caution, there were three NWS RAOBs, and
two special release RAOBs called CLASS. Comparisons of the two types

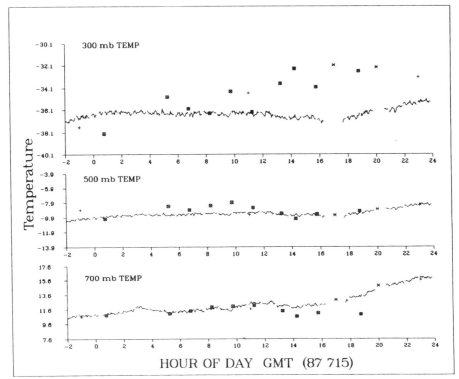

Figure 7.   Time series of 700, 500, and 300 mb temperature derived
            from Radiometric Profiler (solid curve); VAS (⊞); NWS RAOBs
            (+); CLASS RAOBs(✖). Denver, Colorado, 15 July 1987.

of RAOB soundings were not always as close as one would like.) We note that the temperature retrievals of the Profiler are within 0.5 K of RAOBs at 700 and 500 mb, but depart considerably at 300 mb. The VAS retrievals at these levels are also in good agreement with RAOBs, but at 700 mb tend to depart away from the trend after about 1800 UTC. The 300 mb retrievals of both Profiler and VAS differ more from the RAOBs than at the other two levels, but clearly the VAS soundings are more in agreement with the RAOBs. We are now investigating various techniques to take advantage of the continuity of the Profiler data, as updated by VAS data. Here, in Fig. 8, we show only combined retrievals at VAS sounding time. These combined retrievals were determined by Eq. (1).

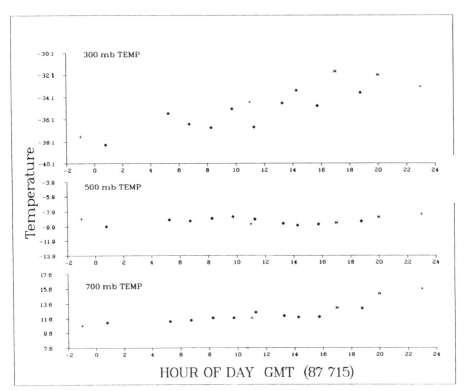

Figure 8.   Combined VAS-Profiler retrievals (◆) for the case shown in
             Fig. 7.

6. PROFILER ADJUSTMENT OF VAS PWV IMAGERY

In addition to the five-channel radiometer at Stapleton Airport, WPL maintains a mesoscale Profiler network in eastern Colorado. This network consists of dual-channel microwave radiometers and both UHF and VHF wind-profiling radars (Fig. 9). The dual-channel radiometers provide almost all-weather soundings of PWV and integrated cloud

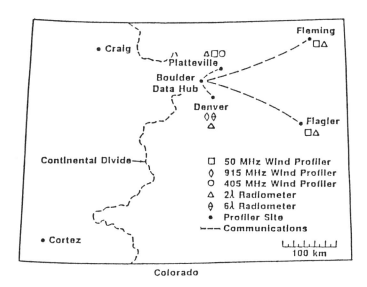

Figure 9.   WPL wind Profiler and radiometer locations - 1985.

liquid.   The accuracy of the PWV determinations should be of the same
quality as the accuracy shown in Fig. 2, since the radiometers have
been routinely calibrated by the tipping curve method.   No in situ
RAOB comparisons are available, however.

    VAS PWV imagery has been available for some time (Chesters et
al., 1983; Smith et al., 1985).   Recently, PROFS generated quantitative imagery for a multi-state region centered over Colorado
(Birkenheuer, 1988).   PROFS is now interested in using this VAS product as a data source for its Local Analysis and Prediction System
(LAPS).   LAPS has a multi-county-scale domain situated over
northeastern central Colorado.   Since the water vapor is computed
through the linear regression of numerous soundings with the VAS
brightness temperature data over a multi-state region, it is not
surprising that the resulting multilinear fit is not exact over a
smaller domain.   In addition, the VAS instrument has only three water
vapor channels, each of which has poor vertical resolution at low
levels; boundary layer moisture is the major contributor to PWV.   We
assume that the VAS moisture soundings err systematically in a manner
correctable by a linear transformation.

    A technique is being investigated that uses Profiler data to
adjust the VAS water vapor field, making the field more suitable for
LAPS analysis.   Profiler PWV measurements are more accurate than VAS
measurements primarily because of the absence of surface effects and
because they are much less sensitive to clouds.   The VAS water vapor
data for the LAPS domain are extracted from the PROFS regional-scale
PWV image and then filtered to reduce high-frequency fluctuations and

interpolated through small-scale data-void regions (Birkenheuer, 1987). Then the VAS field is best fit to the Profiler data. Figure 10 shows the result of one representative test case. The transformed VAS field is computed from

$$V' = aV + b \qquad\qquad (3)$$

where V' is the adjusted VAS PWV (mm) and V is the preanalyzed value of the VAS PWV (mm). The coefficients a and b are determined from

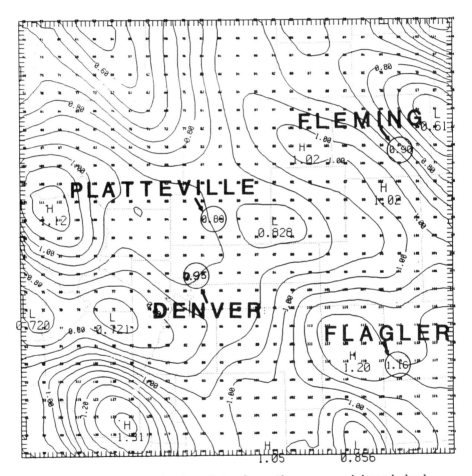

Figure 10. VAS PWV analysis; data from three ground-based dual-channel radiometers are superimposed. The data are for 4 November 1715 UTC; the scale factor is 1.0 and the linear bias is -0.1 mm. In this example, the match between ground-based and VAS data is very close.

optimization such that the difference between V' and Profiler values is minimized with the following physical constraints:

$$a > 0,$$

$$-6 < b < 6,$$

and $\quad\quad\quad\quad 0 < V' < 6$ at all points.

In many cases we have found that the scaling term a is unity and only a bias b is necessary to fit VAS to the Profiler data. This is encouraging since we suspect a bias exists in the VAS data and hope to preserve the gradient structure which would provide information in regions with sparse conventional data. The method described here both detects and corrects such biases. The technique will be expanded in the future to analyze water vapor at various pressure levels to better model moisture advection.

7. CONCLUDING REMARKS

The results presented show that data from ground-based radiometers can be effectively combined with data from both orbiting and geostationary satellites. One reason why this can be done is that the ground-based remote sensors provide data continuously and in almost all weather conditions, and hence are available when the satellite is viewing the region in question. In the case of temperature sensing, a ground-based instrument is continuously operating at only one location and real-time combinations with satellite data are not yet available. However, careful post facto analysis shows that such a combination can provide quite accurate measurements of geopotential heights and thicknesses. If such data were available at a network of stations, the products would have value for synoptic-scale applications. However, vertical resolution of combined temperature retrievals is still not adequate for most mesoscale applications. For water vapor, a limited network of radiometers is available, and the combination of their PWV data with VAS is promising.

ACKNOWLEDGMENT

The authors thank Bob Zamora for his review of the manuscript and for his comments on the accuracy of radiosondes.

REFERENCES

Birkenheuer, D., 1987: Horizontal shape matching applied to VAS data. Proc. Symp. Mesoscale Analysis and Forecasting, Vancouver, Canada, 17-19 August, ESA SP-282, 91-95.
--------------, 1988: PROFS quantitative precipitable water product. Preprints, 3rd AMS Conf. Satellite Meteorology and Oceanography, 31 January - 5 February, Anaheim, CA, (in press).
Chesters, D. C., L. W. Uccellini, and W. D. Robinson, 1983: Low-level water vapor fields from the VISSR Atmospheric Sounder (VAS) "split

window" channels. J. Climate Appl. Meteor., 22. 725-743.

Hogg, D. C., M. T. Decker, F. O. Guiraud, K. B. Earnshaw, D. A. Merritt, K. P. Moran, W. B. Sweezy, R. G. Strauch, E. R. Westwater, and C. G. Little, 1983: An automatic profiler of the temperature, wind, and humidity in the troposphere, J. Climate Appl. Meteor., 22, 807-831.

Rodgers, C. D., 1976: Retrieval of atmosphere temperature and composition from remote measurements of thermal radiation, Rev. Geophys. Space Phys., 14, 609-624.

Smith, W. L., 1983: The retrieval of atmospheric profiles from VAS geostationary radiance observations, J. Atmos. Sci., 40, 2025-2035.

Smith, W. L., G. S. Wade, and H. M. Woolf, 1985: Combined atmospheric sounding/cloud imagery - a new forecasting tool, Bull. Amer. Meteor. Soc., 66, 138-341.

Snook, J. S., 1987: Operational cloud-clearing of VAS radiance data. Proc. Symp. Mesoscale Analysis and Forecasting, Vancouver, Canada, 17-19 August, ESA SP-282, 103-106.

----------, 1988: Operational VAS soundings at PROFS. Preprints, 3rd AMS Conf. Satellite Meteorology and Oceanography, 31 January - 5 February, Anaheim, CA, (in press).

Westwater, E. R., and N. C. Grody, 1980: Combined surface-based and satellite-based microwave temperature profile retrieval, J. Appl. Meteor., 19, 1438-1444.

----------------, W. B. Sweezy, L. M. McMillin, and C. Dean, 1984: Determination of atmospheric temperature profiles from a statistical combination of ground-based Profiler and operational NOAA 6/7 satellite retrievals, J. Climate Appl. Meteor., 23. 689-703.

----------------, Wang Zhenhui, N. C. Grody, and L. M. McMillin, 1985: Remote sensing of temperature profiles from a combination of observations from the satellite-based Microwave Sounding Unit and the ground-based Profiler, J. Atmos. Oceanic Technol., 2. 97-109.

DISCUSSION

Kleespies: I'm glad you brought that point up with the radiosonde because with our comparisons, if you remember our talk yesterday, what we were seeing was about a 1°C bias. I think I would like to talk to you about this further.

Westwater: I think we were very fortunate to be able to have another type of radiosonde operating right beside the National Weather Service sondes. And we do have a 3-month collection of data that we will analyze in detail. These CLASS soundings have a humidity sensor that differs from that on contemporary NWS balloons.

Neuendorffer: You mentioned working under all conditions, although I imagine in Colorado you generally have pretty nice conditions. How adverse would the conditions be that you wouldn't attempt it, like thick clouds, snow, rain?

Westwater: We usually have our quality control algorithm with the 31 GH$_z$ channel; when the brightness temperature in that channel exceeds 100 degrees, which is about 3 millimeters of liquid, we do not have a retrieval. Again, that would be roughly, in terms of 100 degrees brightness temperature, a rain rate of around 5 millimeters per hour. So, in general, we operate up to almost all non-precipitating types of clouds.

SIMULTANEOUS RETRIEVAL OF ATOMIC OXYGEN AND TEMPERATURE IN THE THERMOSPHERE BY AN IR LIMB SCANNING TECHNIQUE[1]

A.S. Zachor
Atmospheric Radiation Consultants, Inc.
Acton, Massachusetts 01720, USA

R.D. Sharma
Air Force Geophysics Laboratory
Hanscom Air Force Base
Bedford, Massachusetts 01731, USA

ABSTRACT

The importance of atomic oxygen in mesospheric/thermospheric processes is the motivation for a study that will determine the feasibility of recovering the O-atom density and translational temperature profiles from a spectrally resolved limb radiance scan in either the 147 or 63 μm line, corresponding to the $^3P_0-^3P_1$ and $^3P_1-^3P_2$ transitions of the oxygen ground electronic state. Described in this paper is an onion-peel retrieval technique that yields simultaneous solutions for the two vertical profiles between approximately 90 km and 300 km altitude. Within each peeled layer the temperature and O-atom density are obtained by solving the least squares normal equations derived by minimizing the rms difference between measured and calculated radiance spectra for the corresponding tangent height. Relationships between the accuracy/stability of the solutions and the available signal-to-noise, available spectral resolution and the desired vertical resolution in the solution profiles are discussed. A preliminary conclusion is that the approach is feasible using a cooled, large-aperture, confocal Fabry-Perot system operating near 147 μm. Briefly described are alternative approaches, one of which, it is hoped, will prove feasible using a less advanced sensor concept.

1. INTRODUCTION

Atomic oxygen plays an important role in chemical and collisional processes in the earth's mesosphere and thermosphere. Current techniques, both in situ and remote, for measuring oxygen atom densities in this regime have produced results with unexplained large disparities. We are investigating the feasibility of recovering vertical profiles of translational temperature and oxygen atom density from measurements of the limb radiance profile near 147 μm and/or 63 μm wavelength, corresponding to the OI transitions ($^3P_0-^3P_1$ and $^3P_1-^3P_2$) of the ground electronic state of atomic oxygen. The assumption that the ^3P fine structure levels are in thermodynamic equilibrium (LTE) with

---

[1]This work was sponsored by the Air Force Systems Command under U.S. Air Force Contract No. F19628-87-C-0053.

RSRM '87: ADVANCES IN
REMOTE SENSING RETRIEVAL METHODS
A. Deepak, H.E. Fleming, and J.S. Theon (Eds.)　　229

the local translational temperature is crucial to the proposed technique, but seems a reasonable one based on the very long radiative lifetimes of the level transitions (see Fig. 1).

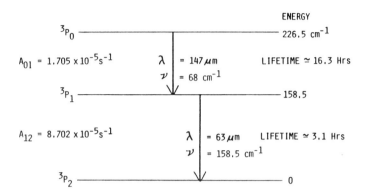

FIGURE 1. Approximate energy-level diagram for the atomic oxygen OI transitions. Values are given for the transition lifetimes, Einstein A-coefficients and upper-state energies.

The two sought-after vertical profiles can, in principle, be recovered from a) a pair of limb radiance profiles representing the total (spectrally integrated) apparent intensities of the two OI lines, or b) a spectrally resolved limb radiance profile for just one of the lines. The use of spectrally resolved data implies an instrument with resolving power greater than 2×10^5, which could be achieved near 63 μm or 147 μm wavelength by a Fabry-Perot system with metal mesh etalons. An IR heterodyne system, if one could be designed to operate near 147 μm, would be more complex but would easily provide the required high resolving power. The physical approach and a Fabry-Perot system concept have been outlined by Sharma, Stair and Smith (1988).

The focus of this paper is on the retrieval technique used in the "spectrally-resolved approach," and on the characteristics of the solutions obtained by applying the technique to synthesized OI limb spectral data. A very similar technique is used in the "non-resolved approach," but evaluation of the latter is still in progress. The results presented include signal-to-noise and spectral resolution requirements for successful recovery of the O-atom density and translational temperature profiles.

2. RETRIEVAL ALGORITHM

We will first summarize the governing radiative transfer equations and describe the retrieval method in general terms. Section 2.2 gives some details of the retrieval algorithm.

2.1 RADIATIVE TRANSFER EQUATION AND RETRIEVAL METHODOLOGY

The transfer of spectral photon radiance N_ν arising from a single transition at wavenumber ν_0, through a slab of thickness Δz along the line-of-sight, can be evaluated by the equations,

$$N_\nu(z+\Delta z) = N_\nu(z)\exp(-\Delta\tau_\nu) + [2c\nu_0{}^2\gamma/(1-\gamma)][1-\exp(-\Delta\tau_\nu)], \qquad (1)$$

$$\gamma = \exp(-c_2\nu_0/T), \qquad (2)$$

$$\Delta\tau_\nu = n \Delta z\, f(\nu-\nu_0)\, S(\nu_0,T), \qquad (3)$$

$$S(\nu_0,T) = (A/8\pi c\nu_0{}^2)[(1-\gamma)/\gamma]\, g_u \exp(-c_2 E_u/T)/Q(T), \qquad (4)$$

$$Q(T) = \exp(-c_2 226.5/T) + 3\exp(-c_2 158.5/T) + 5, \qquad (5)$$

where $T(z)$ and $n(z)$ are temperature and species (O-atom) density, $\Delta\tau_\nu$ is the slab optical thickness, $f(\nu)$ is the normalized shape of the line due to temperature and/or pressure broadening, $S(\nu_0,T)$ is the integrated line intensity and $Q(T)$ is the partition sum; $c_2 = 1.439$ K/cm^{-1} is the second radiation constant and c is the speed of light. The first factor in the second term of Eq. (1) is Planck's function evaluated at ν_0,T. In Eq. (4), A and E_u are the Einstein A-coefficient and upper state energy, respectively; they have one of the values given in Fig. 1, depending on whether $\nu_0 = 68$ cm^{-1} ($\lambda = 147$ μm) or $\nu_0 = 158.5$ cm^{-1} ($\lambda = 63$ μm). The statistical weight g_u of the upper state is unity for the 147 μm line and three for the 63 μm line. Equation 1 is derived by Zachor and Sharma (1985), who give a more general definition of γ that is applicable in the non-LTE case as well.

These equations, together with an expression for the line shape $f(\nu)$, are sufficient for calculating the limb radiance profile spectrally resolved over the 147 μm and 63 μm lines. Above 90 km altitude, $f(\nu)$ is well-approximated by the pure Doppler shape, except in the very far wings of the line. Assuming the Doppler shape is valid, the limb spectral radiance depends solely on the vertical profiles of the O-atom density n and the temperature. Hereinafter these profiles will be referred to as [O] and T, both functions of altitude H. The spectrally resolved limb radiance profile will be denoted $N(H_T)$, where H_T is the tangent height of the line of sight.

For later reference we define the spectrally resolved volume emission rate,

$$\xi_\nu([O],T) = A\cdot[O]\cdot f(\nu-\nu_0)g_u\exp(-c_2 E_u/T)/Q(T), \qquad (6)$$

which is the number of photons emitted (in all directions) per unit wavenumber interval per unit time by a unit volume whose O-atom density and temperature are [O],T. This rate equals $4\pi/\Delta z$ times the second term of Eq. (1) evaluated for $\Delta z \to 0$. Its dependence on temperature derives from the Planck source function, the line intensity S and the Doppler lineshape function f.

The altitude variation of [O] and T from 90 to 300 km altitude is typified by the two models shown in Fig. 2. Figure 3 shows the spectrally resolved limb radiance profile $N_\nu(H_T)$ computed (at infinite

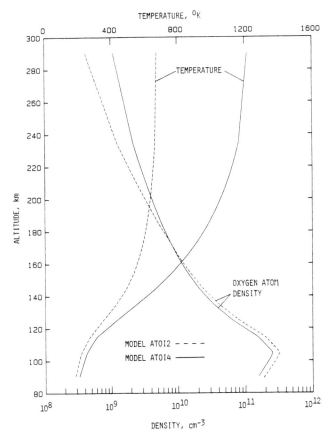

FIGURE 2.   Model profiles for [0] and T representing a cool thermo-
            sphere (ATOI2) and relatively warm thermosphere (ATOI4).

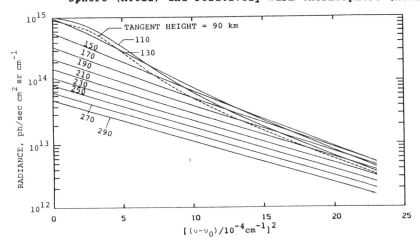

FIGURE 3.   The spectrally resolved limb radiance profile computed
            (for infinite resolving power) for the 147 μm line and
            atmospheric model ATOI4.

spectral resolution) for the 147 µm line using atmospheric model ATOI4
of Fig. 2. Note that the spectral radiances are plotted in the form
$\log N_\nu$ vs. $(\nu-\nu_0)^2$, which results in a straight line when N_ν has the
Doppler shape. For the largest tangent heights N_ν has the Doppler
shape because the tangent path is optically thin and because the
temperature is nearly uniform over the path.

The retrieval technique, basically an "onion-peeling" method,
represents the atmosphere as a series of concentric shells, or layers;
[O] and T are treated as constant within each layer. The retrieval
proceeds downward, layer-by-layer, from some maximum altitude. For the
moment, assume that the sought-after solution is the set of [O]s and Ts
for each of the layers below the maximum altitude, say, 300 km.

Specifically, the "maximum altitude" is defined as the largest
tangent height H_T for which $N_\nu(H_T)$ has been measured, or perhaps the
largest H_T for which the signal-to-noise of the measurement is deemed
adequate. Since the retrieval technique involves a comparison of
measured and calculated limb spectral radiances, it is necessary, or at
least desirable, to include in the calculated values the contributions
of layers above the maximum altitude. This is enabled by an initiation
procedure prior to the onion-peeling process.

The initiation procedure exploits the fact that tangent paths
above $H_T \approx 250$ km are optically thin, and that $N_\nu(H_T)$ tends to fall off
exponentially above this tangent height -- a consequence of the nearly
constant temperature and the exponential decrease in [O] with altitude.
These properties make it possible to analytically invert the
(extrapolated) limb radiance profile via an Abel transform. A closed
form approximation was found that gives the emission rate $\xi\nu$ at 300 km
in terms of $N_\nu(300)$. This rate, and Eq. (6), are used to estimate T
and [O] at 300 km. It was also found possible to develop approximate
functions that allow the net effects of the extrapolated atmosphere to
be included in the $N_\nu(H_T)$ computed for $H_T < 300$ km.

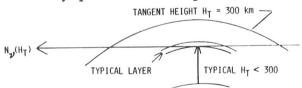

FIGURE 4. **The geometry of the onion-peel retrieval technique.**

The onion-peeling procedure is pictured in Fig. 4 for a
representative layer at some tangent height less than the 300-km
maximum altitude. Previous cycles have provided [O] and T for the
layers above, up to 300 km. If values of [O] and T are assigned to
the layer in question, one can calculate the limb spectral radiance for
this tangent height, including the contributions from the two segments
of the path above 300 km altitude. The solution values are defined as
those for which the rms difference between the calculated spectrum N_ν
and the corresponding measured spectrum is a minimum. No assumption is
made regarding the optical thicknesses of the layers and tangent paths
involved in the onion-peeling procedure.

2.2 MATHEMATICAL AND ALGORITHMIC DETAILS

Let $N_{\nu i}$(CALC) and $N_{\nu i}$(MEAS) denote the calculated and measured limb radiance spectra at wavenumber νi for a particular tangent height H_T. Included in the calculated spectrum is the effect of an instrument line shape (ILS) representing the finite spectral resolution of the measurement. The calculated spectrum depends on the unknown [O] and T at altitude $H = H_T$. Two nonlinear "equations" for these elements of the solution profiles are obtained by minimizing the sum

$$\sum_i \; [N_{\nu i}(\text{CALC}) - N_{\nu i}(\text{MEAS})]^2 \tag{7}$$

with respect to [O] and T. These least-squares normal equations will be denoted X{[O],T} = 0 and Y{[O],T} = 0. As one would expect, X and Y are not expressible in closed form; they represent lengthy function procedures in a computer code. The temperature is obtained from the equation Y = 0 (the one that minimizes the rms spectral difference with respect to T) by Newton's method, starting from a guessed temperature. The [O] corresponding to the guessed T and to each of the iterates in Newton's method is obtained from the other equation, X = 0, which is also solved by Newton's method. All solutions for [O] start from a single initial guess.

The atmosphere must be finely divided to obtain reasonable accuracy in the calculated radiances -- we use a vertical thickness of one km for the layers. Generally, reducing the layer thickness, which reduces each layer's contribution to the limb radiance, would tend to increase the potential for instability in the solution if it consists of independent [O],T values for each and every layer. Thus, we solve for the [O] and T of layers separated vertically by more than one km, and impose the constraint that log[O] and T vary linearly between these "solution layers". The distance ΔH between solution layers is assigned one of several values depending on the altitude (see below). The [O],T values obtained for the maximum altitude in the initiation procedure are used as guesses for the first peeled layer; thereafter, guesses are obtained by linearly extrapolating log[O] and T from the previous pair of solution layers. If X = 0 and/or Y = 0 has no solution, the guesses are systematically revised.

The piecewise-linear temperature solution T(H) is smoothed above H = 130 km, where ΔH is 10 km. The smoothing is performed in each onion peeling cycle, but involves only temperatures obtained in the tw most recent cycles. The smoothing operation eliminates slope discontinuities at the higher altitudes which could adversely affect the stablity of the method.

2.3 CHARACTERISTICS OF THE SOLUTION

Retrievals were performed for the four cases represented by the 63 μm or 147 μm line in combination with synthetic spectral radiance data computed for atmospheric model ATOI2 or ATOI4 (Fig. 2). The synthetic data was degraded in resolution and contaminated by noise. The resolving power, $R \equiv \nu_0/\Delta\nu$, where $\Delta\nu$ is the full-width-at-half-maximum of the ILS used to degrade the data, is used as the measure of resolution. We used an ILS that represents a high-finesse Fabry-Perot system -- it is well approximated by the Cauchy function (Lorentz shape). The noise equivalent spectral radiance (NESR) in the units photons $s^{-1}cm^{-2}sr^{-1}cm$, is the rms of the zero-mean, Gaussian noise added to the spectral radiances. The parameters R and NESR were varied over a wide range. Results obtained using the 63 μm line will not be discussed since those obtained from the 147 μm line for comparable R and NESR were generally superior.

The synthetic 147 μm data for each tangent height consists of samples spaced 8×10^{-5} cm^{-1} apart over half the line. The half-width of the infinitely resolved limb spectral radiance at the higher tangent heights, for model ATOI4, is roughly 2×10^{-4} cm^{-1}, which is equivalent to 2.5 samples. Note that an instrument half-width equal to this value corresponds to resolving power $68/4 \times 10^{-4} = 1.7 \times 10^5$. The spectrally degraded data for the higher tangent heights has the shape of a Doppler line convolved with the Lorentz-shaped ILS, i.e., it has the Voigt shape.

The parameter ΔH, defined earlier, is effectively the "vertical resolution" of the sought-after solution. Initially, retrievals were performed using noise-free, fully resolved ($R = \infty$) data and $\Delta H = 1$ km. The retrieval procedure became very unstable below 135 km. At approximately 130 km altitude it failed, i.e., there was no solution to the normal equations. This behavior, which is explained below, gives some insight into the effects of noise. The final version of the retrieval algorithm uses $\Delta H = 10$ km above 130 km, $\Delta H = 4$ km between 130 and 110 km, and $\Delta H = 2$ km below 110 km. These values are adequate to reproduce the true structure of the [0] and T profiles; they resulted in solution profiles nearly identical to the model profiles over the full range of the limb radiance data (90 to 300 km) for the case $R = \infty$, NESR = 0.

The spacing in tangent height of the data used in the retrievals is the same as the altitude spacing ΔH of the solution layers. The results reported here correspond to an instantaneous field of view (IFOV) subtending roughly one km at the tangent altitude; IFOV smearing effects were not considered. Synthetic data for tangent heights greater than 220 km was generated for one-km intervals and then smoothed to obtain the spectral radiances for $\Delta H_T = 10$ km. A given NESR refers to the noise level before smoothing.

The solutions of the nonlinear normal equations inevitably contain some error, even if the data has no noise. Errors in [0],T propagate downward as these values are used to compute the limb spectral radiance N_{ν}, and then [0] and T, for a lower tangent height (altitude). Also, as the retrieval proceeds downward, the temperature at the tangent altitude changes, and the sensitivity of N_{ν} to this temperature is spectrally redistributed. The contribution to N_{ν} of a unit volume at the tangent altitude is proportional to the volume's spectrally resolved emission rate $\xi\nu$ times the spectral transmittance of the path between the tangent point and the observing instrument. Figure 5 shows $\xi\nu(T)$ for $\nu = 68$ cm^{-1} (the center of the 147 μm line), and the

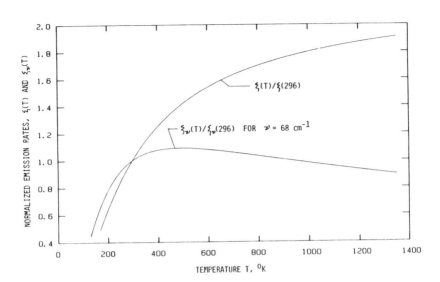

FIGURE 5.   The volume emission rate $\xi_\nu(T)$ at the center of the 147 µm
line ($\nu = 68$ cm^{-1}), and $\xi(T)$ for the 147 µm line. Both are
normalized to their values at T = 296 K. The spectrally
resolved rate ξ_ν is given by Eq. (6).

spectrally unresolved rate $\xi(T)$ for the 147 µm line. The line center
emission decreases with temperature both through Doppler broadening and
the line intensity S(T), but increases through the Planck function.
These competing temperature effects result in the broad relative
maximum in $\xi_{68}(T)$ near T = 500K. For atmospheric model ATOI4 this
temperature occurs at approximately 130 km altitude. Thus, N_ν for $H_T =$
130 km provides temperature information about altitude H = 130 km
principally in wings of N_ν. But a significant portion of the wing
radiance is due to higher layers, which are warmer (have broader ξ_ν due
to Doppler broadening) than the 130-km layer. Errors propagated via
the computation of wing radiances eventually become much larger than
the change in wing radiance due to a reasonable change in temperature
at the tangent altitude, resulting in the failure mentioned above.
Obviously, noise in the data can cause similar error propagation and
instability.

It was found that the solutions tend to fall into three
categories: For relatively low resolving power and high noise, the
retrieval procedure fails at approximately 130 km altitude, but
provides an accurate [0] solution and a usable, albeit noisey,
temperature solution above this altitude. Over a rather wide range of
higher R and lower NESR, the procedure fails at approximately 104 km,
but yields accurate [0] and T values for H > 104 km. This altitude
coincides with the peak of the O-atom density profile. Still higher
and/or lower NESR results in very accurate retrieved profiles for the
entire altitude range 90 to 300 km.

The failure of solutions at the altitude of the [0]-profile peak is due to error propagation in combination with rapidly increasing atmospheric opacity. The transmittance from the tangent point to the observing instrument, at the center of the 147 μm line, is less than ~0.01 for tangent heights less than 104 km. The 63 μm line gave poorer results than the 147 μm line, particularly for the lower altitudes, because of its higher opacity. The 63-μm transmittance to the tangent point is less than ~0.01 for tangent heights less than 120 km.

2.4 SUMMARY OF RESULTS (for 147 μm LINE)

The noise levels and resolving powers corresponding to the three solution categories are defined in Fig. 6. For example, any combination of R and NESR corresponding to a point between the two lowest solid curves results in accurate [0] and T solutions between 300 km altitude and the peak of the [0] profile, near 104 km. Figure 6 is a composite of two such plots obtained separately for the two model atmospheres. Thus, it may not accurately characterize the solution when [0] or T are very different from those of the models.

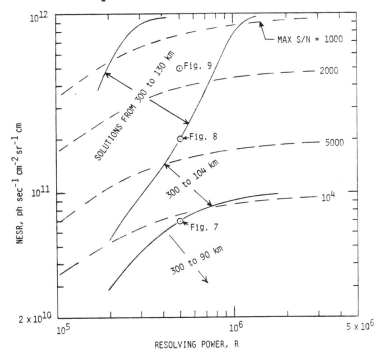

FIGURE 6. Noise-equivalent photon spectral radiance (NESR) and resolving power corresponding to the three solution categories (for the 147 μm line).

Shown in Fig. 6 are lines of constant maximum signal-to-noise, defined as the maximum of the measured $N_\nu(H_T)$ for $\nu = \nu_0 = 68$ cm^{-1}, divided by the NESR. The maximum signal occurs at $H_T \approx 120$ km for a given resolving power. As R is increased the maximum signal becomes

larger, and a given maximum S/N corresponds to larger NESR. The
signal-to-noise is considerably lower than the maximum in the wings of
the line and at the higher tangent heights, which is to say that $N_\nu(H_T)$
has a very large range.

Figures 7 through 9 show retrieved [0],T profiles representative
of the three categories. The point symbols in Fig. 6 identify the NESR
and figure number for each case; the resolving power is 5 x 10^5.
Figures 7 and 9 correspond to the model representing a cool
thermosphere; Fig. 8 corresponds to the other model.

Figures 8 and 9 show clearly that the [0] profile can be recovered
to high relative accuracy even when there are large errors, e.g., 100K,
in the retrieved temperature profile. This is true at least for the
higher altitudes. Generally, the temperature errors decrease from high
to low altitude, due presumably to the increase in S/N, provided the
solution doesn't fail. When solutions can be obtained down to 90 km,
the temperature errors at the lower altitudes (90 to ~104 km) become
very small, but the corresponding errors in [0] can be large relative
to those at high altitude.

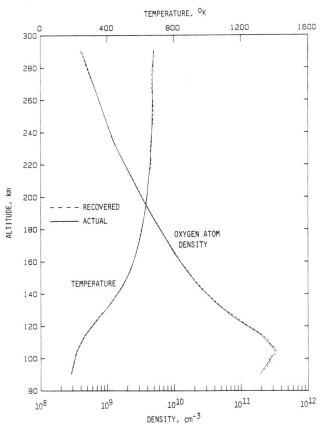

FIGURE 7.   Actual and retrieved [0] and T profiles for resolving power
            5 x 10^5. The NESR is 7 x 10^{10} ph $s^{-1}cm^{-2}sr^{-1}cm$.

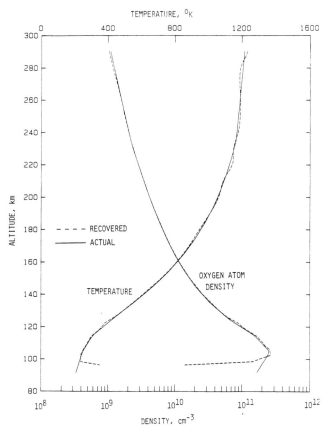

FIGURE 8. Same as Fig. 7, except NESR is 2×10^{11} ph s^{-1}cm^{-2}sr^{-1}cm.

3. CONCLUSIONS

The described technique will yield the translational temperature and O-atom density profiles from a limb radiance scan spectrally resolved over the 147 μm line. The level of noise in the data and the spectral resolution have, of course, an effect on the accuracy of the retrieved profiles, but more importantly, they determine the altitude at which the onion-peel retrieval technique becomes unstable. Very accurate profiles can be recovered from ~300 km down to 90 km altitude if, for example, the resolving power is 5×10^5 and the noise-equivalent photon spectral radiance is less than 7×10^{10} ph s^{-1}cm^{-2}sr^{-1} cm. With half this resolving power and ten times the noise it is possible to recover an accurate O-atom density profile and a rough approximation to the temperature profile above ~130 km altitude. Above 130 km the instrument need only resolve approximately 10 km at the tangent altitude.

A resolving power of 5×10^5 and photon NESR of 7×10^{10} ph s^{-1}cm^{-2} sr^{-1}cm at λ = 147 μm work out to a required noise-equivalent radiance

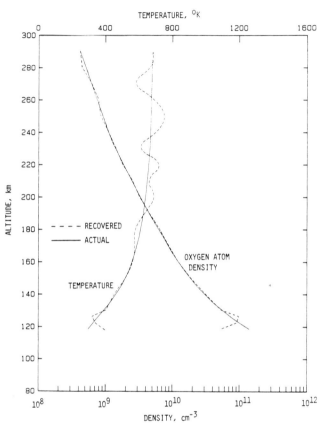

FIGURE 9. Same as Fig. 7, except NESR is 5×10^{11} ph $s^{-1}cm^{-2}sr^{-1}cm$.

of 2×10^{-14} W $cm^{-2}sr^{-1}$. The measurements would have to resolve
approximately two km at the tangent altitude (for retrievals down to
90 km). These signal requirements were translated to basic design
requirements for a Fabry-Perot system operating at orbital altitude.
It was found that the required high performance could be theoretically
achieved only in a confocal design, i.e., with spherical mesh etalons.
Assuming a stressed Ge:Ga detector with NEP of 5×10^{-17} W $Hz^{-1/2}$ and a
one-Hz noise bandwidth, the system would need foreoptics of roughly
80 cm diameter, cooled to 10K. Thus, retrievals down to 90 km by the
described technique imply a somewhat advanced sensor.

As mentioned, the spectrally non-resolved variant of the technique
is still under study. It is described, together with results obtained
from noise-free synthetic data, by Sharma, Harlow and Riehl (1988). We
conclude this discussion by briefly outlining a different alternative
approach that we hope will prove feasible using a less advanced sensor
concept.

Spectrally resolved limb radiance measurements in the fundamental
($\Delta v = 1$) band of nitric oxide near 5.3 μm can be used to retrieve a

portion of the translational (rotational) temperature profile using an algorithm described by Zachor, et al. (1985). It is assumed that the NO rotational levels are in equilibrium with the local translational temperature, which is reasonable up to approximately 200 km altitude. Nitric oxide is optically thin for all tangent heights. The observed limb radiance is due to excited NO, whose density profile peaks at approximately 130 km altitude. Spectral resolution no greater than ~0.1 µm is needed to define the band contour. Depending on the S/N of the spectral data it may be possible to recover the temperature (as well as the NO excited density) between 90 and 200 km, or perhaps 90 to 250 km. If the temperature can be obtained up to 250 km, the O-atom density profile can be recovered from a spectrally non-resolved limb radiance profile for the 147 µm line, since the temperature is nearly constant above 250 km. If the 5.3 µm data yields the temperature to only 200 km, then the 147 µm data for tangent heights greater than 200 km would have to be spectrally resolved, but relatively low resolving power (and low tangent height resolution) would suffice.

REFERENCES

Sharma, R. D., H. B. Harlow, and J. P. Riehl, 1988: Determination of atomic oxygen density and temperature of the thermosphere by remote sensing, accepted by publication in Planet. Space Sci.

Sharma, R. D., A. T. Stair, Jr., and H. A. Smith, 1988: A space-borne passive infrared experiment for remote sensing of the atomic oxygen density and temperature, and total density in the upper atmosphere, accepted for publication in Adv. Space Res.

Zachor, A. S., and R. D. Sharma, 1985: Retrieval of non-LTE vertical structure from a spectrally resolved infrared limb radiance profile, J. Geophys. Res., 90, 467-475.

Zachor, A. S., R. D. Sharma, R. M. Nadile, and A. T. Stair, Jr., 1985: Inversion of a spectrally resolved limb radiance profile for the NO fundamental band, J. Geophys. Res., 90, 9776-9782.

DISCUSSION

Neuendorffer: I was curious, I've never worked around where the peak of the oxygen is. You are going to have a lot of dissociation taking place and I was wondering if the oxygen temperature at that point is representative of the overall temperature or whether it is biased because of the dissociation affects.

Zachor: The translational temperature there is well-defined.

INVERSION TECHNIQUE FOR SAGE II DATA

W.P. Chu
NASA Langley Research Center
Hampton, Virginia 23665, USA

ABSTRACT

A brief description of the operational algorithm used for the processing of SAGE II satellite data is provided in this paper. The inversion technique involves the separation of different contributions in the multi-spectral channel measurements due to aerosol, ozone, water vapor, and nitrogen dioxide, and the retrieval of the vertical profiles for each of these species. The results show that the vertical profiles of each species can be retrieved from the satellite measurements with high accuracy. Diagnostics study based on the approach developed by Rodgers (Rodgers, 1988) has been applied to the SAGE II algorithm for ozone retrieval. The results show that high vertical resolution profiles of ozone can be retrieved from the SAGE II measurements with small interference from aerosol and nitrogen dioxide.

1. INTRODUCTION

The SAGE II instrument is one of the operational satellite instruments using the solar occultation approach to provide measurements of vertical profiles of stratospheric constituents.(Mauldin et al., 1985) It is a seven spectral channel photometer operating in the wavelength region extending from the near UV to the near IR. The objective of SAGE II is to provide near global stratospheric profiles of aerosol extinction properties at different spectral wavelength region, together with concentration profiles of ozone, nitrogen dioxide, and water vapor. Inversion of the solar occultation measurements have been discussed in association with the previous for spectral channel SAGE experiment(Chu, 1977; Chu and McCormick. 1979). However. the previous SAGE data inversion scheme had to be modified in order to accommodate the additional channel data from the SAGE II instrument. This paper will briefly describe the inversion technique used in the inversion of SAGE II data. The discussion will be restricted to the retrieval of aerosol and ozone. In addition, a diagnostic study of the SAGE II algorithm using the approach recently developed by Rodgers (1988) for analyzing the characteristics of the retrieved ozone profiles will be presented.

2. MEASUREMENT APPROACH

SAGE II measurements are obtained during spacecraft sunrise and sunset events. Attenuation of solar radiation by the earth's atmosphere can then be

RSRM '87: ADVANCES IN
REMOTE SENSING RETRIEVAL METHODS
A. Deepak, H.E. Fleming, and J.S. Theon (Eds.)   243

determined from the solar irradiance measurements at the different spectral channels. The instantaneous irradiance measured by the instrument at time t is given by:

$$H_\lambda = \int_{\Delta\lambda} \int_{\Delta\omega} W_\lambda(\theta,\phi) F_\lambda(\theta,\phi,t) T_\lambda(\theta) d\Omega d\lambda \qquad (1)$$

where W is the radiometer's field of view function. ϕ is the azimuthal angle, Ω is the solid angle, T is the transmittance of the atmosphere as a function of view angle θ, and F is the extraterrestrial solar radiance for wavelength λ. By ratioing all irradiance measurements to values obtained when the atmospheric attenuation is insignificant, corresponding to the situation when the sun is high above the atmosphere, the mean transmittance profiles can be obtained which are related to the atmospheric species through the Bouguer law as follows.

$$T_\lambda(h_t) = \exp\left[-\int \sigma_\lambda(h) d\rho_\lambda(h)\right] \qquad (2)$$

where σ_λ is the total extinction coefficient of the atmosphere as a function of tangent altitude h and wavelength λ, and ρ is the geometric slant path through the atmosphere corrected for atmospheric refraction.

Figure 1 shows the typical extinction contributions in the atmosphere at a tangent height of 18 km between the 300 to 1200 nm wavelength region.

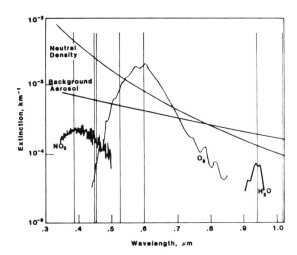

FIGURE 1. Atmospheric constituent extinction versus wavelength at 18 km with the location of the 7 SAGE II channels indicated by vertical lines.

The spectral locations of the seven SAGE II channels are indicated by the vertical lines. They are centered at 1020, 940, 600, 525, 453, 448, 385 nm respectively. The 940 nm channel is for water vapor sensing. The two narrow spectral channels at 453 and 448 nm are for nitrogen dioxide sensing using the differential absorption technique. The other four channels are for aerosol and ozone sensing. The 600 nm channel is located at the peak of the ozone Chappuis absorption band and it is, therefore, most sensitive to ozone absorption. Inspection of figure 1 indicats that measurements at each of the spectral channel location consists of contributions from different species, and the inversion procedure has to decouple the various contributions in order to retrieve the vertical profiles for each species separately.

3. INVERSION APPROACH

The SAGE II measurements can be reduced to seven spectral channel slant path optical depth δ versus tangent height profiles following the procedure described before (Chu and McCormick, 1979). After removal of the Rayleigh component using the NWS (National Weather Service) temperature versus altitude data, and neglecting the water vapor channel at 940 nm, we are left with six channel data which can be expressed as,

$$\delta_\lambda(h_i) = \int (\sigma_\lambda^{ozone} + \sigma_\lambda^{aero} + \sigma_\lambda^{NO_2}) \, d\rho(h) \tag{3}$$

We have six simultaneous equations at each height which we must solve in order to determine the vertical profiles for aerosol, ozone, and nitrogen dioxide. The approach taken in the SAGE II operational algorithm is first to separate the different species contributions independently at each height level. After this procedure is applied through the complete altitude range, we obtain the slant path optical depth profiles for aerosol at the different wavelengths, and for ozone and nitrogen dioxide at thier particular reference wavelength. Finally, the different slant path optical depth profiles can be inverted to produce aerosol extinction, and ozone, and nitrogen dioxide concentration vertical profiles.

The procedure for species separation is performed by using the following steps:

i) At each height level, a nitrogen dioxide value is estimated from the differential channels at 453 and 448 nm by assuming the aerosol contribution to be small and estimating the ozone contribution using the 600 nm channel data.

ii) By removing the nitrogen dioxide contribution from the other five wavelength channels, we are left with five optical depth values at wavelengths of 1020, 600, 525, 453, and 385 nm with contribution from aerosol and ozone only.

iii) The ozone and aerosol contributions at the five wavelength channels are then separated using a method which will be described later resulting in ozone optical depth values at 600 nm, and aerosol optical depth values at 1020, 525, 453, and 385 nm.

iv) Steps (i) through (iii) are then repeated one more time with the updated values of aerosol and ozone. the reason for additional updating is to minimize the first guess influence on the nitrogen dioxide retrieval.

v) The above procedure is applied to each height level of the slant path optical depth data. As a result, we reduce the data into slant path optical depth profiles separately for aerosol, ozone, and nitrogen dioxide.

The resulting retrievals can be expressed for the different species as follows.

$$\delta_\lambda^{species}(h_i) \quad = \quad \int \sigma_\lambda^{species}(h) \; d\rho(h) \tag{4}$$

Assuming the atmosphere to be spherically symmetric, and using the mean extinction values at 1-km thick layers for each species, the integral equations in Equation (4) can be reduced to a system of simultaneous equations of the form,

$$\delta_i \quad = \quad \sum_j P_{ij}\sigma_j \tag{5}$$

where P_{ij} is the j-path length element for the i-level. Equation (5) is inverted using the Twomey's modification of Chahine's nonlinear inversion algorithm (Twomey, 1975). The solution at each tangent height level is updated as follows.

$$\sigma_j^{n+1} \quad = \quad \sigma_j^n \left[1 \; + \; (r_i \; - \; 1)P_{ij}/P_{ii} \right] \tag{6}$$

where r_i is the ratio between the measured optical depth to the computed optical depth value at the n-iteration. The iterations are terminated when the differences between the measurements and the calculated values approach the estimated noise level. In addition, a 5-km vertical smoothing during each iteration cycles is incorporated in the algorithm for retrieval at higher altitude levels where the signal to noise level is usually low. These height levels correspond to about 50 km for the ozone retrievals, and about 24 km for the aerosol retrievals.

4. SEPARATION OF SPECIES CONTRIBUTIONS

Inspection of equation (3) indicates that there are six unknowns to be solved for from the five equations at each height level. The five equations correspond to the slant path optical depth data at 1020, 600, 525, 453, and 385 nm wavelength channels. The six unknowns consist of ozone and the five aerosol optical depth values at each of the wavelength channel location. The nitrogen dioxide values are assumed known and are removed. Since the aerosol extinction versus wavelength behavior is determined from the aerosol size distribution and its optical properties, it is natural to look at way of either expressing the aerosol extinction versus wavelength behavior as a polynomials or performing interpolation of the extinction values to the ozone wavelength at 600 nm. The SAGE II algorithm has adapted the approach of interpolating the aerosol extinction values at 1020, 525, 453, and 385 nm to a value at 600 nm. Therefore the aerosol extinction at the four wavelengths will be treated as the four aerosol unknowns, while the aerosol extinction value at 600nm could be expressed as a known function of the extinction values at the other wavelengths.

To arrive at the desired interpolation scheme, the problem is viewed from the perspective of inverting the aerosol size distribution from the four aerosol extinction values. It should be understood that in our case we are considering the column extinction values corresponding to the optical depth values. Assuming the columned aerosol size distribution is given by the m discrete size array $\mathbf{x} = (\mathbf{N_1}, \mathbf{N_2}, \ldots, \mathbf{N_m})$ and the aerosol extinction cross section similarly computed from the Mie formalism to be $\mathbf{K_i} = (\beta_{i1}, \beta_{i2}, \ldots, \beta_{im})$, then the aerosol optical depth values are given by

$$\delta_i = \sum_{j=1}^{m} \beta_{ij} N_j \quad i = 1, \ldots, 4 \tag{7}$$

which can be put into the matrix form.

$$\delta_i = \mathbf{K_i x} \tag{8}$$

The above equation in principle can be inverted for aerosol size distribution information even though the inverse problem is ill-conditioned. Adapting the Twomey's linear constraint solution (Twomey, 1963), we have,

$$\mathbf{x} = \mathbf{K^T}(\mathbf{KK^T} + \mathbf{\Gamma})^{-1}\delta_i \tag{9}$$

Using the inverted size information, we can calculate the aerosol column extinction value at any other wavelength as,

$$\delta_\lambda = \mathbf{K_\lambda x} = \mathbf{K_\lambda K^T}(\mathbf{KK^T} + \mathbf{\Gamma})^{-1}\delta_i = \sum_{i=1}^{4} \alpha_i \delta_i \tag{10}$$

Therefore the aerosol extinction value at any other wavelength location can be expressed as a linear combination of the extinction values at the four specific wavelengths with coefficients α_i. The α_i coefficients can be pre-computed for the algorithm. As a result, the total number of unknowns in this case will be equal the number of equations, and the exact solution can be obtained for ozone and the four aerosol extinction values.

5. DIAGNOSTICS

The SAGE II retrieval algorithm as described above has been analyzed using the diagnostic approach developed by Rodgers (Rodgers, 1988). Two aspects of the retrieval algorithm have been carefully examine. They consist of the averaging kernel and the covariance of the measurement noise. Only the analyses on the ozone retrieval will be discussed in this paper. A very brief summary of the diagnostic approach will be presented here. If we define our forward model as Equation (3) which describes the measurements as a function of the extinction parameters. The inverse model would be the algorithmic procedure as described before for inverting the measured optical depth data into species extinction profiles. Then a transfer function T can be defined relating the true extinction parameters to the retrieved extinction parameters as below.

$$\hat{\sigma} \;=\; T(\sigma, c, d) \;+\; \epsilon_\sigma \tag{11}$$

where c and d are model parameters for the forward and inverse models respectively, and ϵ_σ is the random noise in the retrieved extinction due to measurement noise.

Following the typical procedure of linearization at some mean states denoted by $\bar{\sigma}, \quad \bar{c}, \quad \bar{d}$, we obtain the expansion form,

$$\hat{\sigma} \;=\; T(\bar{\sigma}, \bar{c}, \bar{d}) \;+\; A_\sigma(\sigma - \bar{\sigma}) \;+\; A_c(c - \bar{c})$$

$$+ \; D_d(d - \bar{d}) \;+\; D_\delta \epsilon_\delta \tag{12}$$

where A_σ is the A matrix where rows are the averaging kernels and columns are the delta function response, the D matrix are the contribution functions, and ϵ_δ is the measurement noise.

The averaging kernels for the ozone retrieval for the SAGE II algorithm are obtained by using a midlatitude mean ozone profile as the linearization points and by perturbing the forward model. Sensitivity analyses have also been performed with the contributions of aerosol and nitrogen dioxide to the ozone retrieval. The nitrogen dioxide contribution was estimated by using a mean sunset, midlatitude nitrogen dioxide profile for linearization and by perturbing the forward model. The aerosol contributions are estimated by using a mean 1985, midlatitude aerosol vertical profile assuming the size distribution for stratospheric aerosols to be lognormal as described by Rosen (Rosen, et al., 1975) with refractive idex of 1.43. There are a total of three parameters for the aerosol description; the total number, the mean radius, and the spread of the distribution. The sensitivity of the aerosol forward model is obtained by perturbing the three aerosol size parameters separately and the three final contributions added to give the overall aerosol sensitivity to the ozone retrieval.

Figure 2 shows the averaging kernels at every 5-km altitude interval for the ozone retrieval with contributions from ozone and aerosol. The contribution from nitrogen dioxide was found to be insignificant over the altitude range being considered here. The ozone averaging kernels show the typical sharp peak behavior with half width of about 1-km, indicating the high vertical resolution nature of the SAGE II measurement and retrieval process. For averaging kernels centered below 20-km, the widths are generally broader due to the fact that the ozone retrieval is below the ozone peak and subsequently the retrieved profile would be slightly smeared out. The aerosol contributions for the ozone retrieval are predominately negative, and are about 4 to 5 percent at altitudes below 20-km where the aerosol concentration is at it's peak.

As discussed by Rodgers (Rodgers, 1988), the random errors in the retrieval would consist of two components contributed by the measurement noise and the null space errors. However, SAGE II is probably the only instrument that can provide an ozone profile with such fine vertical resolution. There does not exist any ozone data base with higher vertical resolution that can be used to provide a covariance description of the true ozone field. Therefore, we have not attempted to estimate the null space error for the SAGE II retrieval due to the lack of information to construct the covariant matrix for the true ozone field. Instead, we have looked at the measurement error component by analyzing the error patterns for the measurement noise. Figure 3 shows the first three normalized eigenvectors of the measurement covariant matrix. The predominated error pattern can be seen to be the lower noise level for altitudes above 30-km, and an increasing random fluctuation below 30-km. This is typical of the SAGE II measurements where sun scans are overlapped together to generate the atmospheric transmission profiles. The noise levels are general higher at lower altitudes due to the increased attenuation at lower altitudes

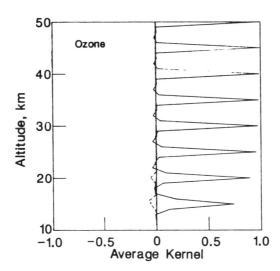

FIGURE 2. Averging kernels centered at every 5 km for ozone retrieval with ozone (solid lines) and aerosol (dashed lines) contributions.

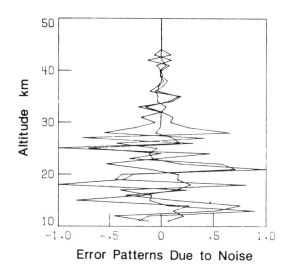

FIGURE 3. Error patterns for the ozone retrieval showing the first three normalized eigenvectors of the noise covariant matrix.

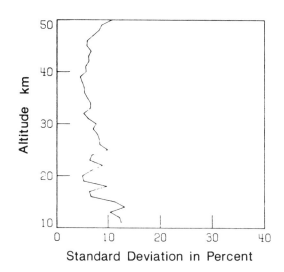

FIGURE 4. Residual standard deviation for the ozone retrieval.

together with the increase of atmospheric refraction correction. Figure 4 shows the square root of the diagonal component of the covariant matrix which is an indication of the noise magnitude due to measurement errors. It can be seen that the ozone retrieval can produce profiles with random errors less than 10 percent.

6. SUMMARY

A brief description of the SAGE II operational inversion algorithm is presented here. Diagnostic analyses on the ozone retrieval indicates that SAGE II can provide vertical profiles of ozone with a vertical resolution of about one km. The interferences from other species such as aerosol and nitrogen dioxide are found to be small and relatively insignificant. The random errors on the SAGE II ozone retrieval due to measurement noise are found to be less than 10 percent.

REFERENCES

Chu, W. P., 1977: The inversion of stratospheric aerosol and ozone profiles from spacecraft solar extinction measurements, in A. Deepak, (Ed.), Inversion Methods in Atmospheric Remote Sounding, Academic Press, New York, 505-516.

Chu, W. P. and M. P. McCormick, 1979: Inversion of stratospheric aerosol and gaseous constituents from spacecraft solar extinction data in the 0.38 - 1.0 μm wavelength region, Appl. Opt., 18, 1404-1414.

Mauldin III, L. E., N. H. Zaun, M. P. McCormick, J. H. Guy, and W. R. Vaughn, 1985: Stratospheric Aerosol and Gas Experiment II instrument: A functional description, Optical Eng., 24, 307-312.

Rodgers, C., 1988: A general error analysis for profile retrieval, in the current issue of this proceding.

Rosen, J. M., D. T. Hofmann, and J. Laby, 1975: Stratospheric aerosol measurements II: The worldwide distribution. J. Atmos. Sci., 32, 1457-1462.

Twomey, S., 1963: On the numerical solution of Fredholm integral equation of the first kind by the inversion of the linear system produced by quadrature, J. Assoc. Comput. Mach., 10, 99.

Twomey, S., 1975, Comparison of constrained linear inversion and an iterative nonlinear algorithm applied to the indirect estimation of particle size distribution. J. Comp. Phys., 18, 188.

DISCUSSION

NO QUESTIONS

THE SIMULTANEOUS RETRIEVAL OF AEROSOL PROPERTIES
FROM THE WAVELENGTH DEPENDENCE OF EXTINCTION MEASURED
BY THE SAGE II EXPERIMENT

Glenn K. Yue, M.P. McCormick, and W.P. Chu
Atmospheric Sciences Division
NASA Langley Research Center
Hampton, Virginia 23665, USA

P. Wang
Science and Technology Corporation
Hampton, Virginia 23666

E. Chiou
ST Systems Corporation
Hampton, Virginia 23666

ABSTRACT

The SAGE II satellite system has seven radiometric channels which measure the intensity of solar radiation traversing the earth's limb at wavelengths ranging from 0.385 μm to 1.02 μm. By assuming stratospheric aerosols are supercooled sulfuric acid droplets, the retrieval of aerosol composition from the water vapor channel information and the corresponding ambient temperature data is elucidated and the associated uncertainties in the retrieved values are discussed. By assuming the aerosol size distribution is lognormal, several methods of retrieving parameters in the size distribution expression from the wavelength dependence of aerosol extinction are discussed. Since at lower altitudes, aerosol extinction at shorter wavelengths is not available due to instrument dynamics range limitations, a slightly different method is applied for retrieval at these altitudes. From the SAGE II results, aerosol properties are retrieved with the suggested techniques. Some examples of the retrieved aerosol properties are presented.

1. INTRODUCTION

The SAGE II (Stratospheric Aerosol and Gas Experiment II) satellite system was designed to monitor the global distribution of minor constituents such as aerosols, O_3, NO_2, and H_2O in the stratosphere and upper troposphere. It is aboard the Earth Radiation Budget Satellite (ERBS) which was launched by the space shuttle Challenger on October 5, 1984.

The SAGE II instrument is a sunphotometer designed to measure solar irradiance through the earth's atmospheric limb during each sunrise and sunset event encountered by the spacecraft as it orbits the earth. The instrument locks onto the centroid of the sun's disc and scans vertically across the disc with a field-of-view of about 0.5 arc

RSRM '87: ADVANCES IN
REMOTE SENSING RETRIEVAL METHODS
A. Deepak, H.E. Fleming, and J.S. Theon (Eds.)

253

minutes in the vertical and 2.5 arc minutes in the horizontal.
Measurements are made in seven radiometeric channels centered at wavelengths of 1.02, 0.94, 0.6, 0.525, 0.453, 0.448, and 0.385 μm. The
0.94 μm channel provides water vapor concentration profile measurements, while the 0.60 μm channel located at the peak of the ozone
Chappius absorption band provides ozone concentration profile measurements. The 0.448 μm and 0.453 μm channel are designed to monitor NO_2
concentration by differential absorption at these wavelengths. The
measurements during each sunrise or sunset event are self-calibrated
relative to the brightness of the sun disc. The small field-of-view
of SAGE II provides about one kilometer vertical resolution for each
observation. A more detailed description of the design and performance of the SAGE II experiment is given by Mauldin et al. (1985) and
McCormick (1987).

In order to separate overlapping contributions at each wavelength
channel due to the presence of aerosol, ozone, nitrogen dioxide, water
vapor, or Rayleigh scattering, the SAGE II raw data are inverted by a
two-step inversion algorithm. The path transmission values at different wavelengths are converted to vertical extinction profiles using a
homogeneous atmospheric shell model. A detailed description of the
inversion scheme is given by Chu (1986, 1988). The useful products of
SAGE II for scientific analysis are vertical profiles of O_3, NO_2, H_2O
concentration and aerosol extinction at 1.02, 0.525, 0.453, and 0.385
μm wavelengths. However, due to instrument dynamic range limitations,
aerosol extinction at shorter wavelengths are not available at the
lowest altitudes. In general, the lower altitude limit for the aerosol extinction at 1.02 μm is nearly the ground or to cloud top, the
lower altitudes limit for the aerosol extinction at 0.525 μm, 0.453 μm
and 0.385 μm are 6.5 km, 10.5 km, and 14.5 km, respectively, if they
are above cloud top.

In this paper, we will discuss several methods of simultaneously
retrieving aerosol size distribution from the aerosol extinction at
four wavelengths provided by the SAGE II experiment. In addition, the
method of retrieving aerosol composition from the SAGE II water vapor
channel and the temperature information provided by the National
Meteorological Center of NOAA will be discussed. Some examples of the
retrieved aerosol properties from the results of the SAGE II experiment will also be presented.

2. THE RETRIEVAL METHODS

2.1 THE RETRIEVAL OF AEROSOL COMPOSITION

In this study, we assume that aerosol particles in the stratosphere and upper troposphere are a supercooled sulfuric acid and water
solution. This assumption is justified since in situ measurements
show that most of the stratospheric aerosols are aqueous sulfuric acid
droplets (Toon and Pollack, 1973, 1976; Rosen 1971; Woods and Chuan
1983). A paper by Turco et al. (1982) reviewed the controversy of

whether ammonium sulfate is present in the stratospheric aerosol and concluded that even if ammonium ions are present, their amount must be very small. Generally speaking, in order for aqueous sulfuric acid droplets to remain in equilibrium, both the H_2O and H_2SO_4 partial vapor pressure above the surface of the aerosol particle must equal to the ambient vapor pressure of H_2O and H_2SO_4. However, since the concentration of water vapor in the stratosphere is orders of magnitude higher than that of H_2SO_4 molecules, the partial vapor pressure of H_2O rather than the partial vapor pressure of H_2SO_4 is in equilibrium with the ambient vapor pressure of that species.

The partial vapor pressure of H_2O above the surface of an aqueous sulfuric acid solution is given by:

$$p(x,T) = \exp \left[A(x)\ln \frac{298.15}{T} + \frac{B(x)}{T} + C(x) + D(x) \, T \right], \qquad (1)$$

where x is the weight percentage of H_2SO_4 in the solution, T is the ambient absolute temperature, and $A(x)$, $B(x)$, $C(x)$ and $D(x)$ are coefficients that are functions of x only. Numerical values of the coefficients for 36 values of x are given by Gmitro and Vermeulen (1964). A plot of the value of $p(x,T)$ in units of ppmv at 20 km as a function of x and T is shown in Figure 1. The aerosol composition x can easily be retrieved from the measured water vapor concentration and the ambient temperature T. As discussed in a paper by Yue et al. (1986), if the ambient temperature is $-55°C$, an uncertainty of 10% in the water vapor pressure will only produce an uncertainty of about 3% in the estimated H_2SO_4 weight percentage. Similarly, for an uncertainty of $2°K$ in ambient temperature, the uncertainty in the estimated H_2SO_4 weight percentage is at most about 3%. Thus, in general, the composition of the aerosol particle can be quite accurately retrieved. However, if the concentration of water vapor is relatively high and the ambient temperature is very low, a slight change of ambient temperature will upset the equilibrium balance and the composition of aerosol particles will be dramatically changed due to the release or absorption of a large amount of water molecules. In this case, the value x is very low and there is a large uncertainty in the retrieved aerosol composition.

2.2 THE RETRIEVAL OF AEROSOL SIZE DISTRIBUTION

In this study, we assume that the size distribution of aerosol particles in the stratosphere and upper troposphere can be expressed by the following lognormal formula:

$$n(r) = \frac{N_0}{(2\pi)^{1/2}\ln\sigma} \cdot \frac{1}{r} \exp \left(- \frac{(\ln r/r_g)^2}{2 \ln^2\sigma} \right) \qquad (2)$$

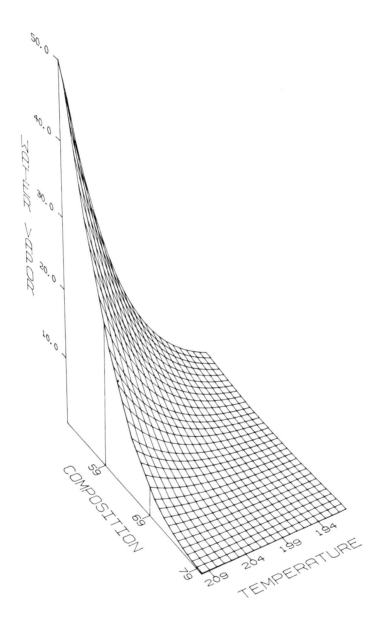

FIGURE 1.   Three-dimensional plots of the water vapor pressure in
units of ppmv at 20 km as a function of temperature and
weight percentage of H_2SO_4 in the aqueous aerosol
particle.

where n(r) is the number of aerosol particle per unit volume of air with radii in the range from r to r + dr in units of $cm^{-3}\mu m^{-1}$, N_0 is the total number of particles per unit volume of air, r_g is the mode radius, and σ is the width of the lognormal distribution. The lognormal size distribution is chosen because it has three adjustable parameters, r_g, σ, and N_0, and it is probably the most widely used analytical expression for background stratospheric aerosols. After a major volcanic eruption, aerosol particles in the stratosphere may be expressed by the sum of two lognormal size distributions to reflect larger sizes. But the number of unknown parameters in such an expression is more than the number of aerosol extinction values at different wavelengths provided by SAGE II. In this study we are limited to study background aerosol particles and the retrieved parameters r_g, σ, and N_0 should be considered as the parameters in the equivalent lognormal size distribution that produce the same wavelength dependence of the aerosol extinction measured by the SAGE II experiment.

In this study we further assume that the aerosol particles are spherical and the extinction coefficient can be calculated by the formula:

$$\beta(\lambda) = \int_{r_1}^{r_2} n(r)\; Q\left(\frac{2\pi r}{\lambda},\, m\right)\; \pi r^2 dr, \qquad (3)$$

where $\beta(\lambda)$ is the volume extinction coefficient of solar radiation at wavelength λ. $Q\left(\frac{2\pi r}{\lambda}, m\right)$ is the Mie extinction efficiency, which is a function of size parameter $2\pi r/\lambda$ and the refractive index m, and r_1 and r_2 are, respectively, the lower and upper limits, of the radii of aerosol particles under consideration. Since n(r) in Eq.(3) is given by Eq.(2), $\beta(\lambda)$ is a function of r_g, σ and N_0. In order to eliminate the factor N_0, we define aerosol extinction ratios R_1, R_2 and R_3 by the following expressions:

$$R_1 = \frac{\beta(\lambda = 0.385)}{\beta(\lambda = 1.02)} = f_1(r_g,\, \sigma) \qquad (4)$$

$$R_2 = \frac{\beta(\lambda = 0.453)}{\beta(\lambda = 1.02)} = f_2(r_g,\, \sigma) \qquad (5)$$

$$R_3 = \frac{\beta(\lambda = 0.525)}{\beta(\lambda = 1.02))} = f_3(r_g,\, \sigma) \qquad (6)$$

Once r_g and σ are retrieved, the effective radius, r_e, given by the following expression can be calculated:

$$r_e = r_g \exp\left(\frac{5}{2} \ln^2 \sigma\right) \qquad (7)$$

this effective radius is a measure of the aerosol radiative property.

The value of N_0 can be obtained by taking the ratio of the measured $\beta(\lambda=1.02)$ to the value of $\beta(\lambda=1.02)$ calculated from Eq.(3) with N_0 in the size distribution expression being set to 1.

Some investigations (Brogniez and Lenoble, 1987) prefer to use the Angstrom coefficient A_i deduced from the relation.

$$R_i = \lambda_i^{-A_i}$$

(8)

Where λ_i equals 0.385, 0.453, and 0.525 for i=1,2,3, respectively.

Several methods can be used to retrieve r_g and σ from the values of R_1, R_2, and R_3 measured by SAGE II. These methods will be discussed in the following subsections:

2.2.1. THE NUMERICAL METHODS

Standard numerical methods for solving the three simultaneous equations (4), (5), and (6) with two unknowns r_g and σ are available. Two common methods are the Newton-Raphson Method (Issacson and Keller, 1966) and the Levenberg-Marquardt method (More et al., 1980). Since only two simultaneous equations are required to solve for the two unknowns, two more accurate measurements of the extinction ratios can be chosen to solve for r_g and σ in applying the Newton-Raphson method. Basically these two methods require an initial guess of the solution. The second guess of the solution is calculated by substituting the values of the first guess into a formula which depends on the method used. The iteration process is continued until a convergence condition is met.

An advantage of the numerical methods is that computer algorithms are already available. However, their success may depend on the initial guess. In applying the Levenberg-Marquardt method, if one value of the extinction ratios has a larger error bar, solution to this set of equations may not be obtained. The greatest disadvantage of these numerical methods is that Mie calculations have to be repeated many times for each set of extinction ratios measured by SAGE II. In order to analyze large quantities of aerosol extinction data measured by SAGE II, a simple, fast and operational method should be used.

2.2.2. THE ANALYTICAL METHODS

Two types of analytical methods can be used to retrieve r_g and σ from the wavelength dependence of aerosol extinction measured by SAGE II. The first type is to derive an analytical relationship between r_g and σ for each set of values of the extinction ratios. The second type is to derive an analytical relationship between extinction and wavelength λ for each set of values of r_g and σ.

In the first method, values of R_i where i=1,2,3 are calculated for possible ranges of r_g and σ. These values are shown as three dimensional plots in Figures 2 (a), (b), and (c). As discussed in detail in a paper by Yue et al. (1986), each contour plot in the r_g versus σ plane for a given value of extinction ratio can be accurately described by a quadratic equation:

$$\ln \sigma = C_0 + C_1 r_g + C_2 r_g^2 \qquad (9)$$

Consequently, there are three analytical relationships between r_g and σ for the three measured values of extinction ratio. The values of r_g and σ can be retrieved by solving these simultaneous quadratic equations.

The advantage of this method is that once the quadratic expressions for each contour are obtained from the Mie calculation, the tedious Mie calculation need not be repeated for each extinction ratio measured by SAGE II. In addition, the variation of the retrieved values of r_g and σ corresponding to a variation of the measured values of extinction ratio can easily be analyzed with this method. As discussed by Yue et al. (1986), the retrieved σ is more reliable than r_g.

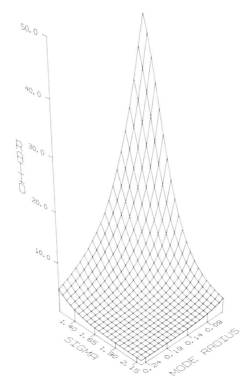

FIGURE 2. Three-dimensional plots of the aerosol extinction ratio as a function of sigma (σ) and mode radius (r_g) in the aerosol size distribution: (a) the ratio is $\beta(0.385)/\beta(1.02)$; (b) the ratio is $\beta(0.453)/\beta(1.02)$; (c) the ratio is $\beta(0.525)/\beta(1.02)$.

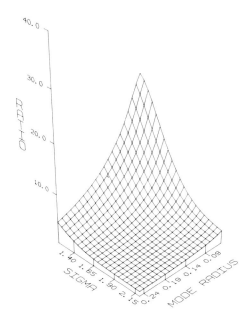

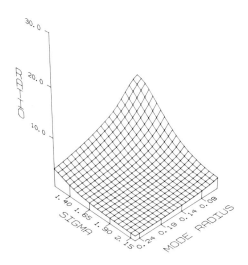

FIGURE 2.   (Continued)

Since two simultaneous equations are required to solve the two unknowns, only two extinction ratios out of the three values are required in this method. We may choose a set of extinction ratios with which more reliable results can be deduced, or we may take the average values of r_g and σ retrieved from two sets of extinction ratios.

The second analytical method is to express the extinction ratio as a function of λ for each set of r_g and σ values. A plot of extinction ratio versus wavelength for different sets of values of r_g and σ are shown in Figures 3 (a) and (b). As can be seen from these figures, the values of R or log R can be quite accurately expressed as a quadratic function of λ or log λ. The coefficients a_1, and a_2 in the following quadratic function:

$$\log R = a_1(1.02-\lambda) + a_2(1.02-\lambda)^2, \tag{10}$$

for several sets of values of r_g and σ are listed in Table 1. Once the coefficients corresponding to a possible range of r_g and σ are

TABLE 1. COEFFICIENTS a_1 AND a_2 IN EQ. (10) FOR DIFFERENT VALUES OF r_g AND σ

r_g (μm)	σ	a_1	a_2
0.05	2.0	0.907	0.216
0.07	2.0	0.804	0.0202
0.09	2.0	0.716	0.172
0.12	2.0	0.598	-0.303
0.10	1.6	1.19	0.167
0.10	1.8	0.923	-0.122
0.10	2.0	0.675	-0.226

generated, the r_g and σ can be retrieved by first calculating the coefficients a_1 and a_2 from the experimental results of wavelength dependence of the extinction ratio, followed by matching the calculated coefficients with those tabulated values. Contour plots of coefficients a_1 and a_2 as a function of r_g and σ are shown in Figures 4(a) and (b), respectively.

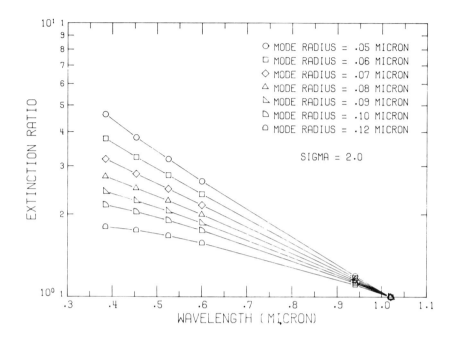

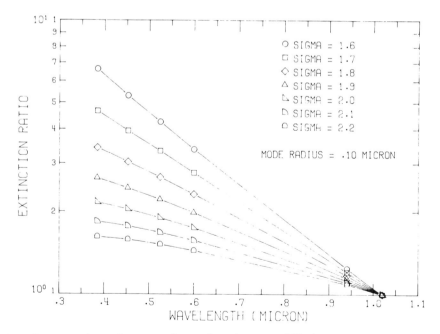

FIGURE 3.   Ratio of aerosol extinction coefficients versus wave-
            length for several values of width σ and mode radius
            r_g.   (a)   σ=2.00 and r_g varies from 0.05 to 0.12 μm
            (b) r_g = 0.10 μm and σ varies from 1.60 to 2.20.

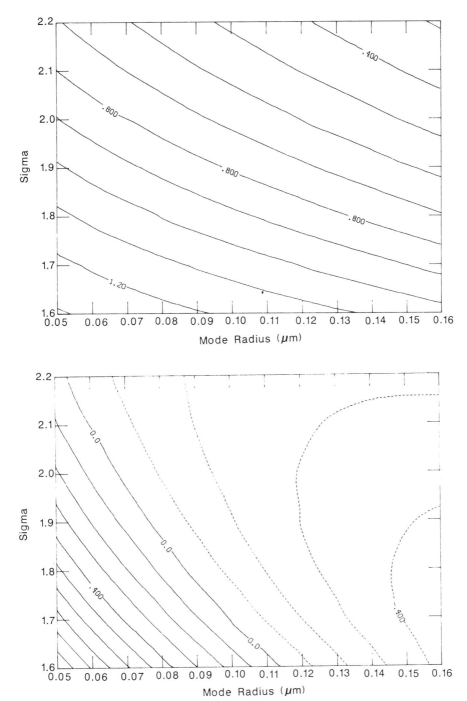

FIGURE 4.   Contour plots of coefficients a_1 and a_2 in Eq. (10) as a
function of r_g and σ.   (a) for a_1; (b) for a_2.

Since the index of refraction of sulfuric acid aerosols is a function of composition and temperature, the coefficients in the analytical expression derived by Mie calculation will be changed if there is a significant change in ambient water vapor and temperature. More accurate results will be obtained if the effect of water vapor and temperature on the coefficient is considered.

2.2.3. THE MODELING METHODS

Basically these methods are the same as the analytical methods described in the previous subsection. The only difference is that instead of using analytical expressions, the modeling results of relationship between r_g and σ for each extinction ratio R_i (i = 1,2,3) or the relationship between extinction ratio and wavelength for each pair of r_g and σ are stored as numerical values. We can interpolate these numerical values to match the experimental results of SAGE II. The corresponding values of r_g and σ are the retrieved parameters in the assumed lognormal aerosol size distribution.

3. RESULTS AND CONCLUDING REMARKS

The modeling method was applied to retrieve aerosol properties from some early results of the SAGE II experiment. The profiles, averaged over 10° latitude bands around 65°S, of the extinction ratio for the periods of December 1984 and December 1985 are presented in Figures 5(a) and 5(b), respectively. It should be noted that the lower altitude limits are different for different ratios as mentioned previously. The retrieved profiles of effective radius and aerosol composition for the months of December 1984 and December 1985 are shown in Figure 6(a) and 6(b), respectively.

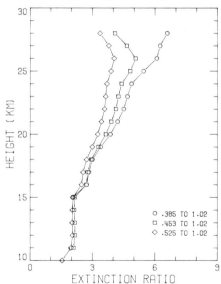

FIGURE 5.   Monthly average profiles of aerosol extinction ratios at 65°S latitude band measured by SAGE II experiment.   (a) For the month of December 1984, (b) for the month of December 1985.

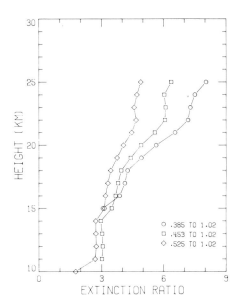

FIGURE 5.   (Continued)

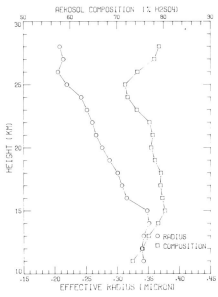

FIGURE 6.   The retrieved profile of effective radius and composition
             at 65°S latitude band.   (a) for the month of December
             1984; (b) for the month of December 1985.

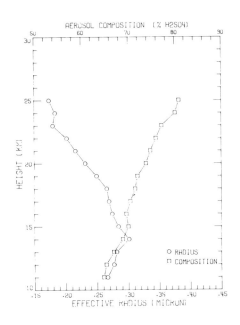

FIGURE 6. (Continued)

        Since at altitudes below 10.5 km only extinction ratio R_3 is
available, we need to assume the value σ and retrieve the mode radius
r_g. At altitudes from 10.5 km to 14.5 km, both ratios R_2 and R_3 are
available. The contours corresponding to the measured values of R_2
and R_3 are first determined by finding the intersection of the planes
with ratio equal to R_2 and R_3 and the corresponding mesh surfaces
shown in Figures 2(b) and (c), respectively. The values of σ and r_g
can be retrieved by determining the intersection of these two
contours. At altitudes above 14.5 km, in general, we can choose any
two of the three ratios to retrieve aerosol size distribution. We
prefer to use R_1 and R_3, since the difference between them is larger
than the difference between R_1 and R_2. The point of intersection can
be more accurately determined if the slopes of the two contours are
quite different from each other.

        The limited scope of this paper prevents us from discussing the
retrieved aerosol properties in more detail. However, it can be seen
from Figures 6(a) and 6(b) that in general the effective radius
decreases gradually with altitude and the percentage of H_2SO_4 in the
aerosol particle increases with altitude. The range of the percentage
of H_2SO_4 in the aerosol particle is between 60 and 80, which is in
good agreement with the generally accepted value of 75 (Rosen, 1971)
and the more recent measurements of the sulfuric acid weight percent
in the stratospheric aerosol after the El Chichon eruption (Hofmann

and Rosen, 1984). When we compare the profile of effective radius for the month of December 1984 and the profile 1 year later, we can see the decrease of aerosol size with time at the latitude under consideration.

In this paper, we have discussed the method of retrieving aerosol composition from the SAGE II water vapor channel and the corresponding ambient temperature information. We have also discussed several methods of retrieving aerosol size distribution from the wavelength dependence of aerosol extinction measured by the SAGE II experiment. As examples of the retrieval methods, some profiles of aerosol properties deduced from the SAGE II experimental results are presented. They were shown to be consistent with accepted values for stratospheric aerosols.

REFERENCES

Brogniez, C., and J. Lenoble, 1987: Modeling of the stratospheric background aerosols from zonally average SAGE profile, J. Geophys. Res., 92, 3051-3060.

Chu, W. P., 1986: Inversion of SAGE II measurements, Sixth Conf. on Atmospheric Radiation, AMS, May 13-16, 1986, Williamsburg, VA. (Available from Am. Meteorol. Soc., 45 Beacon St., Boston, VA 02108)

Chu, W. P., 1988: Inversion techniques for SAGE II data, Advances in Remote Sensing Retrieval Methods, A. Deepak Publishing.

Gmitro, J. I., and T. Vermeulen, 1964: Vapor-liquid equilibria for aqueous sulfuric acid, Amer. Inst. Chem. Eng. J., 10, 740-746.

Hofmann, D. J., and J. M. Rosen, 1984: Measurement of the sulfuric acid weight percent in the stratospheric aerosol from the El Chichon eruption, Geofisica Internacional, 23, 309-320.

Issacson, E., and H. B. Keller, 1966: Analysis of numerical methods, John Wiley and Son, New York, pp. 113-119.

Mauldin, L. E. III, N. H. Zaun, M. P. McCormick, J. H. Guy and W. R. Vaughn, 1985: Stratospheric Aerosol and Gas Experiment II instrument: A functional description, Optical Engineering, 24, 307-312.

McCormick, M. P., 1987: SAGE II: An overview, Adv. Space Res., 7, 219-226.

More, J. J., B. S. Garbow, and K. E. Hillstrom, 1980: User guide for MINPACK-1, Rep. ANL-80-74, Argonne National Laboratory, 261 pp.

Rosen, J. M., 1971: The boiling point of stratospheric aerosol, J. Appl. Meteor., 10, 1044-1046.

Toon, O. B., and J. B. Pollack, 1973: Physical properties of the stratospheric aerosols, J. Geophys. Res., 78, 7051-7059.

Toon, O. B., and J. B. Pollack, 1976: A global average model of atmospheric aerosol for radiative transfer calculations, J. Appl. Meteor., 15, 225-246.

Turco, R. P., R. C. Whittan, and O. B. Toon, 1982: Stratospheric aerosols: Observation and theory, Rev. Geophys. Space Phys., 20, 233-279.

Woods, D. C., and R. L. Chuan, 1983: Size-specific composition of aerosols in the Chichon volcanic cloud, Geophys. Res. Lett., 11, 1041-1044.

Yue, G. K., M. P. McCormick, and W. P. Chu, 1986: Retrieval of composition and size distribution of stratospheric aerosols with the SAGE II satellite experiment, J. Atmos. and Oceanic Technology, 3, 371-380.

DISCUSSION

Gille: Have you looked at the effect of having condensed nitric acid in the aerosols in high latitudes?

Yue: The effect of having condensed nitric acid in the aerosol is to increase the size. So, in general, aerosols will have larger sizes. I think the presence of nitric acid won't affect the aerosol size that much unless the temperature is very low.

Gille: Some of the theories of the Antarctic ozone hole are based on having nitric acid condense into the particles. Have you investigated whether the optical effect of nitric acid in the particles could be detected by SAGE II?

Yue: In this study we assume aerosols are background aerosols of supercooled sulfuric acid droplets only.

OPEN DISCUSSION

SESSION C.3 - CLIVE RODGERS

Rodgers: My major comment is that these have always has been very good meetings, full of interesting papers and well worth coming to. One point that Larry McMillin made over breakfast was there has been much less controversy this time than there was in previous meetings of this group and I was wondering if this is perhaps because we are beginning now to know how to solve the problem and we are all doing it more or less right. One encouraging thing that has begun to become apparent here is that we are now tackling much higher dimensionality of problems. There were a couple of topics that I was surprised to see that we didn't really represent in the session. One of them was sequential estimation. Derek Cunnold gave us a paper on mapping but that seemed to be the only one apart from a brief mention from Grace on filtering. And the other thing I was very surprised not to see any papers on was classification of statistics. In my own session perhaps the most interesting paper was Bill Smith's paper. That seems to me to be a very major breakthrough in sounding the troposphere. The way that vertical resolution has apparently increased enormously and it will be very interesting to understand just how that has happened.

Crosby: I was sort of curious that on the military sounders they were doing it entirely with microwaves. I guess there wasn't any indication that even through the year 2000 they are not going to be using any infrared instruments at all. Does anyone have any comments on that?

Kleespies: The AF a number of years ago made a commitment just to go pure microwave. And I think the reason for that had to do with the buzz word "all weather." The concept was, whether you agree with it or not, that microwaves can sense through clouds. I believe there is one SSH package which they have been talking about flying off and on, and when I left the NEPRF I left the software in place to process the data. But whether or not they do process it, I have no idea.

Rodgers: Will we get a decent high resolution vertical sounder using microwaves if we look at microwave water? Has anyone any feeling about that?

Neuendorffer: I've looked a little bit into the microwave water vapor. There are a lot of problems with it. For one thing the line we are using in the microwave isn't as strong as one would like it to be. You can't see really up high. Once you start looking above 7 kilometers or so, you are near the center of the line and you don't get the pressure effect anymore; this is a big disadvantage. In the other direction, once you start getting near the surface the effects of water vapor reverse: normally the less water vapor you have, the warmer the radiance but near the surface less water vapor is cooler. Consequently, there is going to be a null region above the boundary layer where you can't sound.

McMillin: I would like to make one comment on classification. I think a paper on classification was scheduled but it was withdrawn. This technique has been tried by several people, a lot of them outside the U.S. The experience with it has been rather consistent in that the people that have tried it have gotten rather substantial improvements over the other methods that have been tried. People tend to favor either regression or physical methods but they all tend to give about the same accuracy. The one standout is the result from classification.

Johnson: This problem with just using microwaves to sense water vapor brings up again the point I made before. Single techniques aren't the way to go. I am reminded of something I saw a couple of years ago, a glossy brochure for all the defense contractors, about what the U.S. Army wanted: more and more capability and smaller and smaller single-technique packages. In essence, they wanted a silver bullet that would answer all of their questions. Not only is that non-physical and excessively naive, it is completely wrong. It is exactly the opposite direction in which to go. While microwaves may not sound water vapor better than microwaves under conditions appropriate for the infrared, it is clear that they work a lot better than infrared in the presence of clouds, consequently one should be able to utilize both techniques. My biggest point here: you must have a broad perspective of what you can measure and how you could measure it.

Kleespies: To put my previous comment in historical perspective, in April of 1980 there was a technical exchange group meeting at Scott AF Base and they were representing the Navy and AF as well as NASA there. There was much discussion as to the relative merits of microwave and IR or combinations thereof. And at the end of the meeting, there was a very strong written statement made for a combined system. And so we made that strong statement and passed it up the line, and the AF decided to fly only microwave.

Neuendorffer: I concur with all this talk and I think it is very important and it is very scientific to do it that way. One of the reasons probably we don't have an emphasis on classification is the need for people who feel that there should be a silver bullet, and for the moment the silver bullet happens to be physical retrievals. So you've got to go one direction and whether that will be the best method or not will remain to be seen. But I think a lot of science is suppressed when we don't pursue the methods you are talking about in very broad terms.

A COMPARISON OF RETRIEVAL METHODS: OPTIMAL ESTIMATION, ONION-PEELING, AND A COMBINATION OF THE TWO

Brian J. Connor
NASA Langley Research Center
Hampton, Virginia 23665, USA

Clive D. Rodgers
University of Oxford
Oxford, England

ABSTRACT

We have compared two well-known retrieval methods, and a hybrid method, by applying them to a simulation of the planned ozone measurement by the ISAMS instrument. The methods were tested by retrieval from simulated observations and by a formal error analysis. All three are equivalent in accuracy in the limit of high signal-to-noise ratio; for lower signal-to-noise, optimal estimation yields more accurate results. Onion-peeling takes much less computation time than optimal estimation; the hybrid method, "onion-peeling by optimal estimation," is both fast and accurate.

1. INTRODUCTION

Limb scanning measurements made from satellites are among the most powerful tools available for probing the Earth's middle atmosphere. Several highly successful experiments of this type have been flown in recent years, including the Limb Infrared Monitor of the Stratosphere (LIMS), the Stratospheric and Mesospheric Sounder (SAMS), and the Stratospheric Aerosol and Gas Experiment II (SAGE II). The Upper Atmosphere Research Satellite (UARS), now under construction, will include several limb scanning experiments.

We are presently developing retrieval algorithms for use with one of the UARS limb scanning experiments, namely the Improved Stratospheric and Mesospheric Sounder (ISAMS). Two of the retrieval methods most widely used with earlier limb scanning measurements are "onion-peeling" (Russell and Drayson, 1972) and "optimal estimation" (Rodgers, 1976). In the former, the retrieval is performed one layer at a time, starting from the top of the atmosphere. At each level, the values retrieved from higher levels are left fixed, and a value for that level is found which reproduces the measured radiance within experimental error. In the latter, the measured radiances are statistically combined with an a priori atmospheric profile; the resulting profile is then tested for statistical consistency with the measurement. Proponents of each method have tended to take different conceptual, almost philosophical, approaches to the retrieval problem, with the difference focused on the use of a priori information. Some hold that the a priori information may bias the retrieval either in a way that obscures real information

RSRM '87: ADVANCES IN
REMOTE SENSING RETRIEVAL METHODS
A. Deepak, H.E. Fleming, and J.S. Theon (Eds.)   271

in the measurement, thereby degrading it, or in a way that overstates the real information. Others stress that the use of an a priori profile adds real information to that available in the measurement, resulting in more information in the retrieval.

Because of this background, we have made a direct comparison of the two approaches as part of the development of retrieval algorithms for ISAMS. In the process, we developed and tested a hybrid algorithm which combines features of both. It is our hope that this study may not only advance our work on ISAMS, but also inform other workers involved in limb scanning measurements of the middle atmosphere.

2. PROCEDURE

2.1 OVERVIEW

We chose to use a model of the ISAMS ozone measurement as the vehicle for the comparison. The reasons for this choice were several. First, we wanted to focus on the retrieval rather than the forward calculation, and so wanted to use as simple a forward model as possible, while still making a fairly realistic simulation of an actual measurement. Most of the ISAMS channels will employ pressure modulated radiometry, while the ozone measurement will be a simpler broad-band measurement. Second, we wanted to employ many real atmospheric profiles as a test bed for the retrieval methods, and measurements of the ozone profile are readily available, though this is not so for some other species. Third, particular concern has often been expressed that optimal estimation would perform poorly on a high signal-to-noise ratio measurement. The ISAMS ozone measurement has a design signal-to-noise ratio of ~ 3000 in the mid-stratosphere, and so is a good choice to test this idea. The ISAMS ozone band extends from 990-1010 cm^{-1}; for simplicity we have assumed that the filter transmission is unity within the band and zero elsewhere. Similarly, we have assumed that the instrument field of view is infinitesimal. We performed the forward radiance calculation using the emissivity growth approximation (Gordley and Russell, 1981).

We selected two sets of about 50 ozone profiles each, measured by SAGE II. It was arranged that each set was roughly uniformly distributed in latitude; otherwise each profile was randomly selected from a collection of several hundred profiles. One set, referred to as the "statistics profiles," was used to aid the development of the algorithms; the second, referred to as the "blind profiles," was reserved for testing.

We performed the retrieval comparison by adopting two parallel approaches. The first approach is to compare the performance of the methods on retrieval from simulated observations. We assume the atmospheric profile, calculate radiances from it, and add assumed experimental error. The resulting radiance profile is the simulated observation. We then perform retrievals with each method, compare the results to the assumed profile, and compare the accuracy with which the different methods reproduce the assumed profile. We do this for each of the blind profiles and then compare the performance of the methods statistically.

The second approach is to apply the formal error analysis of Rodgers (1987). The columns of the matrix $\mathbf{D}$ of contribution functions for each retrieval method are calculated by perturbing a standard radiance profile, performing a retrieval, and comparing the result to the profile retrieved from the unperturbed radiances. Similarly, the columns of the matrix of averaging kernals $\mathbf{A}$ are calculated for each method by perturbing a standard mixing ratio profile, calculating radiances from the perturbed profile, performing a retrieval, and comparing the result to the profile retrieved in the unperturbed case. The matrices $\mathbf{D}$ and $\mathbf{A}$ are then used to derive the covariance matrices for measurement, null space, and total error, and orthogonal error patterns calculated from these.

The parallel approaches are complementary. In the first place, the total error predicted by the formal analysis may be directly compared to the actual error in the simulations, so that the two validate each other. Also, the simulations provide the most direct and easily understood comparison, while the formal error analysis provides a much more complete characterization of the errors as well as insight into the differences among the methods.

2.2 THE RETRIEVAL METHODS

The onion-peeling and optimal estimation methods are described in detail elsewhere, and their formulations will not be given here. In addition to the references given in the Introduction, Rodgers et al. (1984) discuss the use of optimal estimation with the SAMS experiment and Gille and Russell (1984) and Remsberg et al. (1984) show examples of onion-peeling applied to the LIMS measurements. We will confine ourselves to a brief description of the methods as applied here and to defining the hybrid method.

The usual optimal estimation approach, which we shall call vector optimal estimation, has two important features. First, the solution is a weighted average of the measurement and a "virtual measurement," namely the a priori profile. The weights depend on the inverse variances of the two quantities. Second, correlations between atmospheric layers (and also correlated measurement errors) are explicitly allowed for. As applied here, the entire mixing ratio profile is updated from each radiance, taken sequentially from the top down. We refer to this method as "vector" optimal estimation because mixing ratio always appears as a vector quantity.

Onion-peeling, on the other hand, treats each layer as independent and uses one radiance to find one mixing ratio value, which is then fixed. The onion-peeling algorithm used here has an additional feature. It begins the retrieval at an altitude well above the region of measurement sensitivity. After retrieving the two highest altitudes, it compares the values. If they differ by more than a pre-determined amount, the difference is assumed to be a spurious oscillation and the retrieval is rejected. Then the two top layers are combined into a single layer, and the process begun again. This continues into the region where the measurement is fully sensitive, after which this "stability check" is not performed.

Our hybrid algorithm retains the first feature of optimal estimation, namely the use of the a priori profile, but proceeds by using one radiance to find one mixing ratio starting from the top, i.e. by "onion-peeling;" hence we call it "onion-peeling by optimal estimation" or, since all quantities appear as scalars, "scalar optimal estimation."

The scalar optimal estimation algorithm proceeds iteratively in two stages. Starting from $u_{n,j}$, the mixing ratio value at level j after n iterations, a linear estimate of the exact solution is made:

$$u_{x,j} = u_{n,j} + (L_j - L_{n,j}) \frac{du_j}{dL_j}|_{u_{n,j}} \qquad (1)$$

where L_j is the measured radiance and $L_{n,j}$ is the radiance calculated using $u_{n,j}$. The variance of $u_{x,j}$ is calculated from

$$\sigma^2_{x,j} = \sigma^2_{L_j} \frac{du_j}{dL_j}|_{u_{n,j}} \qquad (2)$$

where $\sigma^2_{L_j}$ is the measurement variance. Combining $u_{x,j}$ with the a priori mixing ratio $u_{apr,j}$ and its variance $\sigma^2_{apr,j}$ yields the next iterate:

$$u_{n+1,j} = \frac{u_{x,j}/\sigma^2_{x,j} + u_{apr,j}/\sigma^2_{apr,j}}{1/\sigma^2_{x,j} + 1/\sigma^2_{apr,j}} \qquad (3)$$

Convergence is reached when successive iterates differ by less than some fraction, typically 10%, of $\sigma_{x,j}$. The final result must then be checked for statistical consistency with the data. If it passes this test, it is accepted as a valid retrieval. If it fails, then either the a priori or the measurement is in error; the solution is rejected and an error message written.

3.   RESULTS

3.1 RETRIEVAL SIMULATIONS

We performed two sets of retrieval simulations with each retrieval method. In the first, we assumed errors appropriate to the design goals for the ISAMS ozone measurement. To the synthetic radiances we added noise and an offset, each with a standard deviation of 10^{-4} W m^{-2} ster^{-1}, and we multiplied the synthetic radiances by an assumed scale error of 1.01. In the second set, we simply increased the noise by a factor 10, leaving the offset and scale errors unchanged.

3.1.1 Accuracy

The results of the simulations are illustrated in Figures 1 and 2. Figure 1 shows sample simulations, while Figure 2 shows the rms errors in the full set of retrievals. In each panel of Figure 1, the mixing ratio profile marked by plus signs is the assumed ozone profile. The radiance profile marked with plus signs is the simulated observation, that is, the radiances calculated from the assumed ozone profile plus assumed experimental error. The retrieval from the simulated observation is the mixing ratio profile given by the solid line. The solid line radiance profile is that calculated from the retrieved ozone. For the optimal

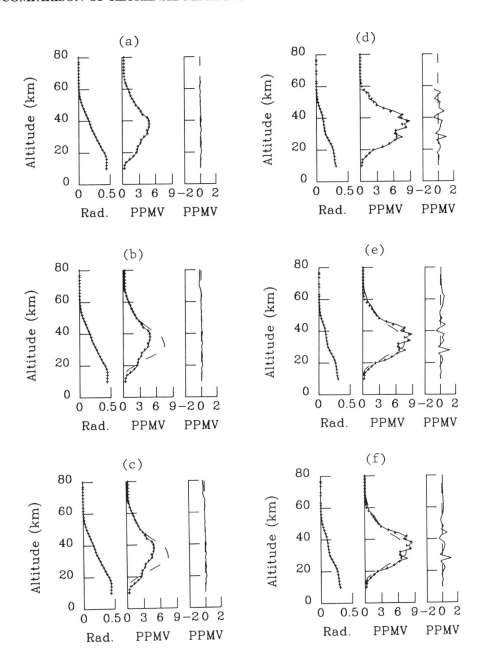

FIGURE 1. Retrieval simulations. Panels a, b, and c are from onion-peeling, scalar optimal estimation, and vector optimal estimation, respectively; the assumed radiance errors are the ISAMS design errors. Panels d, e, and f are corresponding, except for a different assumed ozone profile, and assumed noise 10 times higher.

estimation examples, the a priori profile is shown as the dashed mixing ratio profile. Finally, the graph at the right of each panel shows the difference between the assumed and retrieved ozone profiles.

Panels a, b, and c of Figure 1 show a typical result from the set of simulations which assumed noise of 10^{-4} W m^{-2} ster^{-1}. All three methods give excellent, in fact nearly identical, results. The optimal estimation methods perform well despite the fact that the assumed profile is very different from the a priori. The onion-peeling retrieval stops at 66 km because the solution at higher altitudes was too oscillatory; however, that is not a significant fault. The intrinsic sensitivity of the measurement is falling off rapidly above 60 km, and is effectively nil above 70 km. Therefore, there is nothing to choose between the methods, based on Figure 1. Note that the assumed 1% scale error causes all the retrieved profiles to be slightly larger than the assumed profile over most of the altitude range, just as one would expect.

Panels d, e, and f of Figure 1 are typical of the simulations with assumed noise of 10^{-3} W m^{-2} ster^{-1}. The three methods again yield very similar, quite good results, up to about 40 km. Above that, the optimal methods perform better than onion-peeling, exhibiting less noise in the retrieval. Of the two optimal methods, the vector method does somewhat better near 60 km. It is worth pointing out in this case that all the methods do a good job of reproducing the fine structure in the mid-stratosphere.

Figure 2a shows the rms error in the full set of 52 retrievals for the case of 10^{-4} W m^{-2} ster^{-1} noise. The performance of all three methods is very similar. Figure 2b shows the result for 10^{-3} W m^{-2} ster^{-1} noise. In this case the optimal methods perform significantly better than onion-peeling between 40 and 60 km, and especially between 50 and 60 km. Of the two optimal methods, vector optimal

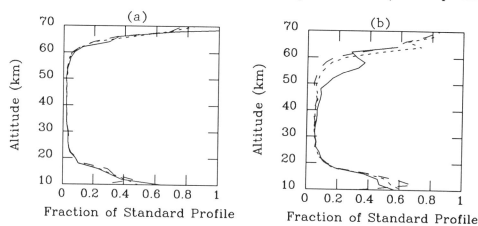

FIGURE 2. rms errors for simulated retrievals by onion-peeling (solid line), scalar optimal estimation (short dashes), and vector optimal estimation (long dashes). Panel a: the assumed errors are the ISAMS design errors. Panel b: assumed noise 10 times higher.

estimation does somewhat better at these altitudes, but the difference between them is small compared to the difference between both of them and onion-peeling. Note that only values accepted as valid by the algorithms are included in Figure 2. For onion-peeling, the stability check rejects many of the solutions above about 56 km in the higher noise case and above about 64 km in the lower noise case, resulting in an improved rms error for the remaining profiles.

3.1.2 Computation Time

The computation times required for the three retrieval methods, as applied here, are in the approximate ratios 2:1:8, for onion-peeling, scalar, and vector optimal estimation, respectively. These ratios should not be taken as fundamental to the methods as it would likely be possible to reduce the time required by both onion-peeling and vector optimal estimation. However, we are confident the ordering (i.e. scalar optimal estimation fastest, etc.) would not change. While it is to be expected that vector optimal estimation is the slowest of the three methods, it is at first surprising to find that scalar optimal estimation is significantly faster than simple onion-peeling. It is so because the onion-peeling may perform several sets of calculations preliminary to the radiance calculation before it reaches the altitude where the retrieval is sufficiently stable, whereas optimal estimation performs the preliminary calculations only once.

3.2 FORMAL ANALYSIS

We have applied the formal error analysis of Rodgers (1987) to the three retrieval methods. The covariance matrix assumed correct for the real atmosphere was calculated from the combined sets of statistics and blind profiles. For purposes of the formal analysis, we assumed that the only experimental error was noise.

A modified version of onion-peeling aided the performance of the analysis. This is actually a simpler algorithm which accepts the solution at all altitudes irrespective of oscillations. The retrieval errors with the modified onion-peeling only differ from the standard version at altitudes where the solution is usually rejected as too oscillatory. Therefore, the error analysis should be valid at lower altitudes for both programs.

3.2.1 Accuracy

Figure 3 shows the predicted rms error in the retrievals for both cases, noise of 10^{-4} and 10^{-3} W m^{-2} ster^{-1}. (These are the square roots of the diagonal elements of the total error covariance matrix.) Note that this Figure is not quantitatively comparable with Figure 2 because the simulations on which Figure 2 is based assumed scale and offset errors as well as noise. However, simulations performed with noise errors only yield results essentially identical to Figure 3.

Figure 3 does qualitatively agree with Figure 2. Optimal estimation does better than onion-peeling above about 40 km in the higher noise case, with vector optimal estimation somewhat better than scalar. In the lower noise case, optimal estimation does somewhat better near 60 km, where the signal-to-noise ratio is no longer so high.

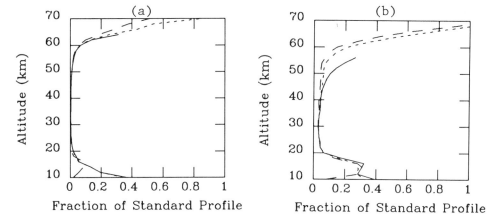

FIGURE 3. rms errors predicted by formal analysis, for noise errors only, for onion-peeling (solid line), scalar optimal estimation (short dashes), and vector optimal estimation (long dashes). Panel a: noise as in ISAMS design. Panel b: noise level 10 times higher.

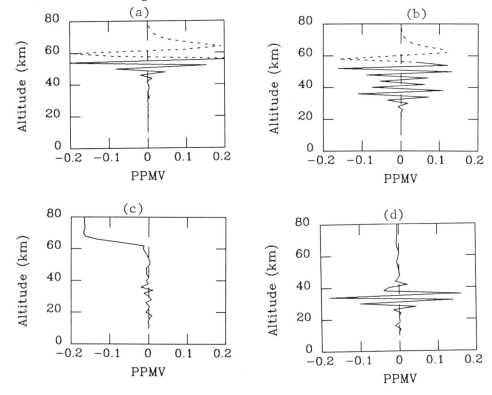

FIGURE 4. Selected error patterns for onion-peeling (a, b) and optimal estimation (c, d), for assumed noise of 10^{-3} W m^{-2} ster^{-1}. The error analysis does not apply above 56 km to the onion-peeling algorithm used for the simulations, so the patterns are shown dashed above that altitude.

3.2.2 Retrieval Diagnostics

The diagnostic tools of the formal analysis enable us to explore the differences between the methods in more detail. In the remainder of this section, we confine our attention to the case with noise of 10^{-3} W m^{-2} ster^{-1} since it is only in that case that the methods give very different results. Also, for simplicity, we compare vector optimal estimation with onion-peeling only; scalar optimal estimation is intermediate between the two, although more like vector optimal estimation.

Figure 4 shows two of the largest error patterns for the two methods (these are eigenvectors of the total error covariance matrix, multiplied by the square root of the corresponding eigenvalue). Optimal estimation exhibits a high correlation between layers at the highest altitudes, with oscillation confined to below about 40 km. Onion-peeling displays highly oscillatory errors, increasing in amplitude with increasing altitude up to the highest altitude used.

One of the most readily understood of the diagnostics is the **A** matrix. The rows of **A** are the averaging kernals which show how the real profile is smoothed in the retrieval. Naively, the closer the averaging kernals are to δ-functions, the better. In fact, narrow averaging kernals are desirable, but Figure 5 shows that the width of the averaging kernals is not the whole story. The onion-peeling averaging kernals are δ- functions at all altitudes, while the optimal estimation averaging kernals become gradually broader and of lower amplitude as altitude increases. This shows that as the signal-to-noise ratio decreases, optimal estimation begins to smooth the solution, smoothing it more as the ratio becomes less, and is one way of conceptually accounting for the lesser noise sensitivity of optimal estimation.

An alternate conceptual view invokes explicitly the contribution of the a priori to the retrieval. The eigenvectors and eigenvalues of **A** can be used to explore that contribution. In particular, eigenvectors which correspond to eigenvalues $\lambda \sim 0$ are poorly measured, and such components of the retrieval come

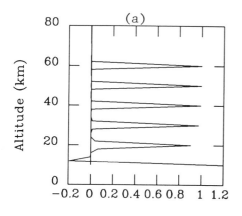
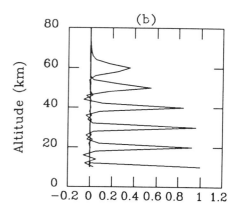

FIGURE 5. Selected averaging kernals for onion-peeling (a) and optimal estimation (b), for assumed noise of 10^{-3} W m^{-2} ster^{-1}.

almost entirely from the a priori, while eigenvectors for which $\lambda \sim 1$ are well measured and receive no contribution from the a priori. Figure 6 shows four of the eigenvectors of $\mathbf{A}$ for vector optimal estimation with noise 10^{-3} W m^{-2} ster^{-1}. Figure 6a shows a component with fine-scale oscillations near 50-60 km, which must come primarily, though not entirely, from the a priori. Figures 6b and 6c show that fine structure between 30 and 50 km and broad structure above 50 km come mostly from the measurement, with some contribution from the a priori. Finally, Figure 6d shows an example of a component which is due almost entirely to the measurement.

4. DISCUSSION

A comparison of Figures 2 and 3 shows the effect of scale and offset errors on the retrievals to be surprisingly large. Since all experiments are prone to such errors, their influence on the various retrieval methods should be further investigated.

Otherwise, the simulations and the formal error analysis lead to clear, consistent conclusions which may be summarized as follows. In the limit of high signal-to-noise ratio, all three methods perform equally well, producing, in fact, almost identical results, although vector optimal estimation is much slower than the other methods. As the signal-to-noise ratio decreases, optimal estimation

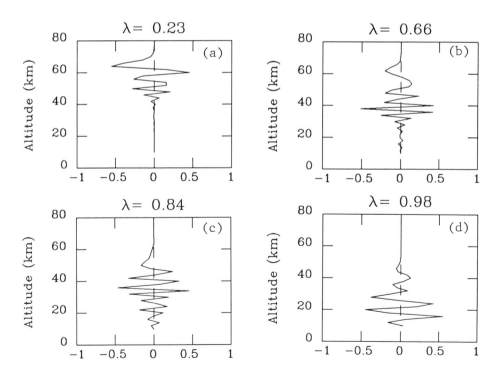

FIGURE 6. Selected eigenvectors of the averaging kernal matrix for optimal estimation, for assumed noise of 10^{-3} W m^{-2} ster^{-1}.

begins to out-perform onion-peeling, its greater stability derived from the contribution of the a priori information. The hybrid method, onion-peeling by optimal estimation, may be a good choice for operational processing; it is both insensitive to noise and computationally efficient.

ACKNOWLEDGMENTS

We thank J.M. Russell III and K.V. Haggard for their assistance and encouragement in this work, and L.L. Gordley, R.E. Thompson, and B.T. Marshall for providing the software and data base for the forward model.

REFERENCES

Gille, J.C., and J.M. Russell III, 1984: The Limb Infrared Monitor of the Stratosphere: Experiment Description, Performance, and Results, *J. Geophys. Res., 89*, 5125-5140.

Gordley, L.L., and J.M. Russell III, 1981: Rapid Inversion of Limb Radiance Data Using an Emissivity Growth Approximation, *Appl. Optics, 20*, 807-813.

Rodgers, C.D., 1976: Retrieval of Atmospheric Temperature and Composition from Remote Measurements of Thermal Radiation, *Rev. of Geophys. and Space Phys., 14*, 609-624.

Rodgers, C.D., 1987: A General Error Analysis for Profile Retrieval, these proceedings.

Rodgers, C.D., R.L. Jones, and J.J. Barnett, 1984: Retrieval of Temperature and Composition from NIMBUS 7 SAMS Measurements, *J. Geophys. Res., 89*, 5280-5286.

Remsberg, E.E., J.M. Russell III, J.C. Gille, L.L. Gordley, P.L. Bailey, W.G. Planet, and J.E. Harries, 1984: The Validation of NIMBUS 7 LIMS Measurements of Ozone, *J. Geophys. Res., 89*, 5161-5178.

Russell, J.M. III, and S.R. Drayson, 1972: The Inference of Atmospheric Ozone Using Satellite Horizon Measurements in the 1042 cm^{-1} Band, *J. Atmos. Sci., 29*, 376-390.

DISCUSSION

Westwater: Does the onion-peeling method take into account explicitly the noise level of the observations?

Connor: The one that we used for the simulations does, yes.

Westwater: So, you do need to know just the noise level but there is no information about the profile itself.

Connor: The only way in which it uses the noise level is that after each iteration it checks to see if it has matched the radiance within that amount. And if it has, it stops.

OPEN DISCUSSION

SESSION C.4 - ED WESTWATER

The first paper in my session was by Art Jordan. It should remind us that there is a huge community of people out there who do work on inverse problems in other fields. Art addressed the electromagnetic inverse problem of measuring the field distribution on some area and inferring the properties of the medium within. Art himself has led several inverse problem symposiums which relate to electromagnetic inverse, optical inverse, ocultation. There is just a huge community of people working on inverse problems and they have techniques which perhaps could be brought to bear on some of the problems that we are more intimately faced with. The second paper, as I understand it, concluded that the best technique was really a combination of the two competing inverse techniques. Thus, we should be emphasizing how techniques can be combined to get the optimum combination. I think in terms of the entire conference I was quite impressed with some of the breakthroughs that have been made, particularly in the multispectral scanning techniques that Fleming and Smith are pioneering. I think one absence, however, was representatives of the user community that will ultimately use the remote sensing data to put into forecasts. I believe that they could have contributed substantially. And finally, as a comment, I believe that many people know that the U.S. will be covered in the next four or five years by a system, of ground-based meteorological radars which will measure horizontal winds. These winds will be measured with a time resolution of perhaps once every hour, or perhaps even less than that. I believe one of the real challenges will be to assimilate data like these. Since the temperature and the winds are connected through dynamical equations of motion, I believe it is the combined aspects of the complimentary sources of data that will lead to the tremendous improvements in the future of numerical weather forecasting.

Wahba: Maybe this isn't the place to make a comment I wanted to make before, however, narrowing down on the particular problem of numerical weather forecasting, one should think that ultimately data from satellites should be used to update the state variables of the model.

Kleespies: I am not a dynamicist, but from the NWP classes that I have taken, the indications are that the atmosphere tends to be more sensitive to the temperature field in the planetary waves and for the synoptic scale waves the analysis tends to like the wind field more. Now, with that in mind, I agree with your earlier statement and I strongly support it. The next time we get together we should have some modelers here to tell us what they really want.

Goroch: I would like to add that the new models that are being developed will use moisture more than the current models. Models are tuned to the data input and it is in the next few years that the retrievals of moisture that we are providing to the modelers will be used to tune the forecast models. It is very important to get all the modelers in here while the next generation of models is being developed.

Neuendorffer: It seems like it is just as important to try to take the radiances as they are and provide the information that we can provide and try to encourage people in figuring what to do with the information.

Kleespies: Basically the models have two generic components: the dynamics which are determined by the atmospheric state, and the physics which has to do with diabatic processes, and, in particular, the moisture. In the past, the models were not initialized with moisture and, as Andy Goroch stated, the new models will be initialized with moisture. But going back to the winds, the only way you are going to get winds from satellites is either from cloud tracking or going to an active sensor like a spaceborne laser. And that is a little down the road. We are really more interested in what we can do right now with passive sensors. I think we really need to give more consideration to the water retrieval and be very sensitive to what the modelers want in terms of how the water vapor retrieval can be initialized into the models.

A GENERAL ERROR ANALYSIS FOR PROFILE RETRIEVAL

Clive D. Rodgers
University of Oxford
Oxford, England

ABSTRACT

The error analysis of profiles retrieved from remote measurements of the atmosphere presents conceptual problems, particularly concerning inter-level correlations between errors, the smoothing effect of remote sounding and the contribution of *a priori* information to profile. This paper attempts to put the error analysis on a reasonably straightforward but rigorous basis, and gives an intuitive description of the nature of the errors.

A formal error analysis for profile retrieval is developed which is independent of the nature of the retrieval method, provided that the measurement process can be modelled adequately. The error separates naturally into three components, (a) random error due to measurement noise, (b) systematic error due to uncertain model parameters and inverse model bias, and (c) null space error due to the inherent finite vertical resolution of the observing system. A recipe is given for evaluating each of the components in any particular case.

Most of the error terms appear as covariance matrices, rather than simple error variances. These matrices can be interpreted in terms of 'error patterns', which are statistically independent contributions to the total error. They are the multidimensional equivalent of 'error bars'.

An approach is described which clarifies the relation of *a priori* data to the retrieved profile.

1. INTRODUCTION

The error analysis of profiles retrieved from remote measurements of the atmosphere has presented conceptual problems for many workers in the subject, particularly concerning inter-level correlations between errors, the smoothing effect of remote sounding and the contribution of *a priori* information to profile. The result has been that it is often difficult to make objective comparisons between different sounding systems, based on the published literature.

The purpose of this paper is to put the error analysis on a reasonably straightforward but rigorous basis, and to give an intuitive description of the nature of the errors. The analysis is carried out in such a way that it can be applied to any kind of approach to the retrieval problem, not only statistically optimum methods. Retrieval methods as such are not discussed, except for the purpose of illustration. For a review of inverse theory in general, see for example Menke (1984). For a

RSRM '87: ADVANCES IN
REMOTE SENSING RETRIEVAL METHODS
A. Deepak, H.E. Fleming, and J.S. Theon (Eds.)   285

review with an atmospheric bias, see Rodgers (1976).

The total error is separated into components due to (a) systematic errors in the forward model and the inverse model, (b) measurement noise in the instrument, (c) components in the profile that the observing system cannot see. A method is described whereby the effect of errors that are correlated between different levels can be readily appreciated.

The context of the discussion is in atmospheric remote sounding, but the concepts are sufficiently general that they may be applied to any type of profile retrieval from remote measurements.

2. DEFINITIONS

The *observing system* is the combination of the measurement and the profile retrieval method.

The *state vector* x is a vector of unknowns to be estimated from the measurements, describing the state of the atmosphere. Usually it will be a profile of some quantity, given at a finite number n of levels, where n is large enough to represent the possible atmospheric variations adequately. However it may in principle comprise any set of relevant variables, such as coefficients for a representation of the profile, see section 6 below. Unless otherwise stated, the discussion below will assume that x is a profile.

The *measurement* y is a vector of m measured quantities. Usually $m \ll n$, and the inverse problem is formally ill-posed. Measurements are made to a finite accuracy, with *measurement error* ϵ_y assumed to be normally distributed with mean zero and known error covariance S_ϵ.

The *Forward Model F* characterises the measurement y, describing how it depends on the state vector x. It may be an algebraic or algorithmic description

$$y = F(x, b) + \epsilon_y$$

where b is a vector of model parameters (such as spectral line data, calibration parameters, etc) that are not perfectly known to the observer. They are a possible source of systematic differences between calculated and measured values of y. We assume that the forward model represents the physics of the measurement accurately; characterising forward model errors that cannot be represented in terms of model parameters is beyond the scope of this discussion. Note that in some works on inverse theory, the term *model parameter* is used to denote the quantity called *state vector* here. In fact the distinction between them is vague: the measurement depends on the model parameters, so they could be regarded as quantities to be retrieved.

For the purpose of error analysis we may linearise the forward model about some reference state $(\bar{x}, \bar{b})$ which may be thought of as the true state x, the retrieved state $\hat{x}$, an ensemble mean, or any arbitrary state, as required.

$$y = F(\bar{x}, \bar{b}) + \frac{\partial F}{\partial x}(x - \bar{x}) + \frac{\partial F}{\partial b}(b - \bar{b}) + \epsilon_y$$

We define the *Weighting Function* $\mathbf{K}_x$ as $\partial F/\partial \mathbf{x}$, so called because in some cases it is normalised to unit area, and $\mathbf{y} - \bar{\mathbf{y}}$ is then a weighted mean of $\mathbf{x} - \bar{\mathbf{x}}$ with weights given by the rows of $\mathbf{K}_x$. Weighting functions give a broad idea of the information content of a set of measurements, in that they show the part of the profile that is represented by each measurement. However even if they have well defined peaks, their width does not necessarily give a good indication of the vertical resolution, as this depends on both the 'width' of the individual peaks and their spacing.

We also define $\mathbf{K}_b$ to be the sensitivity $\partial F/\partial \mathbf{b}$ of the measurement to the forward model parameters. If the derivatives in the linearisation cannot be found algebraically, they may be evaluated numerically by perturbing the forward model.

The *Inverse Model I* describes how the retrieved state $\hat{\mathbf{x}}$ is obtained from the measurement:

$$\hat{\mathbf{x}} = I(\mathbf{y}, \mathbf{b}, \mathbf{c})$$

where $\mathbf{c}$ is a vector of parameters used in the retrieval that do not appear in the forward model, i.e. are unconnected with the measurements, but may reasonably be varied, e.g. *a priori* data. The inverse model may be an explicit algebraic function, or may be defined algorithmically.

For the purpose of error analysis we also linearise the inverse model:

$$\hat{\mathbf{x}} = I(\bar{\mathbf{y}}, \bar{\mathbf{b}}, \bar{\mathbf{c}}) + \frac{\partial I}{\partial \mathbf{y}}(\mathbf{y} - \bar{\mathbf{y}}) + \frac{\partial I}{\partial \mathbf{b}}(\mathbf{b} - \bar{\mathbf{b}}) + \frac{\partial I}{\partial \mathbf{c}}(\mathbf{c} - \bar{\mathbf{c}})$$

and define the *contribution function* $\mathbf{D}_y$ as $\partial I/\partial \mathbf{y}$, so called because each row of $\mathbf{D}$ is the contribution to the solution due to a unit change in the corresponding element of $\mathbf{y}$. We also define $\mathbf{D}_b$ and $\mathbf{D}_c$ to be the sensitivities $\partial I/\partial \mathbf{b}$ and $\partial I/\partial \mathbf{c}$ of the inverse model to the model parameters.

In the error analysis, we must take care when assigning errors to the model parameters. For example, $\mathbf{b}$ takes its exact value as far as the observing system is concerned, but takes an estimated value when it appears in the inverse model, or when the forward model is being evaluated by the observer. The identification of model parameters may not be straightforward. For example, calibration coefficients are forward model parameters, the observing system knows their true values; the observer uses estimated values. This is easily overlooked if the forward model is formulated to give scientific units, and the measurement $\mathbf{y}$ is calibrated and also presented in scientific units.

3. CHARACTERISATION OF CORRELATED ERRORS

Error bars on scalar quantities are easy to understand. When errors in different quantities such as the elements of $\mathbf{x}$ are correlated, as expressed by an error covariance matrix $\mathbf{S}$, it becomes more difficult to understand the implications. The diagonal elements of $\mathbf{S}$ are the error variances of the elements of $\mathbf{x}$, but if the off-diagonal elements show that the errors are highly correlated, then we have more information about $\mathbf{x}$ than is the case if the off-diagonal elements are small. One way of understanding the error covariance matrix is to diagonalise it, i.e. find its eigenvalues λ_i and eigenvectors l_i, so that $\mathbf{S}l_i = \lambda_i l_i$. As $\mathbf{S}$ is symmetric it can be decomposed as

$$\mathbf{S} = \sum_i \lambda_i l_i l_i^T = \sum_i e_i e_i^T$$

The orthogonal vectors $e_i = \lambda_i^{\frac{1}{2}} l_i$ can be thought of as *error patterns*, the multi-dimensional equivalent of error bars. The error vector ϵ_x in $\mathbf{x}$ is a sum of these error patterns, each multiplied by a random factor a_i having unit variance:

$$\epsilon_x = \sum_{i=1}^{n} a_i e_i$$

4. EXAMPLES

To illustrate, the concepts being presented, we will apply them to the SBUV ozone sounding instrument on NIMBUS 7 (Heath *et al.*, 1975). A set of weighting functions $\mathbf{K}_x$ are shown in Fig. 1, which depend on an assumed mean ozone profile as a linearisation point. Only this one linearisation point has been studied. The state vector $\mathbf{x}$ consists of logarithms of layer amounts of ozone in a set of closely spaced layers, and the measurement vector $\mathbf{y}$ consists of the logarithm of the measured albedo in 10 spectral intervals, together with a measurement of total ozone that is obtained from a separate non-linear retrieval from three spectral intervals. Curve A is the weighting function for the total ozone measurement.

The retrieval approach analysed for the illustration is not that used by the SBUV team. It is a simple minded maximum likelihood retrieval (Rodgers 1976, Eq. (21)), chosen for algebraic convenience. It is assumed that the measurement noise $\mathbf{S}_\epsilon = 10^{-4}\mathbf{I}$, i.e. 1% in all channels, and the a priori covariance $\mathbf{S}_a$ is of the form:

$$S_{zz'} = (S_{zz} S_{z'z'})^{\frac{1}{2}} \exp(-(z - z')^2/l^2) \tag{1}$$

with $l = 5$ km and $S_{zz}=0.04$, i.e. a 20% variance in ozone layer amount. This form is qualitatively reasonable, but it has been chosen for illustration only, and should not be used for serious retrievals without validation.

The contribution functions $\mathbf{D}_y$ for this retrieval may be written as

$$\mathbf{D}_y = \mathbf{S}_a \mathbf{K}_x^T (\mathbf{K}_x \mathbf{S}_a \mathbf{K}_x^T + \mathbf{S}_\epsilon)^{-1} = (\mathbf{S}_a^{-1} + \mathbf{K}_x^T \mathbf{S}_\epsilon^{-1} \mathbf{K}_x)^{-1} \mathbf{K}_x^T \mathbf{S}_\epsilon^{-1}$$

They are given in Fig. 2, showing that each channel contributes in a complicated way to the overall profile, although there is a slight tendency for the information to be put into the profile in the same general altitude region as the peak of the corresponding weighting function.

5. ERROR ANALYSIS

A retrieval is unlikely to be identical to the unknown profile. Without regard to errors, we can relate them by a *Transfer Function*, T:

$$\hat{\mathbf{x}} = I(F(\mathbf{x}, \mathbf{b}), \mathbf{b}, \mathbf{c}) = T(\mathbf{x}, \mathbf{b}, \mathbf{c})$$

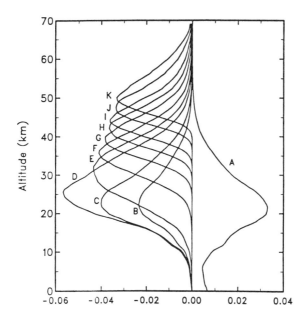

FIGURE 1. Weighting functions for the SBUV ozone sounding instrument. Curve A corresponds to the total ozone constraint.

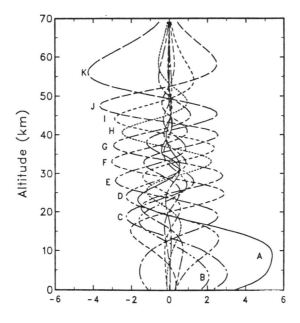

FIGURE 2. Contribution functions for a maximum likelihood retrieval method for SBUV. The alphabetical labeling corresponds to Fig. 1.

One way to understand the nature of retrieved state vector is to study this transfer function. For the error analysis, we include measurement error ϵ_y, and note that the inverse model uses our best estimates $\hat{b}$ and $\hat{c}$, whilst the measurement system 'knows' exact values of the model parameter b:

$$\hat{x} = I(F(x,b) + \epsilon_y, \hat{b}, \hat{c})$$

Linearising about $(\bar{x}, \hat{b}, \hat{c})$ gives some insight into the nature of the transfer function:

$$\hat{x} = T(\bar{x}, \hat{b}, \hat{c}) + D_y K_x (x - \bar{x}) + D_y K_b (b - \hat{b}) + D_y \epsilon_y \qquad (2)$$

Linearisation about the point (x, b, c) might be more appropriate, but as it is not known it would not be possible to evaluate anything there.

5.1. THE AVERAGING KERNEL

Components of the solution corresponding to departures of the profile from $\bar{x}$ are smoothed by the operator

$$A = \frac{\partial T}{\partial x} = D_y K_x$$

Following Backus and Gilbert (1970), we call rows of this matrix the *Averaging Kernel*. The retrieval at any location is an average of the whole profile weighted by this row. Columns may be thought of as the *δ-function response* of the observing system. A δ-function disturbance in the real profile will be reflected in the retrieval as the shape of a column of A. For many retrieval methods, this may be the best way of evaluating A. Both rows and columns would be expected to represent peaked functions, with the width of the peak being a qualitative measure of the resolution of the observing system. Menke (1984) calls A the *model resolution matrix*.

Fig. 3 shows averaging kernels, for the SBUV example, illustrating how the retrieved profile is a smoothed version of the true profile. For clarity, these are not plotted at every level. The width of the primary peak of each curve is a measure of the resolution of the observing system. Note that the retrieval at the top two levels shown (for 62 and 69 km) does not correspond to the true profile around these levels as there is little information in the measurements here. Similarly the retrieval below about 19 km (curves for 6 and 13 km) does not correspond well to the true ozone below 19 km for the same reason.

5.2. THE TOTAL ERROR

From Eq. (2), the total error in a retrieval becomes

$$\hat{x} - x = [T(\bar{x}, \hat{b}, \hat{c}) - \bar{x}] + (A - I)(x - \bar{x}) + D_y K_b \epsilon_b + D_y \epsilon_y \qquad (3)$$

where I is a unit matrix. We have written ϵ_b for $b - \hat{b}$, representing the uncertainty in our knowledge of these quantities. Uncertainty in our knowledge of c only

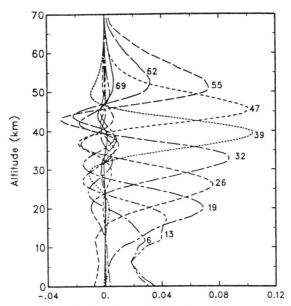

FIGURE 3. SBUV averaging kernels for selected altitudes. The curves are labeled with altitude in kilometers.

appears in the term $T(\bar{\mathbf{x}}, \hat{\mathbf{b}}, \hat{\mathbf{c}})$ at this stage. Each term has its own interpretation, as follows.

5.2.1 Null-space error.

The departure of the transfer function from a δ-function leads to a component of error that we will call the *null-space error*, as it corresponds to those portions of profile space that cannot be measured by the observing system. For any one profile, the null-space error is $(\mathbf{A} - \mathbf{I})(\mathbf{x} - \bar{\mathbf{x}})$. The statistics of this error can be estimated only if we have independent estimates of the statistics of $\mathbf{x} - \bar{\mathbf{x}}$, and in particular the covariance $\mathbf{S}_x$ of its variability. The null-space error covariance is then

$$\mathbf{S}_N = (\mathbf{A} - \mathbf{I})\mathbf{S}_x(\mathbf{A} - \mathbf{I})^T$$

It may be difficult to obtain a good value for $\mathbf{S}_x$ in any particular case, as its calculation requires measurements of an ensemble of profiles with significantly better quality than is obtainable with the observing system under discussion. Often a realistic estimate of the diagonal elements S_{zz} (the variances at each level) will be available, and if the off-diagonal elements (the inter-level covariances) are unknown, a conservative assumption is to set them to zero. This would give a maximum entropy estimate of the ensemble probability density function. If some idea of the length scale l of correlations is available, then an expression such as Eq. (1) can be used. If an artificial covariance matrix is constructed in this way, it must be positive definite.

Fig. 4 presents the nine largest null-space error patterns for the example

SBUV retrieval, showing orthogonal structures that the observing system cannot see. In this case we assume that $S_a = S_x$, when they are proportional to the eigenvectors of

$$S_N = (S_a^{-1} + K_x^T S_\epsilon^{-1} K_x)^{-1} S_a^{-1} (S_a^{-1} + K_x^T S_\epsilon^{-1} K_x)^{-1}$$

Errors in the profiles due to this source will be a linear combination of these functions, with random coefficients having unit variance. These vectors tend to have larger values at the top and bottom of the plot, where there is little information about the true profile in the measurements. Further, they all have a fairly short scale vertical structure, somewhat shorter than the width of the averaging kernels, but not very short. The *a priori* covariance matrix chosen has little variance on scales much shorter than 5 km, so very fine scales do not appear.

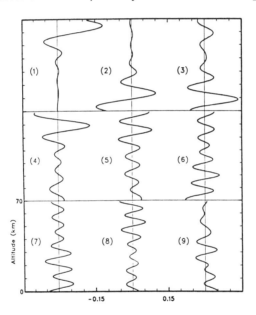

FIGURE 4. The nine null space error patterns with largest eigenvalue for the example SBUV retrieval.

5.2.2 Measurement Error.

As seen in Eq. (3), the contribution of measurement error ϵ_y directly to error in the retrieval is $D_y \epsilon_y$. This has error covariance

$$S_M = D_y S_\epsilon D_y^T$$

Example error patterns for the contribution of measurement noise to the overall error are shown in Fig. 5. In this case, the measurement error covariance is

$$S_M = (S_a^{-1} + K_x^T S_\epsilon^{-1} K_x)^{-1} K_x^T S_\epsilon^{-1} K_x (S_a^{-1} + K_x^T S_\epsilon^{-1} K_x)^{-1}$$

Patterns with broader scale structure generally correspond to smaller errors, as there is information in the measurements to determine these components. Patterns with short vertical scale also give small error, as the inverse model does not attempt to determine these, information on this scale coming from the *a priori*. Larger errors are associated with intermediate scale patterns, on a scale comparable with the weighting function width.

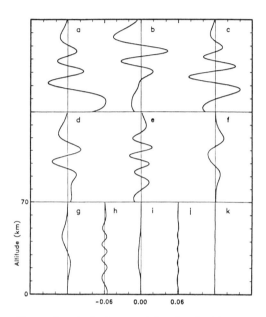

FIGURE 5. Error patterns due to instrumental noise for the example SBUV retrieval.

The total random error of the retrieval at each level are given in Fig. 6. Curve B is the total residual standard deviation, or the total error ignoring inter-level correlations. This is simply the square root of the diagonal of

$$\mathbf{S}_M + \mathbf{S}_N = (\mathbf{S}_a^{-1} + \mathbf{K}^T \mathbf{S}_\epsilon^{-1} \mathbf{K})^{-1}$$

Curve A gives the component due to measurement noise, $\mathbf{S}_M$. These curves are simply the root-sum-squares of the corresponding error patterns, which clearly give more detailed information.

5.2.3 Systematic Errors

If the value of ϵ_b does not change randomly from one measurement to the next, the term in b in Eq. (3) corresponds to systematic error owing to uncertain model parameters. Otherwise, it represents random errors, and could be included with ϵ_y. The error covariance due to this source of error is given by

$$\mathbf{S}_S = \mathbf{A}_b \mathbf{S}_b \mathbf{A}_b^T$$

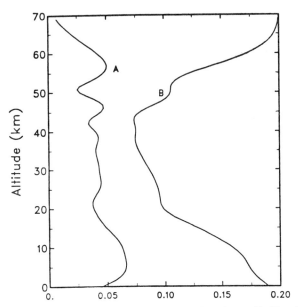

FIGURE 6. A: R.m.s. error due to instrumental noise alone. B: Total r.m.s. error.

where $\mathbf{S}_b$ is the errror covariance of our knowledge of the model parameters.

Systematic error due to bias in the inverse models relative to the forward model is given by $T(\bar{\mathbf{x}}, \hat{\mathbf{b}}, \hat{\mathbf{c}}) - \bar{\mathbf{x}}$. This will clearly not necessarily be zero for any particular $\bar{\mathbf{x}}$, but the inverse model can be redefined where appropriate by subtracting this quantity from it.

The overall error in the retrieval thus comprises three elements, (1) Null space error (2) Measurement error and (3) Systematic error. To compare different observing systems we must compare all these, and note especially that the total random error is the sum of the null-space and measurement errors.

6. EFFECT OF A PRIORI DATA.

Particular concern is often expressed about the contribution of *a priori* data to the retrieval, so a few words are in order here to illustrate how this may be understood. *A priori* may be taken to mean many things, for example the starting point of an *ad hoc* iteration, which may or may not affect the outcome, a constraint expressed by a representation of the profile wih finite number of parameters, or an independent estimate of the state obtained from sources other than the direct measurement (a *virtual* measurement). For the purpose of the error analysis, we only consider the latter type.

A helpful approach is to note that the solution may be put into the form:

$$\hat{\mathbf{x}} - \bar{\mathbf{x}} = \mathbf{A}(\mathbf{x} - \bar{\mathbf{x}}) + \mathbf{D}_a(\mathbf{x}^a - \bar{\mathbf{x}}) + \text{ random terms} \qquad (4$$

where the 'random terms' are from independent error sources. Any reasonabl inverse model should reproduce the *a priori* when given it as input. In these cases we would expect to find

$$\mathbf{x}^a - \bar{\mathbf{x}} = \mathbf{A}(\mathbf{x}^a - \bar{\mathbf{x}}) + \mathbf{D}_a(\mathbf{x}^a - \bar{\mathbf{x}})$$

hence $\mathbf{D}_a = \mathbf{I} - \mathbf{A}$. Thus the retrieval will be a weighted mean of the true state vector and the *a priori*, with $\mathbf{A}$ and $\mathbf{I} - \mathbf{A}$ as weights, plus contributions from other sources of error, that will be ignored for this discussion. To understand how this weighting affects the solution it is helpful to use eigenvectors, as before. If the matrix of eigenvectors of $\mathbf{A}$ is $\mathbf{U}$, so that $\mathbf{AU} = \mathbf{U\Lambda}$ and $\mathbf{U}^{-1}\mathbf{A} = \mathbf{\Lambda U}^{-1}$, then premultiplying Eq. (4) by $\mathbf{U}^{-1}$ gives

$$\mathbf{U}^{-1}(\hat{\mathbf{x}} - \bar{\mathbf{x}}) = \mathbf{\Lambda U}^{-1}(\mathbf{x} - \bar{\mathbf{x}}) + (\mathbf{I} - \mathbf{\Lambda})\mathbf{U}^{-1}(\mathbf{x}^a - \bar{\mathbf{x}})$$

If we write $\mathbf{u} = \mathbf{U}^{-1}(\mathbf{x} - \bar{\mathbf{x}})$, then we obtain $\hat{\mathbf{u}} = \mathbf{\Lambda u} + (\mathbf{I} - \mathbf{\Lambda})\bar{\mathbf{u}}$, which can be separated into components, showing that the elements of $\hat{\mathbf{u}}$ are weighted means of the corresponding elements of $\mathbf{u}$ and $\mathbf{u}^a$:

$$\hat{u}_i = \lambda_i u_i + (1 - \lambda_i)u_i^a$$

Thus we can decompose the state vector into patterns, some of which (those with $\lambda_i \sim 1$) will be well reproduced by the measurement system, and others (with $\lambda_i \sim 0$) that will come mainly from the *a priori* vector. The patterns themselves will the columns of $\mathbf{U}$, i.e. the eigenvectors of $\mathbf{A}$, because we obtain $\mathbf{x}$ from $\mathbf{u}$ according to $\mathbf{x} - \bar{\mathbf{x}} = \mathbf{Uu}$.

Another approach is to estimate the variability in the retrieval that results from possible variations in the *a priori* profile. As mentioned above, *a priori* data can be regarded as one component of $\mathbf{c}$, the inverse model parameters, because they affect the retrieval without entering the forward model. Let the *a priori* state vector estimate be $\mathbf{x}^a$. If we assume that the 'correct' value is $\mathbf{x}^a = \mathbf{x}$, then the contribution to Eq. (3) is $(\partial I/\partial \mathbf{x}^a)(\mathbf{x}^a - \mathbf{x})$. The error covariance due to *a priori* error is therefore $\mathbf{D}_a\mathbf{S}_a\mathbf{D}_a^T$ where $\mathbf{D}_a = \partial I/\partial \mathbf{x}^a$, and $\mathbf{S}_a$ is the error covariance of the *a priori* data. Often $\mathbf{x}^a$ and $\mathbf{S}_a$ will be the same as the $\bar{\mathbf{x}}$ and $\mathbf{S}_x$ referred to in the discussion of null space error, and if $\mathbf{D}_a = \mathbf{I} - \mathbf{A}$ also, then we find that this error covariance is simply the null space error in another guise.

In the example retrieval, the retrieved profile is related to the true profile and the *a priori* by

$$\hat{\mathbf{x}} = \mathbf{x}_a + \mathbf{D}_y\mathbf{K}(\mathbf{x} - \mathbf{x}_a) = \mathbf{Ax} + (\mathbf{I} - \mathbf{A})\mathbf{x}_a$$

which is of the same form as Eq. (4). The largest nine eigenvalues of $\mathbf{A}$ and the corresponding eigenvectors are shown in Fig 7. The first four eigenvalues are essentially unity, implying that the *a priori* data contributes nothing to components of the solution with this vertical structure. The next three or four vectors correspond to vertical structure where both the measurements and the *a priori* contribute.

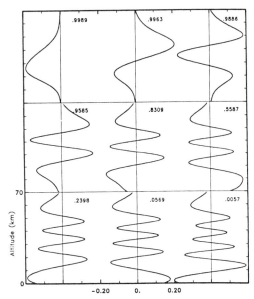

FIGURE 7. First nine eigenvalues and eigenvectors of the averaging kernel matrix.

At scales shorter than this only the *a priori* contributes, the measurements have nothing to say.

7. SUMMARY

A formal error analysis for profile retrieval can be established, which is independent of the nature of the retrieval method, provided that the measurement process can be modelled adequately. The error separates naturally into three components, (a) random error due to measurement noise, (b) systematic error due to uncertain model parameters and inverse model bias, and (c) null space error due to the inherent finite vertical resolution of the observing system. A recipe is given for evaluating each of the components in any particular case.

Most of the error terms appear as covariance matrices, rather than simple error variances. These matrices can be interpreted in terms of 'error patterns', which are statistically independent contributions to the total error. They are the multidimensional equivalent of 'error bars'.

The basic set of diagnostics for characterising errors will consist of (i) the averaging kernels and/or the null space error patterns (ii) the measurement noise error patterns and the level-by-level r.m.s. error (iii) the inverse model bias and (iv) the model parameter error patterns. This set characterises all the terms in the expression for the total error (Eq. (3)).

ACKNOWLEDGMENTS

I would like to thank Bob Hudson of Goddard Space Flight Center for providing the weighting functions for the SBUV, and the 1987 NASA Ozone Trends Panel for providing the motivation for writing up this error analysis.

I would also like to thank the Atmospheric Chemistry Division of NCAR for support, and the opportunity to carry out this work.

REFERENCES

Backus, G.E. and J.F. Gilbert (1970) Uniqueness in the Inversion of Inaccurate Gross Earth Data, *Phil. Trans. Roy. Soc. London, Ser. A,* **266** p 123-192.

Heath, D.F., A.J. Kruger, H.R. Roeder and B.D. Henderson (1975) The Solar Backscatter Ultraviolet and Total Ozone Mapping Spectrometer (SBUV/TOMS) for Nimbus G, *Opt. Eng.,* **14**, p 323-331.

Menke, W. (1984) *Geophysical Data Analysis: Discrete Inverse Theory* Academic Press.

Rodgers, C.D. (1976) Retrieval of Atmospheric Temperature and Composition From Remote Measurements of Thermal Radiation, *Rev. Geophys. and Space Phys.* **14** p 609-624.

DISCUSSION

King: The inversion problem, of course, is notoriously unstable, which is a nice term. For that you need catastrophe theory or certainly non-linear analysis. Am I wrong in inferring that your analysis is of linear analysis and if so, how do you account for these catastrophic things like when your error is too large you get negative temperatures and all of those terrible things?

Rodgers: This analysis is meant to be independent of how the retrieval method works. The only requirement that this is valid is that the inverse method be linear within the error bars of the result. If it is not, then you've got a real problem on your hands and I really can't address that one. It doesn't have to be linear within the variability of the measurements or the solution profile.

Zachor: But aren't you solving for a departure from a mean and assuming you can define a mean that has some relevance?

Rodgers: Not really, I'm just using the mean as some convenient point about which to linearize in order to do the error analysis and it can be chosen such that linearization is valid. The error analysis applies to a single solution. This doesn't mean I've defined an error analysis which applies globally to an inversion method, every solution must have its own error bars. If you are lucky they will be the same on every profile. But if it is non-linear they won't be.

Gille: Could you say something about what it means or what problems might arise if the integral of the averaging kernels isn't unity, that is, if they are not normalized.

Rodgers: There is no reason why they should be normalized. That is not a constraint. There is such a constraint in the original Backus-Gilbert Retrieval Method, but it just doesn't appear in this analysis at all.

Gille: Does it have any implications? Does it mean that there could be some problems?

Rodgers: No, it just means that your retrieval method says that the solution doesn't depend on the measurements in that region. These averaging kernels (Figure 3) are not normalized. Take this one at 69 km. There is a nominal solution at 69 km but, in fact, it is mainly the a priori and it depends only slightly on the profile, but not much.

EFFECT OF CALIBRATION ERRORS ON MINIMUM VARIANCE TEMPERATURE RETRIEVALS[1]

David S. Crosby[2]
National Oceanic and Atmospheric Administration
National Environmental Satellite, Data, and Information Service
Washington, D.C. 20233, USA

ABSTRACT

It is well known that calibration errors can have a significant impact on the accuracy of temperature retrievals. In this paper we analyze satellite temperature retrievals as a linear structural relationship model. By examining the problem as a linear model, it is possible to determine the effects of using an incorrect form of the operator on the variance of the solution vector. In this paper this method is used to determine the effect of calibration errors on the bias and variance of the retrieved temperature. It is found that for certain types of calibration errors it is only necessary to remove the bias error in order to retrieve temperature profiles of good quality.

1. INTRODUCTION

The estimation of atmospheric temperature profiles from satellite measured radiances requires the inversion of the radiative transfer equation. This may be explicit, as in the physical retrieval methods or implicit, as in the regression methods. In this paper we will be concerned with the effects of incorrectly modelling calibration errors on the minimum variance physical retrieval method. This method requires *a priori* knowledge of a number of parameters associated with the atmosphere and the instrument. By examining the radiative transfer equation as a linear structural relationship model it is possible to compute the effect of errors in the specification of these parameters on the variance of the difference between the true temperature profile and the estimated temperature profile.

---

[1]Part of this work was carried out at the Robert Hooke Institute for Cooperative Atmospheric Research, Clarendon Laboratory, Oxford, England.

[2]Permanent affiliation: The American University, Washington, DC.

RSRM '87: ADVANCES IN
REMOTE SENSING RETRIEVAL METHODS
A. Deepak, H.E. Fleming, and J.S. Theon (Eds.)   299

In this paper we will use a technique described in Crosby and Weinreb (1974) and Fleming et al. (1986) to investigate the effect of calibration errors on the retrieved temperature profiles. It is found that for the situation where there are simultaneous measurements of the atmospheric parameters and radiances then these modelling errors are not a serious problem. This is because they affect the retrieval primarily as a bias. This bias is easily removed using the simultaneous measurements. If the calibration of the instrument is changing in an unknown manner, it is found that the modeling errors may affect the retrievals by a small but significant amount.

2. THEORETICAL DEVELOPMENT

2.1 MODEL

We will use the matrix model form of the radiative transfer equation. For details of the development of this model see Fleming and Smith (1972), Rodgers (1976), or Strand and Westwater (1968). In this model we have an equation of the form

$$r = At + \varepsilon \qquad (1)$$

where A is a rectangular matrix of dimension $m \times n$, t and r are temperature and radiance difference vectors respectively and ε is a vector of measurement errors. The elements of the vectors t and r are

$$t(p) = T(p) - \overline{T}(p) \qquad (2a)$$

$$r(\nu) = R(\nu) - \overline{R}(\nu) \qquad (2b)$$

where $T(p)$ is the true atmospheric temperature at pressure p, $R(\nu)$ is the measured radiance at wavenumber ν , $\overline{T}(p)$ and $\overline{R}(\nu)$ are the means of $T(p)$ and $R(\nu)$ over a specified sample.

We let E[ ] represent the expected value operator over the joint distribution of the random vectors t and ε. We then assume

$$E[t] = \emptyset \quad , \quad E[\varepsilon] = \emptyset \quad , \quad E[t\varepsilon^T] = \emptyset \qquad (3a)$$

$$E[tt^T] = \Sigma_t \quad , \quad E[\varepsilon\varepsilon^T] = \Sigma_\varepsilon \qquad (4a)$$

where T superscript represents the transpose and $\emptyset$ represents the zero vector or matrix.

The estimation problem now becomes: given a radiance vector r, estimate a temperature profile vector t. We will assume that the estimate is linear in r. That is we have

$$t^* = C_* r. \tag{5}$$

Under the assumptions of the model the variance-covariance matrix of $(t-t^*)$ is given by

$$\Sigma_{t-t}^* = C_*(A\Sigma_t A^T + \Sigma_\varepsilon)C_*^T - \Sigma_t A^T C_*^T - C_* A\Sigma_t + \Sigma_t \tag{6}$$

The linear least squares estimator is the estimator which uses a C so that the trace of Σ_{t-t}^* is minimized. This C_0 is given by

$$C_0 = \Sigma_t A^T (A\Sigma_t A^T + \Sigma_\varepsilon)^{-1}. \tag{7}$$

We denote this estimate by $\hat{t}$, i.e.,

$$\hat{t} = C_0 r. \tag{8}$$

In this case the variance-covariance matrix of $(t-\hat{t})$ is given by

$$\Sigma_{t-\hat{t}} = \Sigma_t - C_0 A\Sigma_t. \tag{9}$$

When we examine Equation (7), we see that the parameters needed to compute C_0 are Σ_t, A and Σ_ε. Basically the technique used in Crosby and Weinreb (1974), Fleming et al. (1986) and this paper is to examine the effect of incorrectly specifying one or more of these parameters. We evaluate the effect by comparing the diagonal elements of Σ_{t-t}^* and $\Sigma_{t-\hat{t}}$ when an incorrect parameter is used in Equation (7). That is, we then have a non-least squares operator

$$C_* = \Sigma_t^* A^{*T}(A^* \Sigma_t^* A^{*T} + \Sigma_\varepsilon^*)^{-1}. \tag{10}$$

In Equation (10), it is assumed that at least one of the parameters Σ_t^*, A^*, or Σ_ε^* are incorrect. In this paper we will argue that calibration errors will affect the operator C_* by using an incorrect A (A^*) or an incorrect Σ_ε (Σ_ε^*).

2.2 CALIBRATION ERRORS

2.2.1 Fixed Calibration Errors

We will assume the incorrect calibration is linear in its effect. Then we measure

$$R_m(\upsilon)= \alpha(\upsilon)R(\upsilon) + \beta(\upsilon) \tag{11}$$

and not $R(\upsilon)$. That is, in Equation (11) $R_m(\upsilon)$ is the actual measured radiance and $R(\upsilon)$ is the true radiance.

In the arguments which follow, we assume that the proper means are used and that the offset, $\beta(\upsilon)$, is zero. These assumptions will be discussed in a later section.

The true relation between r_m and t is given by

$$r_m = DAt + \varepsilon. \tag{12}$$

where D is a diagonal matrix with α's on the diagonal and ε is assumed to be approximately equal to $D\varepsilon$. Therefore in this case the operator C_*, given by

$$C_* = \Sigma_t A^T (A\Sigma_t A^T + \Sigma_\varepsilon)^{-1} \tag{13}$$

is incorrect. The correct operator is given by

$$C_o = \Sigma_t A^T D (DA\Sigma_t A^T D + \Sigma_\varepsilon)^{-1}. \tag{14}$$

To investigate the effect of this error, we use different D matrices in Equation (14) and compare the diagonals of $\Sigma_{t-t}{}^*$ and $\Sigma_{t-\hat{t}}$ as determined by Equations (6) and (9). The objective of the above development is to transfer the calibration errors from the radiance to the operator.

2.2.2 Variable Calibration Errors

To evaluate the effect of calibration parameters that change with temperature, scan position, etc., we consider the $\alpha(\upsilon)$ and $\beta(\upsilon)$ of Equation (11) to be random variables. We write Equation (12) in the form

$$r_m =(I-(I-D))At + B + \varepsilon \tag{15a}$$

$$r = At -(I-D)At + B + \varepsilon. \tag{15b}$$

In Equation (15), I is an identity matrix and B is a vector of elements of the form $\beta(v)$. We now have a new error term of the form

$$\delta = -(I-D)At + B + \varepsilon. \tag{16}$$

Then we have the equation

$$r_m = At + \delta. \tag{17}$$

This model is considerably more complicated because it is no longer true that $E[t\delta^T]=\emptyset$. However, since the elements of $E[t\delta^T]$ are small relative to the other parameters of the model we will ignore them.

In the usual physical solution it is assumed that Σ_ε is a diagonal matrix. However, because of the fact that the calibration procedures are the same for different channels we would expect that these calibration errors would be dependent. This would imply that Σ_δ will have large off-diagonal elements. Hence, to investigate the effect of these sorts of errors we will examine the effect of using the wrong error covariance matrix, i.e.,

$$C_* = \Sigma_t A^T (A\Sigma_t A^T + \Sigma_\varepsilon)^{-1}. \tag{18a}$$

Note that the correct operator is C_o, which is given by,

$$C_o = \Sigma_t A^T (A\Sigma_t A^T + \Sigma_\delta)^{-1}. \tag{18b}$$

2.3 BIASES

In the above development we have ignored the problem of biases. Writing Equation (11) in matrix form the measured radiance vector is given by

$$R_m = DR + B. \tag{19}$$

We then use the solution

$$T^* = \bar{T} + C(R_m - \bar{R}) \tag{20}$$

where $\bar{R}$ is the mean of R and not of R_m. Then

$$T^* = \bar{T} + C(DR + B - \bar{R}) \tag{21a}$$

$$T^* = \bar{T} + C(R - \bar{R})) + C((D-I)R + B). \tag{21b}$$

If we take the expectation of this estimate, assuming that D and R are independent, we have

$$E[T^*] = \bar{T} + C((D-I)\bar{R} + B). \quad\quad (22)$$

Since the expectation of our estimate should equal $\bar{T}$, the difference $C((D-I)\bar{R}+B)$ is a temperature bias. If we have simultaneous measurements of R and T, then this bias can be removed by a simple adjustment. See Fleming et al. (1986) for a discussion of the problem of bias and its solution. As will be seen , an incorrect calibration affects the bias much more than it affects the variance or standard deviation of the solution.

3. APPLICATION

The technique was applied to data from the Special Sensor Microwave/Temperature (SSM/T) instrument. This is a seven channel microwave sensor and is part of the Defense Meteorological Satellite Program (DMSP). The SSM/T instrument has seven channels in the oxygen band, with one channel serving as a window channel. Channel parameters are given in Table 1. For a more complete description of the instrument see Grody et al. (1985).

TABLE 1. CHANNEL PARAMETERS FOR THE SSM/T INSTRUMENT

Channel number	Frequency (GHz)	Approximate peak weighting function (mb)
1	55.50	surface
2	53.20	800
3	54.35	400
4	54.90	280
5	58.40	10
6	58.825	70
7	59.40	40

For our application we will use channels 2 through 7, giving a six channel sounding instrument. The Σ_t was compiled for the spring-fall in the 30°-60°N latitude region. The soundings were assumed to be in an ocean region and the emissivity of the surface was assumed to be equal to 0.5.

We restrict our study to reasonable model errors to give some insight into the problem. For the study of fixed calibration errors, we assume that the $\beta(\upsilon)$ are zero and that

$$D = \alpha I \quad\quad (23)$$

where α is a constant and I is the identity matrix. For the case of variable calibration errors we assume that

$$\Sigma_\varepsilon = \sigma^2 I \tag{24}$$

and

$$\Sigma_\delta = \sigma^2 I + \psi J \tag{25}$$

where J is a matrix of ones.

4. RESULTS

4.1 BIASES

Before we present the results of using an incorrect operator we will look at the problem of biases and assume that the bias is zero in the remaining sections. The number of different ways that a bias can occur is infinite. We will limit our consideration to one case. We assume that the calibration offsets $\beta(\upsilon)$ are zero and that

$$D= (1.05)I. \tag{26}$$

Then from Equation (23) the bias is $C((D-I)\overline{R}$. The operator C and the mean $\overline{R}$ are computed for the spring-fall for $30°$ to $60°$ N region. The D is assumed to be a constant so that its mean value is equal to itself. Table 2 gives the biases for this situation for several pressure levels of the atmosphere.

TABLE 2. BIAS VECTOR FOR D=1.05I AND $\beta=\varnothing$

Pressure (mb)	Bias (deg C)
10	28.06
50	17.66
150	9.35
300	11.22
400	10.81
500	9.26
700	8.13
850	8.36
920	8.90
1000	9.47

The numbers are quite large and would make the results unusable. In practice, biases of this magnitude are not seen. Since the biases are linear in α, if

$$D = (1.005)I \tag{27}$$

the bias at 1000 mb would be .947. Biases of this order of magnitude or larger have been observed.

4.2 FIXED CALIBRATION ERRORS

We will now examine the effect of using an incorrect operator, i.e., the C_* of Equation (13). The correct operator is the C_o of Equation (14). For the study we assume that in Equations (13) and (14)

$$\Sigma_\varepsilon = (.25)I \tag{28a}$$

$$D = \alpha I. \tag{28b}$$

In each model the covariance matrix of $(t-\hat{t})$ is that of the case where α is equal to one.

Table 3 gives the standard deviation of t and $(t-t^*)$ for different values of α and various pressure levels of the atmosphere. For reference the standard deviation of $(t-\hat{t})$ is given in the column for α equal to one.

TABLE 3. STANDARD DEVIATIONS OF t and $(t-t^*)$ FOR $D = \alpha I$

Pressure (mb)	σ_t	$\sigma_{t-\hat{t}}$ $\alpha=1.00$	σ_{t-t}^* $\alpha=0.95$	σ_{t-t}^* $\alpha=1.05$	σ_{t-t}^* $\alpha=1.10$	σ_{t-t}^* $\alpha=1.20$
10	9.39	2.91	2.95	2.95	3.05	3.42
50	6.15	1.73	1.76	1.76	1.83	2.10
150	6.39	2.67	2.69	2.68	2.74	2.91
300	6.13	2.38	2.40	2.40	2.45	2.63
400	7.54	2.01	2.04	2.04	2.13	2.48
500	8.26	2.02	2.06	2.06	2.17	2.58
700	9.08	2.22	2.26	2.26	2.39	2.83
850	10.15	2.12	2.17	2.18	2.34	2.90
920	10.58	2.01	2.08	2.08	2.27	2.89
1000	11.55	3.71	3.75	3.75	3.86	4.30

It is seen from Table 3 that if the calibration is known to within about five percent the effect on the standard deviation of the retrieval is not large. It does not compare with the magnitude of the bias (see Table 2). Even if the calibration is only known to within about 10 percent the retrievals would still be usable. However, it must be emphasized that unless there are simultaneous measurements of the radiance and the temperature, calibration errors of this order of magnitude will make the retrievals unusable.

4.3 VARIABLE CALIBRATION ERRORS

In this section we look at the effect of incorrectly modelling variable calibration errors. For this situation, we will assume we have

$$C_* = \Sigma_t A^T (A\Sigma_t A^T + \Sigma_\varepsilon) \qquad (29a)$$

$$C_o = \Sigma_t A^T (A\Sigma_t A^T + \Sigma_\delta) \qquad (29b)$$

$$\Sigma_\varepsilon = (.0625)I \qquad (29c)$$

$$\Sigma_\delta = (.0625)I + \psi J \qquad (29d)$$

where J is a matrix of ones. Table 4 gives the standard deviation of $(t-\hat{t})$ and $(t-t^*)$ for two ψ levels. For reference, the standard deviation is given for ψ equal to zero.

TABLE 4. $\sigma_{(t-\hat{t})}$ AND $\sigma_{(t-t^*)}$ FOR VARIABLE CALIBRATION ERRORS

Pressure	$\psi=0.0$	$\psi= 0.5$		$\psi= 1.0$	
(mb)	$\sigma_{t-\hat{t}}$	$\sigma_{t-\hat{t}}$	σ_{t-t^*}	$\sigma_{t-\hat{t}}$	σ_{t-t^*}
10	2.71	3.18	3.21	3.52	3.65
50	1.57	1.91	1.94	2.16	2.25
150	2.27	2.37	2.37	2.44	2.47
300	2.18	2.29	2.30	2.38	2.42
400	1.79	1.89	1.90	1.97	1.99
500	1.71	1.77	1.78	1.82	1.85
700	2.08	2.12	2.13	2.15	2.17
850	1.96	2.01	2.01	2.05	2.07
920	1.50	2.08	2.09	2.15	2.17
1000	3.14	3.19	3.19	3.24	3.25

It is seen from Table 4 that there is not a large difference in the standard deviation of the retrievals using the incorrect operator C_* and the correct operator C_o. However, the difference in the standard deviations at the large values of ψ is of the order of the difference observed between a well modelled physical retrieval technique and regression procedures, the regression procedure being slightly more accurate. For large samples the regression operator converges to the correct linear operator. This suggests that the reason that physical retrievals have never quite attained the accuracy of the regression may be a consequence of this type of calibration error.

5. CONCLUSIONS

The purpose of this paper has not been to answer all possible questions about the effect of calibration errors on satellite temperature retrievals. However, the technique developed is quite general and can be used for any instrument and sounding method which is at least approximately linear.

For the SSM/T it was found that for the situations where it is possible to have simultaneous measurements of the temperature and the radiances, most reasonable calibration errors are not a serious problem. This agrees with the results in Fleming et al. (1986). If the biases can not be removed then calibration errors can make the data unusable.

ACKNOWLEDGMENT

The author wishes to thank Ms. Martha Maiden who provided the data sets which were used in this study.

REFERENCES

Crosby, D. S, and M. P. Weinreb, 1974: Effect of incorrect atmospheric statistics on the accuracy of temperature profiles derived from satellite measurements. *J. Stat. Comput. Simul.*, 3, 41-51.

Fleming, H. E., D. S Crosby, and A. C. Neuendorffer, 1986: Correction of satellite temperature retrieval errors due to errors in atmospheric transmittances. *J. Clim. Appl. Meteor.*, 25, 869-882.

Fleming, H. E., and W. L. Smith, 1972: Inversion techniques for remote sensing of atmospheric temperature profiles. *Temperature: Its Measurement and Control in Science and Industry.* Vol. 4, Part 3, H. H. Plumb (Ed.), Instrument Society of America, 2239-2250.

Grody, N. G., D. G. Gray, C. S. Novak, J. S. Prasad, M. Piepgrass, and C. A. Dean, 1985: Temperature soundings from the DMSP microwave sounder. In A. Deepak, H. E. Fleming and M. T. Chahine (Eds.), *Advances in Remote Sensing Retrieval Methods*, A. Deepak Pub., Hampton, Virginia, 249-265.

Rodgers, C. D., 1976: Retrieval of atmospheric temperature and composition from remote measurements of thermal radiation. *Rev. Geophys. Space Phys.*, 14, 609-624.

Strand, O.N., and E. R. Westwater, 1968: Minimum-RMS Estimation of the Numerical solution of A Fredholm Integral Equation of the First Kind. *SIAM J. Numer. Anal.*, 5, 287-295.

DISCUSSION

Neuendorffer: I was a little curious, you made the remark that you felt certain types of calibration errors were equivalent to cloud errors, where clouds can come and go randomly. I don't quite get the connection.

Crosby: They are equivalent mathematically. Cloud errors can appear at this point in the equations as a calibration error. That is, I can consider calibration errors to be like random calibration. Basically, mathematically they will both fall within the covariance structure of the error matrix.

Falcone: Are there any other questions? If not, I have a question. Why did you call the SSMT a six-channel instrument? It has seven channels.

Crosby: Well, the seventh one sees the surface, I did not use it in the inversions because of the problem of unknown surface emissivity.

VERTICAL RESOLUTION AND ERROR COMPONENTS OF OZONE RETRIEVALS FROM MEASUREMENTS OF INFRARED LIMB EMISSION[1]

John C. Gille and Paul L. Bailey
Naitonal Center for Atmospheric Research[2]
Boulder, Colorado 80307, USA

ABSTRACT

The error analysis approach described by Rodgers (1988) has been used to compare different methods for retrieving ozone from measurements by a broad band limb-scanning radiometer. For onion peeling, smoothed onion peeling, Twomey and maximum likelihood retrievals, the vertical resolution and error components due to measurement noise, uncertainties in input parameters and finite retrieval resolution (null space) are compared. The vertical resolution is uniformly highest for the onion-peeling retrieval, but the maximum likelihood solution leads to the smallest errors, and nearly the same vertical resolution as onion peeling when the signal to noise ratio is high. The ozone errors due to random temperature and registration pressure errors are essentially independent of retrieval method.

1.   INTRODUCTION

The aim of this heuristic study is to clarify the effects of the algorithm on the ozone determination from broad-band infrared limb scanning radiometers. There are two motives for doing this. The first is to gain a better understanding of the way in which various factors in the limb-scanning retrieval algorithms affect the final results. The second is to perform an analysis similar to those being performed on other ozone measuring techniques, and to clarify the extent to which changes in parameters input to the retrieval may affect the results.

In this study, we have followed the methodology outlined previously by Rodgers (1988). For the convenience of the reader, we shall try to use his terminology and symbols as well. The second section describes the method used, and presents some basic information and shows how temperature and pressure errors are considered. Section 3 describes the results of a simple back substitution

---

[1]Supported by NASA Order No. S-10782C and NOAA Order No. NA87AANEG0190
[2]The National Center for Atmospheric Research is sponsored by the National Science Foundation

or ''onion-peeling'' method, while Sec. 4 shows the results of applying a Twomey type constraint. A discussion of the maximum likelihood solution is given in Sec. 5. Section 6 contains some brief conclusions.

2.   OUTLINE OF THE CALCULATIONS

We define a forward model $\mathbf{F}$ that acts on an ozone state vector x to calculate a vector of measured radiances y, making use of other data in a vector b. In particular, using a typical mid-latitude ozone mixing ratio profile as the mean state vector $\bar{x}$, the mean measurement vector is given by $\bar{y} = \mathbf{F}(\bar{x},\bar{b})$ where the vector b includes the parameters of the forward model, such as spectroscopic data or the temperature and pressure profiles.

In this study the temperature and pressure were taken to be the 1976 US Standard Atmosphere mid-latitude temperature profile.

For the forward model in this study, it is necessary to be able to calculate realistic radiances, given initial temperature, pressure and ozone distributions. For simplicity we have assumed that the spacing of the computational levels in the atmosphere is 1.5 km, the same as the spacing of the observations. This is somewhat restrictive, as this sets smaller scales identically equal to zero, and we shall say more about this below. The levels are the boundaries of atmospheric shells, through which atmospheric quantities are assumed to vary linearly with altitude. In this study refraction is neglected; this is reasonable above 25 km, and its omission does not affect the general conclusions reached here.

The transmittance model is based on line-by-line calculations for a range of atmospheric conditions, and includes Voigt effects as well as the temperature dependence of line intensities and half widths. Inhomogeneous paths are treated by the Curtis-Godson method as outlined in Rodgers and Walshaw (1966). Where it has been necessary to consider measurement parameters, we have used parameters similar to those for the LIMS instrument on Nimbus 7, as described by Gille and Russell (1984).

The weighting function is defined as

$$K = dF/dx, \tag{1}$$

i.e., the change in the calculated measurement vector for an ozone change at a particular level. A set of weighting functions is shown in Fig. 1. They show several interesting features. The first is the usual narrow peak typical of limb-viewing experiments. The maximum occurs one level above the tangent height, because a perturbation of that level affects both the tangent shell and the one above it. This results in a slightly larger effect on the outgoing signal than a perturbation of the tangent level, which affects only the tangent shell.

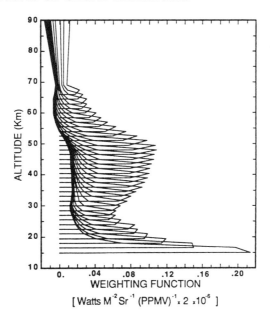

FIGURE 1. The weighting functions. These are the rows of $\mathbf{K}$, defined in Eq. (1).

The second feature is the maximum in weighting function peaks near the 100 mb level. Since response is in units of radiance per ppmv, the changes at that level are likely to be small in absolute units. A final curiosity is the negative values near the mesopause. These show that when ozone is added in this cold region, there is actually a drop in the signal when the tangent height is in a warmer region.

Continuing to follow Rodgers, we denote the inverse model by $\mathbf{I}$, and define the contribution function as

$$D = dI/dy. \tag{2}$$

Then applying the inverse model to the forward model and linearizing about the mean state results in

$$\hat{x} = T(x, \hat{b}, \hat{c}) + A(x - \bar{x}) + A_b(b - \hat{b}) + D\epsilon_y \tag{3}$$

where T is the transfer function, and $A = DK$. The vector c contains parameters that occur in the inverse but not the forward model, and ϵ_y is the measurement noise. The retrieval uses $\hat{b}$ and $\hat{c}$, the best estimates of b and c. The rows of A are referred to as the averaging kernels.

The expression for the total error may then be written

$$\hat{x} - x = [T(\bar{x}, \hat{b}, \hat{c}) - \bar{x}] + (A - I)(x - \bar{x}) + A_b \epsilon_b + D\epsilon_y. \tag{4}$$

Here we assume that $T - \bar{x}$ can be adjusted to be zero (possibly introducing errors through $\hat{b}$). The second term on the right is the null
space error, and the fourth that due to measurement errors. Null
space error arises when the resolution of the retrieval does not allow it to follow small scale variations in the atmosphere. Setting
variations smaller than 1.5 km to zero, as done here, will result in
some underestimation of null space errors.

The third term on the RHS of expression (4) gives the solution
error that results from an error in input parameters, such as temperature or the registration pressure. In this case, temperature
errors enter mainly through the Planck function, but also through
the hydrostatic equation and the temperature dependence of the
transmittances. Similarly, an error in the registration pressure,
usually determined near 10 mb, will change the mass distribution
in the atmosphere, affecting the transmittances and the mixing-
ratios.

Here we shall treat only random errors. In that case, the null-
space error covariance is given by

$$S_N = (A - I)S_x(A - I)^T. \tag{5}$$

The evaluation of this term requires that we have an esti-
mate of S_x, the covariance of $x-\bar{x}$. For this study we have used the
statistics of all the LIMS ozone profiles for January, 1979.

Similarly, the error covariance due to random errors in the
input parameters are given by

$$S_p = A_b S_b A_b^T \tag{6}$$

where S_b is the covariance matrix of the random errors of model input parameters and S_p is the resulting error covariance of the solution. Here, values for the random temperature errors are taken from
Gille et al. (1984). It is very difficult to determine the error
in the pressure determination, but the random error does not seem
likely to be greater than 1.5%.

Finally, S_M, the covariance of the errors due to measurement
noise, is given by

$$S_M = DS_\epsilon D^T. \tag{7}$$

The measurement error at one altitude is assumed to be uncorrelated
with the errors at other altitudes.

These quantities will be calculated to allow a comparison of
the effects of different retrieval methods. To facilitate the presentation of results, the square root of the diagonals of the er-

ror covariance matrices are plotted in the following error plots. This is not strictly accurate, but all of the matrices are found to be strongly diagonal, so this approximation should not be greatly misleading.

3. ONION-PEELING RETRIEVALS

By its geometry, limb scanning results in data for which all levels above the tangent height of the observation contribute to the outgoing radiation, but the levels below do not. Ignoring some effects at the upper boundary, this means that K can be written as a triangular matrix, which greatly facilitates obtaining a formal solution. Onion-peeling, at its most basic, refers to solving first for the topmost or outermost shell from the observations obtained by viewing that shell. This solution is inserted to calculate the effect of the outermost shell when viewing the next lower shell, and so forth down through the atmosphere. This is just a back-substitution in the triangular matrix. The onion-peeling solution is then given by

$$\hat{x} - \bar{x} = K^{-1}(y - \bar{y}) \tag{8}$$

In this case, the averaging kernels for a few levels are shown in Fig. 2. Note that they are delta-functions to the resolution of the system, do not overlap, and all have the same maximum amplitude of 1. These show that the A matrix is diagonal. A change at one level will only affect the solution at that level. The columns of the A matrix, the delta function response functions, will look the same.

The solution in this case (not shown) is jagged and characterized by large high-frequency variations in the vertical. For this retrieval $A - I = 0$, so that the null space error vanishes. The component of retrieval error due to measurement noise indicates small errors at lower altitudes, but large errors, of the order of 1 ppmv, in the upper stratosphere and mesosphere, where the signal to noise ratio (S/N) is low. These errors are much larger than the contributions due to the temperature and pressure uncertainties.

Clearly, this is a solution with very high vertical resolution, but which is (expectedly) unstable to noise. The resolution is paid for with large errors and unrealistic vertical variations.

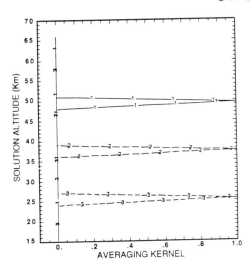

FIGURE 2.   Examples of averaging kernels for the onion-peeling
             solution. These are the rows of **A**, introduced in Eq.
             (3). To the resolution of the calculation they are delta
             functions.

One of the simplest approaches would be to apply a 1-2-1 smoothing to the retrieval. This drastically reduces the size of the oscillations, so now the smoothed solution follows the true profile
much more closely, as indicated by the much reduced ozone measurement errors in the upper stratosphere shown in Fig. 3. The effect
of the smoothing is to double the widths of the triangular averaging kernels. The solution now can not follow the true profile exactly, resulting in the null-space error that is a major component
of the error below 40 km. Uncertainties in the pressure also make a
large contribution, with a peak value about 0.15 ppmv, and a shape
like that of the ozone distribution. The contribution of temperature uncertainties has a similar shape, but a peak magnitude of only
0.1 ppmv.

However, one can argue that this smoothing is too drastic;
the solution at the top suffers from S/N problems, and needs to be
smoothed there, but perhaps not further down. Smoothing at the top
can take numerous forms, many of which do not fit easily into the
general linear framework used here. However, the results should not
depend strongly on the way in which this is done.

4.   USE OF A TWOMEY CONSTRAINT

An alternate approach to smoothing the results is to apply a
Twomey (1963) type constraint. This also turns out to be very instructive. The solution is now written as

$$\hat{x} - \bar{x} = (K^T K + \gamma H)^{-1} K^T (y - \bar{y}) \tag{9}$$

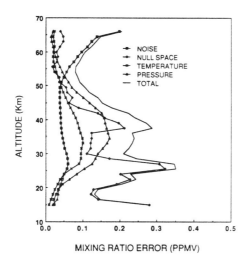

FIGURE 3. Ozone error components for onion-peeling with a 1-2-1 smoothing.

where H is a constraint matrix, and γ is the parameter by which the amount of smoothing can be controlled. In this study we have used the minimum curvature constraint.

If γ is set at 10^{-4}, there is not much smoothing, and the solution again is rather jagged. The error components (not shown) not surprisingly are dominated by measurement error, with only a small null-space error. The total error is about 0.3 ppmv at all altitudes, with little sharp structure.

Increasing γ to 10^{-3} increases the smoothing, resulting in the smoothing of much but not all of the sharp features. The measurement noise (Fig. 4) still dominates above 50 km, and the temperature and pressure errors are not much changed. Now the null space errors are larger in the mid-stratosphere, and clearly dominate around 25 km, where the errors are still about 0.3 ppmv, but at other altitudes up to 60 km the errors are less than 0.2 ppmv.

When γ is increased to 10^{-2}, the solution is smooth except when it comes around the maximum. As Fig. 5 shows, for this γ, the measurement error is now less than the null space error at all levels. Comparison of this with Fig. 4 shows that the noise reduction due to increased smoothing has been more than compensated for by the increase in the null-space error component.

The averaging kernels are illustrated for $\gamma = 10^{-3}$ in Fig. 6. Clearly they are wider and less strongly peaked than for the onion-peeling case, and vary in width with altitude, providing more smoothing at the top where it is needed. As one would expect, the averaging kernels widen as γ is increased.

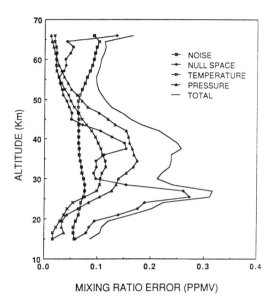

FIGURE 4.   Ozone error components for a solution with a Twomey
             minimum curvature constraint, for $\gamma = 10^{-3}$.

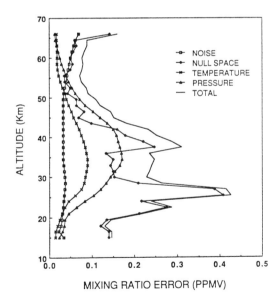

FIGURE 5.   Ozone error components for solution with a Twomey mini-
             mum curvature constraint, for $\gamma = 10^{-2}$.

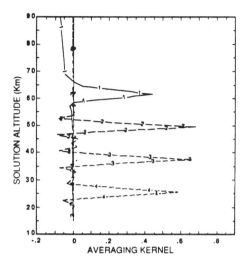

FIGURE 6.   Examples of averaging kernels for solutions obtained
with a Twomey minimum curvature constraint, for $\gamma =$
10^{-3}.

5.   THE MAXIMUM LIKELIHOOD SOLUTION

An alternate way to constrain the solutions is to require that
the solution be the most probable one, given the measurements and
the a priori statistics. Mathematically, this is given by (Rodgers,
1976)

$$\hat{x} - \bar{x} = S_a K^T (K S_a K^T + S_e)^{-1} (y - \bar{y}) \tag{10}$$

where S_a is the covariance matrix of the a priori information, taken
to be S_x.

The use of this solution results in the averaging kernels shown
in Fig. 7. They are broadened considerably at the top, indicating
the expected loss of resolution at these levels where S/N is low.
Below about 45 km they become narrow, but not quite as narrow as for
the onion-peeling solution.

The error components are shown in Fig. 8. For this retrieval,
the total errors are minimized at all levels, and it can be seen that
the measurement and null-space errors are similar in magnitude. The
effects of temperature and pressure uncertainties are not affected
very much. The maximum errors are now less than 0.25 ppmv. Overall,
these errors are similar to those of the Twomey solution with $\gamma =$
10^{-3}, except that the maximum near 25 km is handled better, and the
errors are smaller in the mesosphere.

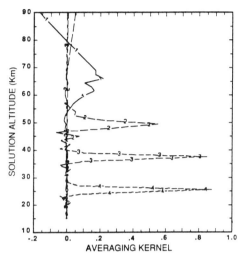

FIGURE 7.   Examples of averaging kernels for the maximum likelihood solution.

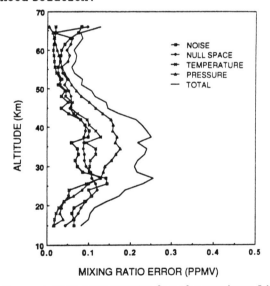

FIGURE 8.   Ozone error components for the maximum likelihood solution.

6.   CONCLUSIONS

These calculations have dealt only with the random errors of a broad-band infrared limb sounder. They show that the onion-peeling solution has the highest vertical resolution, but is unstable to noise, especially at high altitudes, where the signal to noise ratio is low. Simple smoothing, or a Twomey constraint, reduces the error component due to measurement noise, but at the expense of vertical resolution and therefore null space error. The use of the maximum likelihood solution systematically leads to greater smoothing and loss of resolution in regions of low S/N, but does not greatly affect the resolution or solution when the S/N is high. In this case, the effects of measurement noise and null-space error are comparable in size. The effects of random errors in the temperatures and registration pressure are not much affected by the retrieval method, but the latter may be a major source of random ozone error.

However, it must be pointed out that this has been a limited study, and much of the input data is somewhat uncertain. At this time, these results should only be taken as illustrative. It should also be noted that in using the maximum likelihood retrieval, it is necessary to be careful that the a priori statistics do indeed faithfully represent the atmospheric covariances.

REFERENCES

Gille, J. C., and J. M. Russell III, 1984: The Limb Infrared Monitor of the Stratosphere: Experiment Description, Performance and Results. *J. Geophys. Res.*, 89, 5125-5140.

Gille, J. C., J. M. Russell III, P. L. Bailey, L. L. Gordley, E. E. Remsberg, J. H. Lienesch, W. G. Planet, F. B. House, L. V. Lyjak, and S. A. Beck, 1984: Validation of Temperature Retrievals Obtained by the Limb Infrared Monitor of the Stratosphere (LIMS) Experiment on Nimbus 7. *J. Geophys. Res.*, 89, 5147-5160.

Rodgers, C. D., 1976: Retrieval of Atmospheric Temperature and Composition from Remote Measurements of Thermal Radiation, *Revs. Geophys. Space Phys.*, 14, 609-624.

Rodgers, C. D., 1988: A general error analysis for profile retrieval. In A. Deepak and H. Fleming (Eds.), *Remote Sensing RetrievalMethods*, A. Deepak Publishing (this volume).

Rodgers, C. D., and C. D. Walshaw, 1966: The Computation of Infrared Cooling Rates in Planetary Atmospheres. *Quart. J. Roy. Meteor.Soc.*, 92, 67-92.

Twomey, S., 1963: On the Numerical Solution of Fredholm Integral Equations of the First Kind by the Inversion of the Linear System Produced by Quadrature. *J. Assoc. Comput. Mach.*, 10, 97-101.

DISCUSSION

Rodgers: John, I'm not sure whether you said or whether I missed it, what did you use for your covariance matrix in computing the null-space error?

Gille: It was calculated from all the profiles for January, including both Northern and Southern hemisphere data, so it covers a broad range of seasonal and latitudinal variation. In that way, it is probably too pessimistic for a particular location and season.

King: John, I noticed all your averaging kernels were symmetric and if you look at weighting functions, it is true that in limb-viewing they are skinny and very pointed, but they really are not symmetric in the sense that they fall off much more sharply on the downside, right?

Gille: The averaging kernels for the simple onion peeling are symmetric but for the others they are not identically symmetric. The top-most averaging kernel for maximum likelihood showed it most clearly but some of the others have small negative lobes that differ above and below the peak.

King: My comment may not be pertinent then, but, if you have a weighting function that shows the usual behavior of growth, if you go to the limit, asymmetry persists. So the limiting expression really involves not a delta function, which is a symmetric function, but rather the sum of the delta function and the derivative of the delta function. If this is really a factor, which I'm not sure it is, it would show up by not taking into account this asymmetry. There should be some systematic bias in your inference and you indicated that your errors are random. What I am saying is that you should be finding a systematic bias that is too large up at the top and too small at the bottom in your inferencing techniques.

Gille: The weighting functions are shown in Figure 1 of the paper. As you say, they tend to increase with decreasing altitude, reach a peak just above the target weight, and drop off. They are very asymmetric. These are included in the matrix. You must multiply these by Clive's (Rodgers) matrix to get the matrix whose rows are the averaging kernels. The weighting functions and the averaging kernels are two different things. With regard to the second part of your question, the analysis shown here focuses on random errors, in a heuristic. Systematic errors would be shown by different diagnostics. As long as we assume that the measurements, transmittance modeling, and a priori statistics are unbiased, the solutions should not show a systematic bias.

THE FEASIBILITY OF 1-DEG/1-KM TROPOSPHERIC RETRIEVALS

Arthur C. Neuendorffer
National Oceanic and Atmospheric Administration
National Environmental Satellite, Data, and Information Service
Washington, D.C. 20233, USA

ABSTRACT

The next generation of tropospheric sounder will most likely consist of an infrared spectrometer or interferometer with a spectral resolution approaching one part in 1200. The current designation for this type of instrument is Advanced InfraRed Sounder or AIRS. What is the prospect of such an instrument significantly improving upon the current 2 deg RMS sounding accuracy? Can this or any other type of sounder be relied upon for 1 deg RMS accurate soundings as is desired by the meteorological community?

A simple physical analysis leads to the conclusion that the AIRS instrument is adequate for sounding the troposphere to 1 deg accuracy or better. However, AIRS is primarily a dew point depression instrument. Most, though by no means all, of its utility as a tropospheric thermal sound comes indirectly from its remarkable sensitivity for monitoring dew point depression. New and radical methods of retrieval will need to be developed if this powerful new tropospheric sounding device is to be fully utilized.

1. INTRODUCTION

The troposphere is nominally only about 11 km thick with a pressure variation from top to bottom of less than a factor of 5. Moreover, it is permeated with clouds and other aerosols. These two facts greatly limit our capacity to radiometrically "sound" the troposphere. The proposed Advanced InfraRed Sounder (AIRS) would monitor the earth's infrared radiation at a resolution of one part in 1200 and would represent an important advance in the utilization of high horizontal resolution, passive, monochromatic, nadir radiance measurements.

Considering all the difficulties involved in sounding the troposphere, however, what sort of improvements can one realistically expect from such an instrument? In particular, can the "all weather" 2 to 3 degree RMS temperature profiling accuracy of the Advanced Microwave Sounding Unit (AMSU) be enhanced sufficiently to allow for "clear column" 1 degree RMS levels of accuracy? This paper is an attempt to critically evaluate the AIRS concept as it pertains to these issues.

2. MEASUREMENT ACCURACY VS. VERTICAL RESOLUTION

Profiling accuracy requires a combination of measurement accuracy and good vertical resolution. A well-aligned, narrow field of view, low noise instrument with multiple redundant channels should assure good measurement accuracy under most circumstances. While AIRS is designed to have all these characteristics it would be unrealistic and unwise to attribute abnormally low noise values to it. Good vertical

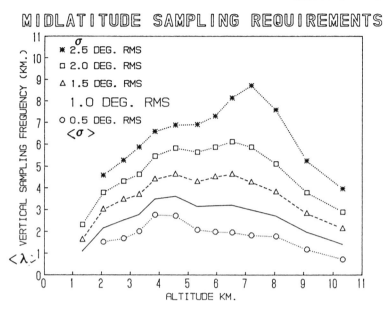

FIG. 1   Anticipated vertical resolution required for given RMS
          retrieval accuracy.

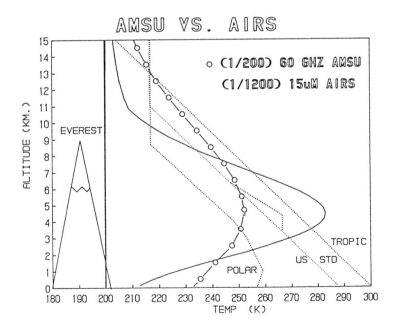

FIG. 2   Typical monochromatic weighting functions for infrared
          and microwave.

resolution is a more concrete issue with which to deal.

The importance of vertical resolution on profiling accuracy is clarified in Fig.1. This figure indicates how coarsely a tropospheric profile may be sampled and still allow for interpolated intervening thermal values of a given RMS accuracy. (A representative all-season ensemble of 400 temperate tropospheric profiles was used for this study.) A crude linear relationship is seen to exist between the sampling rate/scale (λ) and the corresponding RMS accuracy (σ): $\sigma = C\lambda$. with $C = 1$ deg per 3 km in the middle of the troposphere. The linearity is indicative of a variance spectra (of the vertical temperature structure) that falls off cubically with spatial wavenumber:

$$V = \sigma^2 = C^2\lambda^2 = C^2/k^2$$
$$\text{Variance Spectra} = dV/dk = C^2/k^3$$
$$\text{"Amplitude Spectra"} = C/k^{3/2}$$

The rapid high wavenumber drop-off implies that the fine vertical temperature structure is either weak or rare (or both.) Hence, one can simply disregard such structure and still achieve adequate (2 to 3 deg) RMS sounding accuracy. However, we can not continue ignoring fine scale structure indefinitely; some attempt must be made to capture this elusive structure with sensitive IR instrumentation. If narrow field of view but carefully aligned IR channels are used, cloud effects need not be a concern since spectrally "(aligned) cloud noise" falls off even faster (i.e., inverse forth power) than cubically. However, if 1 deg RMS sounding accuracy is required, the sounding instrument must be capable of resolving /retrieving 3-6 km features in accordance with Fig. 1.

3. DRY GAS SOUNDERS

3.1 MICROWAVE / AMSU

IR instruments have a dramatic advantage over microwave instruments in resolving/retrieving 3-6 km temperature features. The reason for this is plain to see in the Fig. 2 weighting function comparison. While 1/200 resolution monochromatic microwave channels have little trouble penetrating cloud features, their vertical resolution is not very good. Microwave can easily distinguish and interpolate between the main tropospheric classes (i.e., tropical/ standard/ and polar) but this can not properly be considered "sounding" the troposphere. Microwave weighting functions are essentially flat within the confines of an 11 km troposphere. Temperature profile features smaller than (9 km) Mt. Everest produce weak, washed out microwave signals which practical inversion algorithms are wise to ignore. If it weren't for the general universality of such tropospheric characteristics as 6.5 deg/km lapse rates and polar inversions, all-microwave tropospheric "soundings" would be useless.

The predictability of such characteristics enables a single microwave tropospheric temperature channel to produce retrievals of 3 deg RMS accuracy or better. Subsidiary information, such as a microwave /classification sounding of the tropopause and a broadband infrared measurement of near surface temperatures (Neuendorffer, 1985), brings "clear column" tropospheric retrievals into the 2 deg RMS accuracy range. However, sensitive monochromatic IR instruments

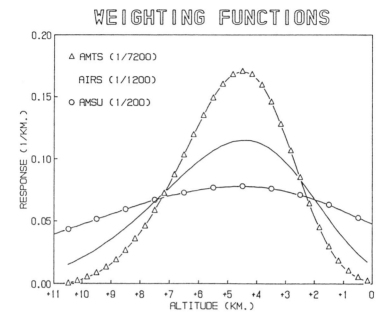

FIG. 3 Potential Infrared improvements in tropospheric
        weighting functions.

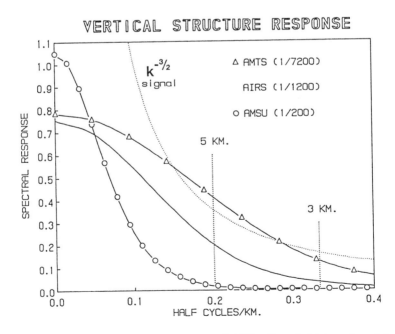

FIG. 4 Amplitude Fourier spectra of FIG. 3 weighting functions.
       3 to 5 km. spacial structure is dramatically suppressed.

will be required in order to break the 2 deg RMS barrier.

3.2 LONGWAVE IR / AIRS

The AIRS instrument will have sufficient spectral resolution (i.e., 1/1200) to resolve the valleys between longwave 15um CO_2 lines. AIRS monochromatic weighting functions are considerably narrower than their microwave counterpart due to 1) a positive temperature sensitivity of the absorption, 2) a moderately strong "Planck effect", and 3) the fact that (unlike 60 GHZ lines) most IR lines are cleanly separated and distinct at tropospheric pressures.

The AIRS longwave channels should have no difficulty in resolving down to the 5 km level and thereby knocking the RMS error down to about 1.5 deg. This is a considerable improvement upon all microwave retrievals. Improvements beyond on this level of accuracy, however, are probably only feasible in highly idealized conditions.

3.3 SHORTWAVE IR / AMTS

AIRS has 4.3 um CO_2/N_2O shortwave channels but none of the tropospheric shortwave lines are close to being resolved. A monochromatic 4.1 um shortwave Advanced Moisture and Temperature Sounder (AMTS) has been proposed (Kaplan et al., 1977) but it would require (in its original pristine form) an additional factor of 6 in resolving power. AMTS weighting functions would be cleaner (vis-a-vis ozone and water vapor absorption) and narrower than AIRS. It is doubtful, however, that signal to noise of a dozen narrow AMTS shortwave channels could be sufficiently enhanced to outperform the multi-channel longwave AIRS system. Certainly, AMTS would work poorly during daytime hours since solar contamination for all but the warmest channels would be horrendous.

3.4 RESOLUTION COMPARISON

It is instructive to compare (Fig. 3) the different weighting functions (as defined by the brightness temperature response to a 1 deg/ 1 km variation about the US Standard atmosphere.) The amplitude Fourier Transforms of these weighting functions (shown in Fig. 4) give some indication of how badly the weak 3-5 km thermal temperature structure is suppressed by convolution with smooth weighting functions:

3-5 km structure is simply obliterated by AMSU.

AIRS and AMTS pick up this structure but at a reduced level. The reduced signal is most likely unrecoverable above their respective noise characteristics (, not shown.)

The real advantage of the dry gas IR systems is in their capacity to retrieve 5 km tropospheric features. 5 km resolution and 1.5 deg RMS accuracy is no insignificant achievement for a tropospheric sounder and would represent a considerable improvement over current sounders. And while there are other possibilities for dry gas IR systems (e.g., another increase of 6 in resolving power would allow one to move up thermally sensitive 2390 cm-1 linewings for a "super AMTS,") there is no convincing evidence for a conceptual breakthrough for any post-AIRS system.

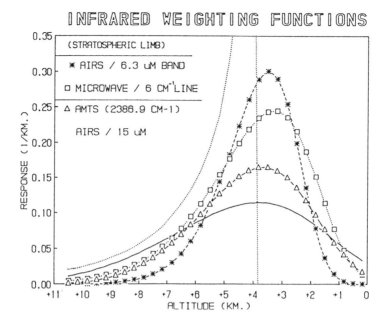

FIG. 5 The narrow thermal response of monochromatic water
        vapor channels.

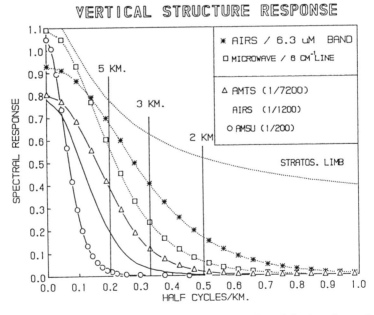

FIG. 6 Amplitude Fourier spectra of FIG. 5 weighting functions.
        3 km. spatial strucure is now "retrievable."

4. WET CHANNELS / AIRS and the 6.3 um BAND

4.1 THERMAL STRUCTURE

Despite the problems inherent in direct tropospheric thermal soundings, I believe that space based 2-5 km retrievals (and thus 1 deg RMS accuracy) are indeed possible; furthermore they are attainable with the AIRS system itself. Most (about 500) of AIRS sounding channels are not in the narrow dry CO_2 bands of 15um and 4.3um but in the broad 6.3 um water vapor band. AIRS spectral resolution of 1/1200 is more than adequate in providing monochromicity for this spectral band. The band has a strong Planck advantage and a good temperature sensitivity.

Above all, 6.3 um channels have the advantage of a rapidly decreasing tropospheric water vapor distribution with height. Because of a strong saturation sensitivity, tropospheric H_2O mixing ratios generally vary as a cubic power of pressure or better. Figs 5 & 6 show that (under nominal conditions) the resulting narrow 6.3 um thermal weighting functions are far more sensitive to 2-5 km temperature structure than any conceivable dry gas sounder. Fig. 6 indicates that over half the 3-5 km thermal signal is retained in the AIRS midwave water vapor band radiances. At the same time, midwave solar and instrumental noise should be minimal.

4.2 DEW POINT STRUCTURE

The problem with using the 6.3 um band, of course, is that its sensitivity to fine thermal structure is matched by its sensitivity to fine water vapor structure: Fig. 7 shows that AIRS' midwave response to (negative) dew point variation is much the same its response to temperature variation. The major difference is that the dew point weighting functions peak about a kilometer higher and are noticeably flatter than their thermal counterparts. If these channels are to provide our sole information on 2-5 km structure how do we determine when the signal corresponds to temperature structure and when to dew point structure?

4.3 DEW POINT DEPRESSION STRUCTURE

Two auxiliary facts clarify the dilemma of interpreting 6.3 band AIRS' radiance signals:

1) 2-5 km dew point structure generally exceeds temperature structure by factors of 3 or more.

2) In most clear column (IR retrievable) situations 2-5 km dew point structure is weakly but distinctly mirrored in the temperature structure.

Hence, despite their somewhat higher thermal sensitivity, AIRS' midwave channels are driven by the stronger dew point structure. More precisely, these channels are sounders of dew point depression with weighting functions close to those for (negative) dew point though somewhat biased towards the thermal. Although AIRS' dew point depression weighting function of Fig. 8 are not quite as sharp as the corresponding thermal one of Fig. 7, this deficit is more than made up for by the enormous amount of strong 2-5 km dew point depression structure

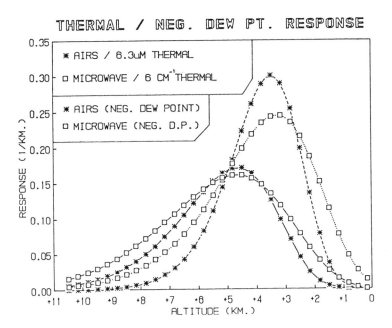

FIG. 7   Dew point depression weighting functions. The two terms
         compliment each other in cases of subsidence.

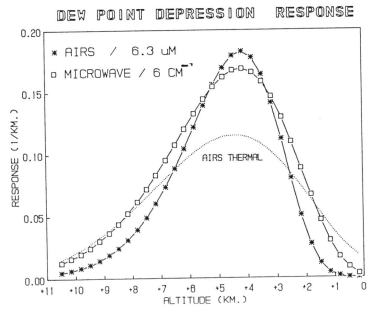

FIG. 8   The dominant water vapor response to dew point features
         still has better resolution than pure thermal channels.

waiting around to be measured. Even the most crudely conceived of possible AIRS systems would provide a wealth of tropospheric dew point depression information. This high vertical resolution data could be used to determine the closely related dew point structure and (to a lesser extent) temperature structure.

The Washington D.C. profile of Fig. 9 (Mastenbrook and Purdy, 1969) illustrates the type of 3 km dew point depression features that AIRS should have no trouble detecting. Two of the three subsidence features (i.e., a & b) display characteristic temperature structure in conjunction with a more easily sensed moisture structure. Predicting such temperature features from the midwave radiances may turn out to be more art than science but it is the type of art that needs to be actively developed. At the very least, the strong, information-rich 6.3 um (dew point depression) signals could be used to tell us when and where to expect fine thermal profile structure so as to provide a priori information for a longwave AIRS dry gas sounding.

It should be noted that a 6 cm⁻¹ (183 GHZ) microwave sounder could also retrieve the central subsidence feature (b) in Fig. 9, but it could not handle the upper or lower features very well. The upper troposphere is too dry for the 6 cm⁻¹ H_2O line to sound and the lower troposphere (just above the mixing layer) is too confused by microwave surface reflections.

CONCLUSION

All the information we need for 1 deg RMS soundings are contained in the AIRS radiances. The tropospheric radiance signal is particularly robust in the mid-wave 6.3 um band. In view of this, it seems whatever limitations AIRS has in sounding the troposphere are less physical than they are conceptual. AIRS is more than capable of monitoring the troposphere from space but it will be up to the ingenuity of the retrieval expert to disentangle the radiance data to produce viable sounding information.

REFERENCES

Kaplan, L.D., M.T. Chahine, J. Susskind, and J.E. Searl, 1977: Spectral band passes for a high precision satellite sounder. Appl. Opt.,16, 322-325.

Mastenbrook, H.J. and D.R. Purdy, 1969: Concurrent measurements of water vapor and ozone over Washington, D.C., During 1967 and 1968. NRL Report 6974, Naval Research Laboratory, Washington, D.C. 20233.

Neuendorffer, A., 1985: Remote sounding of the lower troposphere (700-1000mb). In A. Deepak, H.E. Fleming and M.T. Chahine (Eds.),Remote Sensing Retrieval Methods, A. Deepak Publishing, Hampton, Virginia 23666-1340.

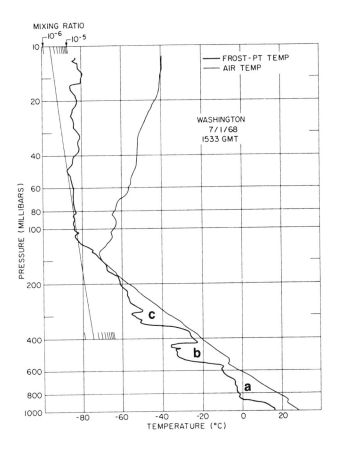

FIG. 9   Summertime midlatitude profile from Mastenbroodk and
Purdy. Three 3km. subsidence features represent the
type of dew point depression structure AIRS is
capable of retrieving.

DISCUSSION

McClatchey: I guess what I have is a comment. The comment is that several years ago I remember looking at the question of water vapor retrieval and concluding two things: one, in order to get water vapor information you had to know the temperature profile first, and two, that the water vapor profiling was far more sensitive to errors than temperature profiling was. And with those two concepts in mind, it sounds to me very much like a bootstrap operation you are proposing. And I guess I am not convinced it can work using the water vapor region to get temperature information without external information from some other part of the spectrum, probably the CO_2 bands to begin with.

Neuendorffer: I concur that you are going to need external information. But it could be that climatological information will be sufficient. I am not saying it is insufficient, I am saying that this sort of a temperature structure you are probably never going to see with the temperature sounder, a dry gas sounder. But, the combination of dry and wet sounders is going to produce a very strong signal; we are talking about an 8-degree signal in the water vapor channels and easily resolvable. That is where the signal is if you can extrapolate it out. I don't see any real problem with determining the water vapor. The problem with water vapor in the past has been that we've only had two or three water vapor channels whereas there is clearly much more information there. So, I think it is going to take at least a dozen or twenty water vapor channels up there, good water vapor channels. Monochromatic preferably, and seeing what you can do with it in a real life situation.

McClatchey: What I recall though, and this is from some work that we did on the blackboard quite a few years ago, is that we concluded that an error in the temperature reflected itself in three times the error in a water vapor inversion in the 6.3 micron region. I can't prove that statement right now, but it seems to me that is provable. If that is the case, you've got to know the temperature or bootstrap your way with the same channels (i.e., the H_2O channels) somehow to get the temperature information.

Neuendorffer: Perhaps, I mean you certainly have to know essentially what the lapse rate is. But I don't see that as being a big problem. The temperature is almost constant compared to dew point and the dew point temperature just dominates everything because the variations in it are so great. You could essentially draw straight lines through that temperature and I think you would do just as good a retrieval on dew point as if you have the actual temperature profile. But I don't know I could be wrong, I haven't done it.

Gille: Have you done any simultaneous temperature and water retrievals, I think that may be what Bob was getting at.

Neuendorffer: Right. No I haven't gotten into that area yet. I am just looking at the feasibility of seeing the structure to begin with and it seems like the structure is in the water vapor channels and if you can't retrieve them you can't retrieve them. But that is where to look. I mean, maybe I'm not going to be able to figure it out or maybe somebody else isn't going to be able to figure it out, but that is where the information is and someday somebody may be smart enough to figure it out.

Speaker: Might that not be where the value of the simultaneous temperature and water vapor retrieval comes in, and that is something that will be included, as I understand it in the new AIRS instrument?

Neuendorffer: Yes, if it is simultaneous in the sense that you actually use the variation of the water vapor to define the variations in temperature. But I don't know what is sufficient. I don't know how far you have to go or how little you have to go to retrieve the temperature information.

Speaker: As I am sure you know, Bill Smith has done some retrievals simultaneously for water and temperature.

Neuendorffer: Yes, I've seen his results and like Henry's results, I am convinced that he is getting his information from the water vapor channels.

Speaker: Yes, and I think he would agree with that.

AN EXTENSION OF THE SPLIT WINDOW TECHNIQUE FOR THE RETRIEVAL OF PRECIPITABLE WATER: EXPERIMENTAL VERIFICATION

Thomas J. Kleespies
Air Force Geophysics Laboratory
Hanscom Air Force Base
Bedford, Massachusetts 01731, USA

Larry M. McMillin
National Oceanic and Atmospheric Administration
National Environmental Satellite, Data, and Information Service
Washington, D.C. 20233, USA

ABSTRACT

A method is presented to estimate the ratio of the transmittances in the two channels of the 11-μm split window. This method uses only the radiances and requires no *a priori* information. A previously published algorithm is then applied to this transmittance ratio to arrive at an estimate of total precipitable water.

1. INTRODUCTION

The "split window" technique was originally derived for the determination of surface skin temperature, specifically sea surface temperature (Anding and Kauth,1969) . The technique makes use of two differentially absorbing channels in the 11- to 12- μm region to remove the contaminating effect of water vapor and thus arrives at an improved estimate of the skin temperature. See McMillin and Crosby(1984) for a detailed discussion of the split window technique and an extensive review of the literature.

More recently the channels used for the split window have been applied to the retrieval of precipitable water (Chesters et al., 1983) , Chesters et al., 1987) . Whereas these methods seemed to produce internally consistent fields of "low level water vapor", they required *a priori* knowledge of the mean air temperature and empirical adjustment of the absorption coefficients in order to bring the results in agreement with in situ observations.

In this paper we present the results of an extension to the split window technique such that precipitable water can be retrieved with a minimum of *a priori* information.

RSRM '87: ADVANCES IN
REMOTE SENSING RETRIEVAL METHODS
A. Deepak, H.E. Fleming, and J.S. Theon (Eds.)

2. THEORETICAL DISCUSSION

Kleespies and McMillin (1984,1986) have presented a theoretical discussion of this extension to the split window technique. Summarized briefly, the upwelling longwave infrared radiance emitted from a plane parallel, non scattering atmosphere in local thermodynamic equilibrium can be expressed as

$$I \ = \ B_s \tau_s + \int_{\tau_s}^{1} B \ d\tau \tag{1}$$

where I is the radiance measured by the satellite, B is the Planck radiance, τ is the transmittance from a given level to the top of the atmosphere, the subscript s refers to the surface of the earth, and the integral is the radiance originating from the atmosphere alone. Equation (1) may also be written as

$$I \ = \ B_s \tau_s + \bar{B}_a (1-\tau_s) \tag{2}$$

where $\bar{B}$ is a weighted average given by

$$\bar{B}_a = \frac{\int_{\tau_s}^{1} B \ d\tau}{\int_{\tau_s}^{1} d\tau} \tag{3}$$

Consider observations of the earth under conditions where the surface contribution to the outgoing infrared radiance varies markedly, but where the atmospheric contribution changes very little. We can now write a set of four equations, one for each of the two channels, and one for each of the different surface observing conditions:

$$I_{11}^1 \ = \ B_{s_{11}}^1 \tau_{s_{11}} + \bar{B}_{a_{11}} (1-\tau_{s_{11}}) \tag{4a}$$

$$I_{12}^1 \ = \ B_{s_{12}}^1 \tau_{s_{12}} + \bar{B}_{a_{12}} (1-\tau_{s_{12}}) \tag{4b}$$

$$I_{11}^2 \ = \ B_{s_{11}}^2 \tau_{s_{11}} + \bar{B}_{a_{11}} (1-\tau_{s_{11}}) \tag{4c}$$

$$I_{12}^2 \ = \ B_{s_{12}}^2 \tau_{s_{12}} + \bar{B}_{a_{12}} (1-\tau_{s_{12}}) \tag{4d}$$

where the superscripts 1 and 2 refer to the viewing conditions and the subscripts 11 and 12 refer to the nominal 11 and 12 micrometer channels in the split window. We can eliminate the atmospheric term $\bar{B}_a$ by differencing to yield two equations

$$\Delta I_{11} = \Delta B_{s_{11}} \tau_{11} \tag{5a}$$

$$\Delta I_{12} = \Delta B_{s_{12}} \tau_{12} \tag{5b}$$

where for compactness we have written the delta quantities as

$$\Delta I_{11} = I_{11}^1 - I_{11}^2 \tag{6a}$$

$$\Delta B_{11} = B_{11}^1 - B_{11}^2 \tag{6b}$$

The ratio of transmittances in the two channels may be formed by dividing Eqs. (5) to yield

$$\frac{\tau_{11}}{\tau_{12}} = \frac{\Delta I_{11} \, \Delta B_{s_{12}}}{\Delta I_{12} \, \Delta B_{s_{11}}} \tag{7}$$

Following the approach of McMillin (1971) , Eq. (7) can be linearized by converting from radiances to temperatures, the ΔB_s become ΔT_s and cancel, and after expanding the delta quantities we are left with

$$\frac{\tau_{11}}{\tau_{12}} = \frac{T_{11}^1 - T_{11}^2}{T_{12}^1 - T_{12}^2} \quad . \tag{8}$$

It has been shown that this ratio can be related to "low level water vapor" i.e., precipitable water (Chesters et al. 1983) .

The observing conditions under which the surface contribution to the upwelling radiances can change markedly but the atmospheric contribution can change very little fall into two general categories; that of variation in time, and that of variation in space. Consecutive observations of a land surface from a geosynchronous satellite during the heating cycle of the day would be one example. Another would be observations from either a geosynchronous or polar orbiting satellite of immediately adjacent land and water surfaces with contrasting skin temperatures.

Kleespies and McMillin (1984) discuss the theoretical application of this extension to the Visible Infrared Spin Scan Radiometer (VISSR) Atmospheric Sounder (VAS), and in their 1986 paper describe, again in theoretical terms, its application to Advanced Very High Resolution Radiometer (AVHRR) split window data. In ensuing sections, application of this technique to these two instruments is demonstrated with real data.

3. APPLICATION TO SATELLITE RADIOMETER DATA

The real test of a retrieval algorithm is to apply it to real data and to somehow verify it with ground truth. However this is fraught with difficulties, including cloud contamination, aerosol problems, collocation inaccuracies, and errors in the satellite instrument and the in situ measurements. In the following sections we apply this technique to measurements made with the VAS and the AVHRR. In all cases of comparison with radiosonde transmittance ratio, the transmittance ratio was computed from collocated radiosondes using the wide-band radiative transfer model described by Weinreb and Hill (1980) .

3.1 APPLICATION TO THE VISSR ATMOSPHERIC SOUNDER

Observations were made with the VAS channels 7 and 8 (12.7- and 11.2- µm respectively) over North America on 25 August 1987. Multispectral imagery were acquired on the AFGL Interactive Meteorological System (AIMS) (Gustafson et al, 1987) at hourly intervals from 11:30 UT to 17:30 UT. This works out approximately from just before local sunrise in the mid-United States to just after local noon on the east coast. In order to achieve the contrasting surface contribution to the outgoing radiance, while minimizing the effects of surface obscuration due to convective cloudiness in the local late morning and early afternoon, a variety of time intervals were tested with the optimal interval subjectively selected to be from 1130-1530 UT. Typical brightness temperature increases due to diurnal heating of the surface during this interval were 5-10 K. Since the VAS oversamples along the scan line by a factor of 2:1 for band 8 and 4:1 for band 7, and the band 7 detector linear dimensions are twice that of band 8, a total of 6 pixels from band 7 were averaged to make one band 7 "retrieval spot" and 6 pixels from each of two lines of band 8 (a total of 12 pixels) were averaged to make one band 8 "retrieval spot". If the average brightness temperature in either channel were less than 273 K at either of the two times, the retrieval spot was assumed to be cloudy and was discarded. Furthermore, if the standard deviation of the brightness temperatures that went in to making the retrieval spot were greater than about 1.5 K for either time, the spot was assumed to be partly cloudy and was discarded.

The transmission ratio was computed for each non-discarded retrieval spot using Eq. (8). Examination of pseudo-imagery of this transmission ratio revealed that while interesting mesoscale features were apparent in this imagery, further spatial averaging was required before quantitative comparisons could be attempted.

Collocations were performed on launch sites of the 1200 UT radiosondes on 25 Aug 1987. Statistics were computed on the non-discarded transmittance ratios from a 9x9 box of transmittance ratios centered on the radiosonde site. Those transmittances outside of one standard deviation from the mean of the ratios were discarded and the mean was recomputed. This filtered mean was compared with the radiosonde only if the the resulting number of retrieval spots exceeded ten. The comparison between the transmittance ratio derived from the VAS and that derived from the radiosonde is presented in Figure 1.

Precipitable water was computed from the mean ratio by adapting the method of Chesters et al. (1987) to this problem

$$PW = 1/\Delta\alpha \left\{ 1/\sec\theta \ \ln[\tau_{11}/\tau_{12}] - \Delta\kappa \right\} \qquad 9)$$

where $\Delta\kappa = 0.051$ and $\Delta\alpha = .136$ are differential absorption coefficients due to mixed gases and water vapor respectively, θ is the local zenith angle and the τ's are the transmittances. The comparison between the precipitable water derived from Eq. (9) and the radiosonde precipitable water is given in Figure 2.

Following the suggestion of Chesters et al. (1983), a qualitative evaluation of the precipitable water product was made. Precipitable water pseudo-imagery were generated from the 25 August 1987 dataset. Equation (9) was applied to each 'sounding spot' and the amount of precipitable water was assigned a particular color. The precipitable water imagery from the 1130-1530 interval and from the the 1330-1730 interval are displayed in Figure 5, accompanied by visible imagery from 1330 UT. It is important to note that the radiances used to generate the precipitable water imagery for these different intervals are separate and independent. The latter interval contained some small scale convective cloudiness which contributed to high frequency noise.

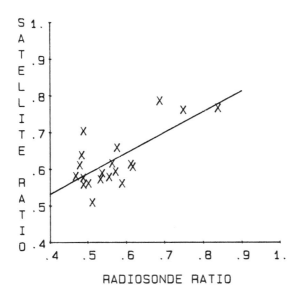

Figure 1. Radiosonde transmittance ratio versus transmittance ratio computed from VAS radiances from 20 collocations over North America. Dimensionless Units. Correlation is .7193. Line of best fit Y=0.3+.56X .

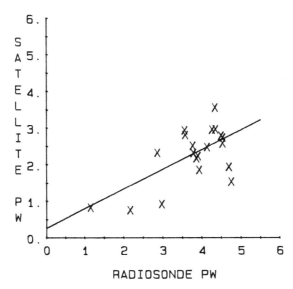

Figure 2. Radiosonde precipitable water versus VAS precipitable water from 20 North American collocations. Units are cm of precipitable water. Correlation is 0.6387. Line of best fit Y = 0.25 + .54X.

Examination of these imagery indicate that there is considerable spatial and temporal coherence to the water vapor features, but that the scale of some of these water vapor features is quite small, on the order of 100-200 km. This leads us to believe that some of the variability in the more objective comparison given in Figure 2 is due to miscollocation of the radiosonde with the the satellite observed features.

3.2 APPLICATION TO THE AVHRR

Global Area Coverage (GAC) data from the AVHRR were collected from NOAA-7 for 11 June 1982. GAC data has a nominal resolution of 4 km and is distinguished from the nominal AVHRR sensor 1 km resolution by the fact that four pixels are averaged along scan and four scans are skipped to make a GAC scan line. Bands 4 and 5 have the nominal wavelength of 10.7- and 11.8- μm respectively which correspond only approximately to those of the VAS (see Figure 3). Nighttime data over North America was used in order to be as close as possible to radiosonde launch time and in order to avoid convective cloudiness. The orbits were from four hours to one hour prior to synoptic time. Cloud free areas were selected at the AIMS workstation by examining 24-bit multispectral imagery created from AVHRR bands 3, 4 and 5 (d'Entremont and Thomason, 1987) . In this imagery opaque clouds appear white, low clouds and fog appear bright red against a brown background, and thin cirrus appears cyan, yielding a fairly unambiguous rendition of clear/cloudy regions. Contrasting surface temperatures were determined by selecting a body of water (lake, river, coastline) which at nighttime was relatively warm compared to the surrounding countryside. A 3×3 array of GAC spots were selected for both the warm water surface and the cooler countryside surrounding it. Since many of these water surfaces did not fill the 3×3 array of GAC pixels, a method was developed to determine the "best" combination of warm and cold brightness temperatures. A comparison of two ensembles of 3×3 arrays yields 81 possible combinations. The brightness temperature differences between these 81 combinations were sorted for each of the two channels. The sum of the rank order of the pixel pairs between the two scenes and the two channels was used as a quality measure, the idea being to maximize the brightness temperature difference between the warm and the cold scenes for both channels. Since there is a danger that sub-pixel cloudiness in one scene but not the other which would contribute to an excessively high quality measure, the transmission ratios computed from Eq. (8) were averaged for the top 10 quality measures. Typical brightness temperature differences were about 5 K.

The method presented by Chesters et al. (1983) for computing precipitable water from transmission ratio applies only to the VAS instrument. Whereas it can be adapted to the AVHRR by recomputation of the differential absorption coefficients, time limitations for publication in these proceedings allow only an indirect comparison with in situ measurements. Rather than compare the radiosonde precipitable water with AVHRR precipitable water using an invalid relationship between precipitable water and transmission ratio, it was deemed most prudent to compare directly between the satellite and the radiosonde transmission ratios.

Radiosondes from 1200 UT on 11 June 1982 over North America were collected from the AFGL McIDAS upper air archive. Collocations were made to the closest radiosonde within 300 km of the satellite observation. The comparison between the modeled transmittance ratio and that computed from Eq. (8) for the AVHRR is given in Figure 4.

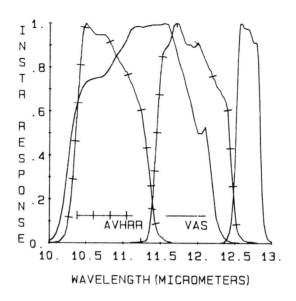

Figure 3. Relative instrument response for VAS and AVHRR split window channels.

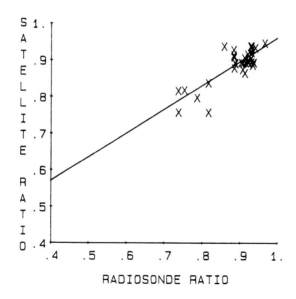

Figure 4. Radiosonde transmission ratio versus AVHRR transmission ratio for 36 North American collocations. Dimensionless units. Correlation is 0.8412. Line of best fit $Y = 0.28 + 0.69X$.

Figure 5a.   Figure 5b.

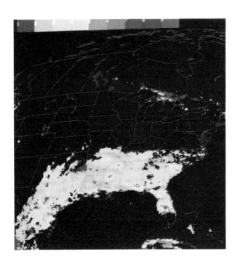 

Figure 5c.

Figure 5. Precipitable water pseudo imagery with accompanying visible
          imagery. Color bar indicates precipitable water amounts in 1 cm
          increments. 5a) Precipitable water estimated from 1130-1530 UT
          radiances. 5b) Precipitable water estimated from 1330-1730 UT
          radiances. 5c) Visible imagery from 1330 UT.

4. DISCUSSION

The correlation between the VAS transmittance ratio and that computed from collocated radiosondes is about 0.72 . The correlation between the AVHRR ratio and the radiosonde ratio is an even more respectable 0.84 . The precipitable water computed from the VAS transmittance ratio had a correlation coefficient of about 0.64 when compared with the radiosonde precipitable water. Chesters et al. (1983) reported a lesser correlation with their initial study, (r=0.43), even though they used ancillary information in the form of the atmospheric temperature averaged over a number of radiosonde sites. In their 1987 paper they substantially improved on this figure by using empirical adjustments to the absorption coefficients and by modifying the atmospheric temperature used in their algorithm. The method presented here requires no ancillary information. However, examination of Figures 1, 2 and 4 indicate that the line of best fit in each figure does not have a zero intercept or a slope of unity. It is clear that empirical adjustments are warranted to help remove the systematic bias which is evident in these figures.

Perhaps the greatest uncertainty in retrieving precipitable water from satellite observations lies in determination of the atmospheric absorption coefficients for these channels. Since these channels are quite broad and water vapor is a significant absorber in this region (the water vapor continuum is not well characterized), absorption uncertainties dominate errors in the retrieval process. The focus of Chesters et al. (1987) is the empirical adjustment of the absorption coefficients to optimize the precipitable water retrievals. Barton (1985) reports similar problems with his sea surface temperature determination.

Another major source of concern in application of this technique has to do with instrument noise. Since Eq. (8) deals with the ratio of a difference in the brightness temperatures, this method is quite sensitive to errors in the measurements. The VAS has a nominal NEΔT of 0.25 for these channels. The AVHRR has a NEΔT of 0.12 for its split window channels. Just on this figure of merit, the AVHRR would seem to be better suited for this purpose since a smaller brightness temperature difference would be required for Eq. (8) in order to minimize the effect of instrumental noise. However, from geometric considerations it is much easier to obtain the desired change in scene brightness temperature by observing diurnal temperature changes from geosynchronous orbit than searching for contrasting skin temperatures from polar orbit. The imager to be flown on GOES NEXT is very similar to the AVHRR, with planned NEΔT even better than the present AVHRR (Koeing,1987) . It is anticipated that with the launch of GOES NEXT, a thorough evaluation of this method can be made.

ACKNOWLEDGMENTS

The authors wish to thank Mssrs. Mike Weinreb and Mike Hill of NOAA/NESDIS for providing the radiative transfer code used in this study, and Dr. Dennis Chesters for providing a pre-publication copy of his 1987 article. The authors also express their appreciation to Mr. Robert d'Entremont of AFGL Satellite Meteorology Branch for his assistance with the AVHRR GAC data.

REFERENCES

Anding, D., and R. Kauth, 1969: Atmospheric Modeling in the Infrared Spectral Region: Atmospheric Effects on Multispectral Sensing of Sea Surface Temperature from Space, *Rep 2676-1-P, Willow Run Lab., Inst. Sci. Technol.*, Univ. Mich. Ann Arbor, Mich.

Barton, I. J., 1985: Transmission Model and Ground-Truth Investigation of Satellite-Derived Sea Surface Temperatures, *J. Clim. Appl. Met.*, Vol. 24, pp. 508-516.

Chesters, Dennis, Louis W. Uccellini and Wayne D. Robinson, 1983: Low Level Water Vapor Fields from the VISSR Atmospheric Sounder (VAS) "Split window" Channels. *J. Clim. Appl. Met.*, Vol 22, No. 5. pp 725-743.

Chesters, Dennis, Wayne D. Robinson and Louis W. Uccellini, 1987: A Note on Optimized Retrievals of Precipitable Water from the VAS "Split Window", *J. Clim. Appl. Met.*, Vol 26, No. 8. pp 1059-1066.

d'Entremont, Robert P. and Larry W. Thomason, 1987: Interpreting Meteorological Satellite Images Using a Color-Composite Technique, *Bull. Am. Met Soc.*, Vol 68, No.7, pp762-768.

Gustafson, G., D. Roberts, C. Ivaldi, R. Schechter, T. Kleespies, K. Hardy, R. d'Entremont, G. Felde and R. Lynch, 1987, The AFGL Interactive Meteorological System, *Preprints, Third International Conference of Interactive Information and Processing Systems for Meteorology, Oceanography, and Hydrology*, 12-16 January 1987, New Orleans, LA.

Kleespies, Thomas J. and Larry M. McMillin, 1984: Physical Retrieval of Precipitable Water Using the Split Window Technique, *Preprints, Conference on Satellite Meteorology/ Remote Sensing and Applications*, Clearwater Beach FL, Am. Met. Soc.

Kleespies, Thomas J. and Larry M. McMillin, 1986: An Extension of the Split Window Technique for the Retrieval of Precipitable Water. *Preprints, Second Conference of Satellite Meteorology/ Remote Sensing and Applications*, Williamsburg VA., Am. Met. Soc.

Koenig, E. W., 1987: GOES I Series Imager and Sounder, ITT Aerospace/Optical Division, 11pp.

McMillin, L. M., 1971: A Method of Determining Surface Temperatures from Measurements of Spectral Radiance at Two Wavelengths, *PhD. Dissertation*, Iowa State Univ., Ames.

McMillin, L. M. and D.S. Crosby, 1984: Theory and Validation of the Multiple Window Sea Surface Temperature Technique, *J. Geoph. Rsch.* Vol 89, No. C3, pp 3655-3661.

Weinreb, Michael P. and Michael L. Hill, 1980: Calculation of Atmospheric Radiance and Brightness Temperatures in Infrared Window Channels of Satellite Radiometers, *NOAA Technical Report NESS 80*, 40 pp.

DISCUSSION

Neuendorffer: The two systems you talked about are AVHRR and GOES. GOES kind of uses water vapor lines whose absorption doesn't increase quite as fast as the amount of water, whereas AVHRR is using the water vapor continuum that is water vapor broadened so it increases faster with water vapor. Do you find an advantage of one system over the other? Do you prefer one system over the other?

Kleespies: I'll tell you, in terms of day-to-day operation, the VAS is easier to deal with just because, as you saw from my slides, I can create precipitable water imagery and look for spatial and temporal coherence. With the AVHRR I have to look at places where the constraints of the observing condition requirements are satisfied and those situations are kind of few and far between. These considerations are purely for geometric reasons. From a spectral consideration, the VAS is better because the absorption by water vapor is stronger in the 13-μm channel. However, the noise of the VAS is much worse than the AVHRR.

Neuendorffer: What about surface emissivity? Is there any problem with that?

Kleespies: I don't think so, surface emissivity is a problem when you're dealing in arid regions. Some silica surfaces have very strong differences; it changes very rapidly in this region. However, I don't think we are going to be looking over deserts for a very strong precipitable water signature.

Smith: When you get around to modeling the absorption coefficients I urge you not to forget the chloroflourocarbons which have strong bands in these regions.

ON THE DYNAMIC ESTIMATION OF RELATIVE WEIGHTS FOR OBSERVATIONS AND FORECAST

Grace Wahba
Department of Statistics
University of Wisconsin-Madison
Madison, Wisconsin 53706, USA

ABSTRACT

We look at the problem of merging direct and remotely sensed (indirect) data with forecast data to get an estimate of the present state of the atmosphere, for the purpose of numerical weather prediction. To carry out this merging optimally, it is necessary to provide an estimate of the relative weights to be given to the observations and forecast. It is possible to do this dynamically from the information to be merged, if the correlation structure of the errors from the various sources is sufficiently different. We describe some new statistical approaches to doing this and quantify conditions under which such estimates are likely to be good.

1. INTRODUCTION

We have been studying various aspects of the problem of simultaneously combining information from various sources, for the purpose of obtaining initial conditions of the atmosphere for numerical weather prediction. By information, we mean data from diverse instruments, information from a forecast, prior information concerning the atmosphere, and physical constraints. See Wahba(1981,1982a, 1985a). Various parts of this what might be called "multispectral" point of view, whereby data from different sources are combined, and several meteorological parameters retrieved simultaneously, constitute a major theme in several papers presented at this conference, in particular, see Isaacs et al. (1988), Smith et al.(1988), Westwater et al. (1988) and Rodgers (1988). See also Lorenc(1986).

This work was sponsored by the National Aeronautics and Space Administration under Contract NAG5-315 and the National Science Foundation under Grant ATM-840373.

In this paper we will first briefly review the variational prescription for combining data from different sources and prior information, and its relation to Gandin (Bayes) estimation. In this prescription are a number of "tuning" parameters, which we will divide into two classes. The first class will be called weighting parameters, and the second smoothing (also known as bandwidth) parameters. The weighting parameters are those which govern the relative weights to be given to various types of observational, forecast, and physical information, while the smoothing parameters control the relative amount of "information" which is to be assigned to "signal" and to "noise". All practical forecast models have many such tuning parameters, and in practice, they are chosen by trial and error, by the use of externally measured data on various sources of error and strength of signal, and in very simple cases, by Kalman filtering, which propagates estimates of covariances forward in the model.

Under certain circumstances certain smoothing parameters can be estimated well from the data at hand (i. e. the data to be analyzed), by cross validation and generalized cross validation(GCV) methods. These methods sit somewhere between "static" estimation methods, with their long response times, and the Kalman filter methods, with their relatively large computational burden, and the possibility that the assumptions of the Kalman filter theory concerning error structure might not be satisfied for model errors in some circumstances.

It is the purpose of this paper to initiate the development of a complimentary theory for the dynamic estimation of certain weighting parameters, which govern the relative weight to be given to different types of observations and forecast. In practice, the estimates resulting from this theory have the potential for being used to decide which of two strongly conflicting sources of information (for example, forecast and satellite radiance data) should be given primary credence. This theory must of necessity have certain subtleties, since, if two instruments (or an instrument and forecast) measure the same quantity, each with a constant bias, the relative sizes of the two biases cannot be discerned from the observations (or observations and forecast). In order to have hope of carrying out this program, we shall see that the spatial (or temporal) error correlation structure from the two different sources has to be sufficiently different. Part of the goal of this theory is to quantify "sufficiently different" in a useful way. We believe that this circumstance of "sufficiently different" occurs in a number of meteorologically important circumstances.

After reviewing the general variational problem, we present the estimates, using 500mb heights from raobs and forecast as a concrete example. Then we give a theorem which quantifies "sufficiently different" and, finally, we describe the general case of both direct and indirect measurements of the same meteorological quantities.

2. VARIATIONAL AND GANDIN OBJECTIVE ANALYSIS

We suppose that $s_1, \ldots, s_p$ are the "state variables" in a global scale numerical weather prediction model, either a grid point or a spectral model. We will assume that IF the best possible values of the state variables were chosen, then the difference between the model atmosphere and the true atmosphere can be treated as "noise". The number p of state variables may be very large.

We will consider two vectors of information (think of these as direct or indirect observational data, or forecast), that are to be combined to obtain an improved estimate of the state variables.

Let $y^{(1)}$ and $y^{(2)}$ be vectors from set 1 and set 2 respectively and suppose

$$y^{(i)} = F^{(i)}(s) + \varepsilon^{(i)}, \quad i=1,2$$

where $y^{(i)}=(y_1^{(i)},\ldots,y_{n_i}^{(i)})$, $s=(s_1,\ldots,s_p)$, $F^{(i)}(s)=(F_1^{(i)}(s),\ldots,F_{n_i}^{(i)}(s))$, $i=1,2$ and we assume that the discrepancies $\varepsilon^{(i)}$ between $y^{(i)}$ and $F^{(i)}$ can be modelled as zero mean random vectors with $E\varepsilon^{(i)}\varepsilon^{(i)'}=\sigma_i^2 Q_i, i=1,2$, and $E\varepsilon^{(1)}\varepsilon^{(2)}=0$, that is, errors in the two data sources are uncorrelated.

$F^{(i)}$ may be a matrix (including the identity matrix) , or a nonlinear operator which models the forward problem arising when a radiometer observes radiant energy remotely, or it may model the relationship between directly measured quantities such as the horizontal wind field, and state variables in the forecast model such as the coefficients in a spherical harmonic expansion of stream function and velocity potential.

We will suppose that a reasonable model for the state variables (the mean having been subtracted off), is

$$Es = 0, \quad Ess' = b\Sigma.$$

See Wahba(1982b), for further references and a discussion as to how Σ may be obtained from historical data.

This approach results in a mandate to find s as the minimizer of the variational problem:

$$(y^{(1)} - F^{(1)}(s))Q_1^{-1}(y^{(1)} - F^{(1)}(s)) + \frac{1}{r}(y^{(2)} - F^{(2)}(s))Q_2^{-1}(y^{(2)} - F^{(2)}(s))$$

$$+ \lambda s'\Sigma^{-1}s \,,$$

where $r=\dfrac{\sigma_2^2}{\sigma_1^2}$ and $\lambda=\dfrac{\sigma_1^2}{b}$. See Wahba (1981, 1982a, 1985a), O'Sullivan and Wahba(1985). We note that physical constraints on the state vector s or functions of s can be inserted by solving the variational problem above subject to these constraints, see Villalobos and Wahba(1982), Svensson(1985). Svensson put constraints on the dry adiabatic lapse rate when solving a variational problem of this form with satellite radiance data to estimate vertical temperature profiles.

In the linear Normally distributed case the estimate of s which minimizes the above variational problem is the Gandin(Bayes) estimate of s, given the prior covariance of s and the covariance matrices of the errors. See, e. g. Kimeldorf and Wahba(1970). Kalman filter theory would give s as the minimizer of the variational problem with the term preceeded by λ absent. Here the inclusion of this term enters apriori (smoothness) information that may have been lost via model error.

If r, Q_1 and Q_2 are known, then λ and some parameters inside Σ may be estimated by the GCV. See, for example, Wahba and Wendelberger(1980), Wahba(1982b), O'Sullivan and Wahba(1985), Merz(1980).

3. DYNAMIC ESTIMATION OF WEIGHTING PARAMETERS

3.1 A SIMPLE EXAMPLE - 500 mb RAOBS AND FORECAST

We first illustrate the method and results by letting $y^{(1)}$ be observed 500mb heights and $y^{(2)}$ be forecast 500mb heights. It is reasonable to take Q_1 to be I for the observations since these measurements can be assumed to be independent, with about the same variance from station to station. Hollingsworth and Lonnberg(1986) and Lonnberg and Hollingsworth(1986) (LH), have recently obtained estimates of r and $Q_f=Q_2$ from three months data from the European Center forecast model. We will use their example and results as an illustration, before going on to the general case.

Figure 1 shows an estimated 500 mb correlation function from LH. Let h_l^o and h_l^f be the observation and the forecast 500 mb height at station l, and ε_l^o and ε_l^f be the observation and forecast errors, (on a particular day). Let

$$\xi_l = h_l^o - h_l^f = e_l^o - e_l^f$$

Here, $y^{(1)}=h^o$ and $y^{(2)}=h^f$; thus both vectors represent the same meteorological quantity: that condition will be relaxed later. Letting τ_{lm} be the distance between stations l and m, LH assumed that

$$E \varepsilon_l^f \varepsilon_m^f = \sigma_f^2 \rho(\tau_{lm})$$

where $\rho(0)=1$. If the e_l^o are independent and identically distributed zero mean random variables then

$$E \xi_l \xi_m = \sigma_1^2 \delta_{lm} + \sigma_f^2 \rho(\tau_{lm})$$

where $\delta_{lm}=1$ if $l=m$ and 0 otherwise. LH collected 90 days of values of ξ_l for each station. Letting j index day, they used sample correlations computed from

$$\frac{1}{90} \sum_{j=1}^{90} \xi_l(j)\xi_m(j)$$

as an estimate of $\sigma_1^2 \delta_{lm}+\sigma_f^2 \rho(\tau_{lm})$. In the figure, sample values of $\dfrac{\sigma_f^2 \rho(\tau)}{\sigma_1^2+\sigma_f^2\rho(0)}$ are plotted, and $\dfrac{\sigma_f^2}{\sigma_f^2+\sigma_1^2}$ is estimated by extrapolating the smooth part of the curve back to the origin by methods described in their paper (quite different than the methods to be discussed here.)

Figure 2 shows a one parameter family of (synthetic) correlation functions, defined by

$$\rho_L(\tau) = \frac{(1-2\theta(L)\cos(\frac{2\pi\tau}{R_o})+\theta^2(L))^{-1/2}-(1+\theta(L))^{-1}}{(1-\theta(L))^{-1}-(1+\theta(L))^{-1}}$$

where $\theta(L)$ is determined by

$$\frac{3}{2}\theta(L) - \frac{1}{2}\theta^3(L) = \cos\frac{2\pi L}{R_O}.$$

Here the parameter L in km is the distance τ for which $\rho_L(\tau)=\frac{1}{2}\rho_L(0)=\frac{1}{2}$, where R_O is the circumference of the earth in km. This family of (isotropic) correlation functions on the sphere has been chosen here partly because of a superficial resemblance to some of the curves obtained by LH and partly for mathematical convenience. We wish to use this family as a moderately realistic example of the estimation of a single (important) parameter in Q_f, namely, the correlation half-distance. Further study is needed to determine if it is appropriate to include other factors, such as anisotropy, variation with latitude, etc. in this correlation function. With this model, letting $\xi=(\xi_1,\ldots,\xi_n)$ be a vector of one day's data, we have that the covariance matrix of ξ is $\sigma_1^2(I + rQ_f(L))$, where the lmth entry of $Q_f(L)$ is $\rho_L(\tau_{lm})$. The GCV estimate of r and L can be shown to be the values of r and L which minimize

$$V(r,L) = \frac{\xi'(I+rQ_f(L))^{-2}\xi}{[\frac{1}{n}Trace(I+rQ_f(L))^{-1}]^2}$$

If ξ is assumed to have a multivariate normal distribution with zero mean, then the maximum likelihood estimates of r and L are the minimizers of

$$M(r,L) = \frac{\xi'(I+rQ_f(L))^{-1}\xi}{[\det(I+rQ_f(L))]^{\frac{1}{n}}}$$

Properties of these estimates are under study, and properties and derivations will appear elsewhere. While the ML estimate has various optimality properties if all of the assumptions of the model are satisfied, the GCV estimate may be more robust to model errors, see Wahba(1985b). These remarks are conjectural at the present time.

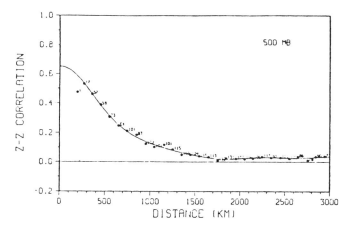

FIGURE 1 (From LH). The correlation of 500mb height forecast errors as a function of station separation.

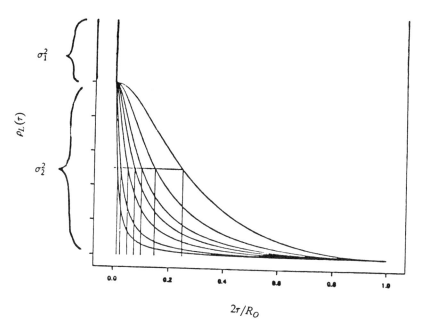

FIGURE 2. Family of synthetic correlation functions.

(Not to same scale as Figure 1.)

3.2 WHEN ARE GOOD ESTIMATES POSSIBLE?

We would like to know whether or not it is reasonable to attempt to obtain an estimate of r and L from the data to be analyzed. In particular, as n becomes arbitrarily large, can we expect that the mean square errors of the estimates will become arbitrarily small?

Fortunately, this question can be answered, by applying the mathematical theory of equivalence and perpendicularity. We first state a rather abstract theorem, adapted from Parzen(1962), then we apply it to our example, and make a few remarks concerning its intuitive meaning.

Theorem:

For each n, let $x=(x_{n1}, \ldots, x_{nn})$ be a zero mean normally distributed random vector with covariance matrix $C=C(\beta)$ where β are some unknown constants. It is desired to estimate β from x. For any two distinct values $\beta^{(1)}$ and $\beta^{(2)}$ of β, let

$$J_n(\beta^{(1)},\beta^{(2)}) = Trace\,[I-C\,(\beta^{(1)})^{-1/2}C\,(\beta^{(2)})C\,(\beta^{(1)})^{-1/2}]^2.$$

If $J_n(\beta^{(1)},\beta^{(2)})$ tends to infinity as n tends to infinity for any two distinct possible pairs $(\beta^{(1)},\beta^{(2)})$, then there exists a sequence of estimates $\beta_n, n=1,2,\ldots$ of β whose mean square error goes to zero as n tends to infinity. If J_n is bounded as n tends to infinity for all possible pairs, then there cannot exist such a sequence of estimates.

In the example here, $C=\sigma_1^2(I+rQ_f(L))$ and $\beta=(\sigma_1^2,r,L)$. It can be shown, that if σ_1^2,r and L are all restricted to be strictly positive, then J_n tends to infinity, and good estimates of these quantities are possible (at least in theory), from one data set, if n is sufficiently large. A deeper analysis of this theorem (and common sense!) suggests that the condition for estimability of r is that $Q_f(L)$, for any L, has eigenvalues that are very large compared to 1 as well as very small. Intuitively, it means that the forecast error has a substantial "low frequency" component.

3.3 THE GENERAL CASE

We now consider the general (linear) case, where

$$y^{(i)} = F^{(i)}s + \varepsilon^{(i)}, \quad i=1,2$$

where $y^{(i)}$ is of dimension $n^{(i)}, i=1,2$ and the covariance of $\varepsilon^{(i)}$ is $\sigma_i^2 Q_i$. To estimate r, we must be able to construct two matrices $B^{(1)}$ and $B^{(2)}$ of dimension $n \times n^{(1)}$ and $n \times n^{(2)}$ respectively, which satisfy $B^{(1)}F^{(1)}=B^{(2)}F^{(2)}$. Let $z^{(i)}$ be defined by

$$z^{(i)} = B^{(i)}y^{(i)}, \quad i=1,2$$

and let w be defined by

$$w = z^{(1)}-z^{(2)} = B^{(1)}y^{(1)}-B^{(2)}y^{(2)} = B^{(1)}\varepsilon^{(1)}-B^{(2)}\varepsilon^{(2)}.$$

The covariance matrix of w is then

$$Ew'w = \sigma_1^2 B^{(1)}Q_1 B^{(1)\prime} + \sigma_2^2 B^{(2)}Q_2 B^{(2)\prime}.$$

Suppose $B^{(1)}Q_1 B^{(1)\prime}$ is of full rank, then we can take the Cholesky decomposition LL' of $B^{(1)}Q_1 B^{(1)}$, where L is lower triangular, and let $\xi=L^{-1}w$. Then the covariance matrix of ξ is

$$E\xi\xi' = \sigma_1^2(I+rQ),$$

where $Q=L^{-1}B^{(2)}Q_2 B^{(2)}L^{-1\prime}$. The ML and GCV estimates are then given by the minimizers of V and M of Section 3.1, and the estimability of r depends on the properties of Q. Loosely speaking, the two vectors $z^{(1)}$ and $z^{(2)}$ need have to have their "energy" at different "wavenumbers".

3.4 COMPUTATIONAL CONSIDERATIONS

Computation of r and L minimizing quantities similar to M and V has been carried out in a relatively straightforward way using matrix decompositions for n up to a few hundred, on a VAX 11/720 in the Statistics Dept. at the University of Wisconsin- Madison, and with n up to about 800 on the Cray XMP at the San Diego Super Computer Center, without much optimization of the code, in under 150 seconds, using GCVPACK (Bates et al. (1985)). Research is continuing on more efficient methods for larger problems. It is probably true that large data sets will be required in practice.

3.5. POTENTIAL APPLICATIONS

A possible important application is to the comparison of satellite-observed radiances to forecast. It is possible to observe certain gross features of the atmosphere in two-dimensional plots of satellite (raw) radiances. Forecast errors frequently tend to be phase errors, with certain spatial features displaced in space. To compare forecast $y^{(2)}$ with satellite radiance data $y^{(1)}$ following the approach in this paper, one should compute from the forecast, radiances that would be seen by the satellite, that is, let $B^{(1)}$ be I and $B^{(2)}$ map forecast into radiance data as would be seen by the satellite. Assuming that realistic Q_1 and Q_2 can be established (remember, Q_2 is in the radiance observational domain), it is likely that r could be estimated, and used to help decide whether to trust the forecast or the radiance data in the event of a major discrepancy.

4. SUMMARY

A study has been initiated into the estimation, from the data to be analyzed, of relative weight to be given to various sources of data, for the purposes of numerical weather prediction. For the estimation of these weights to be successful the various sources have to have sufficiently different error correlation structure. The meaning of "sufficiently different" has been quantified in a theoretical way, and reduction of this theory to practical cases has begun with a study of 500 mb height forecast and observational data. Estimates are proposed and numerical methods for computing them for very large data sets are under study. The method has the potential for determining the relative credence to be given to different sources of information, such as forecast and satellite radiance data.

ACKNOWLEDGMENT

This work has benefitted from numerous helpful discussions with D. R. Johnson

BIBLIOGRAPHY

Bates, D. M., Lindstrom, M. J., Wahba, G., and Yandell, B. (1987), "GCVPACK - Routines for Generalized Cross Validation," *Commun. Statist. Simul. Comput.*, 16, 263-297.

Hollingsworth, A., and Lonnberg, P. (1986), "The Statistical Structure of Short Range Forecast Errors as Determined from Radiosonde Data. Part I: The Wind Field," *Tellus*, 38A, 111-136.

Isaacs, R. G. (1988), "A Unified Retrieval Methodology for the Defense Meteorological Satellite Program Meteorological Sensors," in *International Workshop on Remote Sensing Retrieval Methods*, ed. A. Deepak, Hampton, VA: A. Deepak Publishing.

Kimeldorf, G., and Wahba, G. (1970), "A Correspondance Between Bayesian Estimation of Stochastic Processes and Smoothing by Splines," *Ann. Math. Statist.*, 41, 495-502.

Lonnberg, P., and Hollingsworth, A. (1986), "The Statistical Structure of Short Range Forecast Errors as Determined from Radiosonde Data. Part II: The Covariance of Height and Wind Errors," *Tellus*, 38A, 137-161.

Lorenc, A. C. (1986), "Analysis Methods for Numerical Weather Prediction," *Quart. J. R. Met. Soc.*, 112, 1177-1194.

Merz, P. (1980), "Determination of Adsorption Energy Distribution by Regularizarion and a Characterization of Certain Adsorption Isotherms," *Journal of Computational Physics*, 38, 64-85.

O'Sullivan, F., and Wahba, G. (1985), "A Cross Validated Bayesian Retrieval Algorithm for Non-Linear Remote Sensing Experiments," *J. Comput. Physics*, 59, 441-455.

Parzen, E. (1963), "Probability Density Functionals and Reproducing Kernel Hilbert Spaces," in *Proceedings of the Symposium on Time Series Analysis Held at Brown*, ed. M. Rosenblatt, New York: Wiley, 155-169.

Rodgers, C. (1988), "A General Error Analysis for Profile Retrieval," in *International Workskhop on Remote Sensing Retrieval Methods*, ed. A. Deepak, Hampton, VA: A. Deepak Publishing.

Seaman, R. S., and Hutchinson, M. F. (1985), "Comparative Real Data Tests of Some Objective Analysis Methods by Withholding Observations," *Aust. Met. Mag.*, 33, 37-46.

Smith, W. L., Revercomb, H. E., Howell, H. B., Woolf, H. M., and Huang, A. (1988), "Science Applications of High Resolution Interferometer Spectrometer Measurements from the NASA U2 Aircraft," in *International Workshop on Remote Sensing Retrieval Methods*, ed. A. Deepak, Hampton, VA: A. Deepak Publishing.

Svensson, J. (1985) "A Nonlinear Inversion Method for Derivation of Temperature Profiles from TOVS Data." *Tech. Proc. Second International TOVS Study Conference*, Madison, WI292-307.

Villalobos, M., and Wahba, G. (1987), "Inequality Constrained Multivariate Smoothing Splines with Application to the Estimation of Posterior Probabilities," *J. Am. Statist. Assoc.*, 82, 239-248.

Wahba, G., and Wendelberger, J. (1980), "Some New Mathematical Methods for Variational Objective Analysis Using Splines and Cross-Validation," *Monthly Weather Review*, 108, 1122-1145.

Wahba, G. (1981) "Some New Techniques for Variational Objective Analysis on the Sphere Using Splines, Hough Functions, and Sample Spectral Data." Preprints of the Seventh Conference on Probability and Statistics in the Atmospheric Sciences, American Meteorological Society.

Wahba, G. (1982a), "Variational Methods in Simultaneous Optimum Interpolation and Initialization," in *The Interaction Between Objective Analysis and Investigation*, ed. D. Williamson, Boulder, CO: Atmospheric Analysis and Prediction Division, National Center for Atmospheric Research, 178-185. (Publication in Meteorology 127)

Wahba, G. (1982b), "Vector Splines on the Sphere, with Application to the Estimation of Vorticity and Divergence from Discrete, Noisy Data," in *Multivariate Approximation Theory, Vol. 2*, ed. W. Schempp and K. Zeller, Birkhauser Verlag, 407-429

Wahba, G. (1985a), "Variational Methods for Multidimensional Inverse Problems," in *Advances in Remote Sensing Retrieval Methods*, eds. A. Deepak, H. E. Fleming, and M. Chahine, Hampton, Va: A. Deepak Publishing, 385-410.

Wahba, G. (1985b), "A Comparison of GCV and GML for Choosing the Smoothing Parameter in the Generalized Spline Smoothing Problem," *Ann. Statist.*, 13, 1378-1402.

Westwater, E. R., Falls, M. J., Schroeder, J., and Snook, J. (1988), "Combined Ground- and Satellite-Based Radiometric Remote Sensing," in *International Workshop on Remote Sensing Retrieval Methods*, ed. A. Deepak, Hampton, VA: A. Deepak Publishing.

DISCUSSION

Westwater: Have you looked at different synoptic situations to see how the structure of the forecast error would vary as a function of distance, or is this a single case shot?

Wahba: Well, I haven't, other people have. The picture that I showed there assumes not only that it is isotropic but it is the same throughout the globe. Of course, neither of those assumptions is true. You could make it possible to throw in a few more parameters, for example, to make the forecast error correlation such that the half length varies slowly with the latitude. Now there is going to be some limit to how much you can get out of one chunk of data this way. You could let the half length L depend on latitude in a way that involves only one or two unknown parameters but you can't carry this too far from one chunk of data.

Speaker: Can it be defined by cross-validation, though, for an individual case, what the error correlation is, in the forecast?

Wahba: One parameter in the error correlation PBL/BNP model (the halflength L), can be estimated by cross-validation or maximum likelihood.

Speaker: I don't understand theoretically. I mean, practically it may have to be because of computational limitations but in principle I would think you could determine the statistics of the error in the forecast (your background field) by cross validation.

Wahba: Oh, yes, yes. The point here is that you can determine a lot of things by collecting a lot of history and assuming that there is stationarity or you can just take today's set of data. Now, if you are just going to take today's set of data you can only carry this so far, but probably far enough to be interesting. I know by this theory that I could carry it to the point of estimating two numbers, namely, the ratio of the forecast error to observation error, and the half scale of the correlation. This is for the proposed model for 500 millibar heights right now. Now what you are saying is that we know that these parameters are not really constant over the globe. My answer to that was that I could model them as slowly varying, but I can't expect to get too much out of one day's worth of data. You could combine a few days worth of data, rather than three months. If you are going to do that then you can expect to estimate a few more coefficients.

Gille: Would that same point hold if you include systematic observation errors? I can imagine that if you are using satellite data, your errors may very much depend on the synoptic situation, and you may not be able to make some of the assumptions you did here.

Wahba: I should have said before that, in theory, the more days of data you combine, the more and better you can estimate, provided the situation is stationary. However, to the extent that the situation is not stationary, you will be estimating some sort of average values of the parameters. There will be some tradeoff changes. The goal of the present research is to see (at least from a theoretical point of view) how far one can expect to go on a very short-term basis.

ON INFERRING PLANETARY WAVES
FROM POLAR-ORBITING SATELLITE OBSERVATIONS[1]

D.M. Cunnold and C.C. Wey[2]
Georgie Institute of Technology
Atlanta, Georgia 30332, USA

ABSTRACT

Most techniques for inferring wave structure from observations by polar-orbiting satellites produce estimates of planetary wave amplitudes and phases by fitting stationary waves to the observations. This assumption has mathematical advantages, but when the planetary waves are moving, typically leads to the underestimation of wave amplitudes. An algorithm which fits the observations using moving planetary waves is described. This procedure produces an excellent simulation of both wave amplitudes and phases.

1. INTRODUCTION

The measurement of atmospheric parameters from a polar-orbiting satellite produces a sample of an atmospheric field which varies in four-dimensional space. For many purposes it is useful to interpolate this data to produce a synoptic map of the sampled field. For observations of tropospheric temperatures, this interpolation is often performed with a three-dimensional numerical forecast model. For observations of stratospheric parameters, however, the sampling of the global atmosphere is usually less complete than for the troposphere, and it is doubtful that there is sufficient information with which to initialize a three-dimensional numerical model. The interpolation of stratospheric fields is therefore typically performed based essentially on the simplistic assumption that stratospheric fields remain stationary over 24 hours (or more).

Stratospheric studies, however, indicate the importance of planetary-scale motions and that planetary waves typically possess non-zero phase velocities (with small values for the lowest order modes and larger values for the higher order modes). In this paper two-dimensional aspects of the four-dimensional mapping problem are considered; samples at a prespecified latitude and height (or pressure level) are analyzed. This part of the mapping problem was previously discussed by Salby (1982a and b).

---

[1] This research was supported by contract number NAS5-27264 from the National Aeronautics and Space Administration.

[2] NASA Lewis Research Center, Cleveland, Ohio 44135, USA

RSRM '87: ADVANCES IN
REMOTE SENSING RETRIEVAL METHODS
A. Deepak, H.E. Fleming, and J.S. Theon (Eds.)

2.   THE ESTIMATION PROCEDURE

Linear interpolation of data sampled at a series of longitudes and times will, in the presence of a moving wave, usually result in realistic estimates of the phase of the wave but will typically produce underestimates of the wave amplitude. To remedy this amplitude underestimation problem, our algorithm does not estimate wave parameters based on the stationarity assumption but attempts to estimate parameters which characterize wave movement. Specifically, the following model of atmospheric variations is employed:

$$\chi = A_0 + B_0 t + \sum_{i=1}^{4} (A_i + B_i t) \sin (i\lambda + p_i t + \phi_i) \qquad (2.1)$$

where χ = measured parameter (e.g. ozone mixing ratio)
λ = longitude
$\underline{x}$ = state vector

$\qquad = (A_0, B_0, A_1, B_1, \phi_1, p_1, \ldots)$

t = time (measured in days)

In this study the mathematics are simplified by assuming that there are exactly 16 complete orbits of the satellite in each 24-hour period and we solve for just 4 atmospheric waves (versus the 7 waves for which parameters might be estimated).

The estimation of the parameters in Eq. (2.1) is performed using sequential estimation and Kalman filtering. The state vector of the estimated parameters $(\hat{\underline{x}}_j)$ is updated after each measurement (z_j) according to

$$\hat{\underline{x}}_j (+) = \hat{\underline{x}}_j (-) + \underline{\underline{K}}_j \left[z_j - h_j (\hat{\underline{x}}_j (-)) \right] \qquad (2.2)$$

where $\underline{\underline{K}}_j$ is the Kalman filter gain matrix and $h_j (\hat{x}_j(-))$ is the expected measured value determined from Eqs. (2.1) or (2.2) based on the parameters estimated at the previous timestep. This form of the Kalman filter equation when applied to the estimation of non-linearly interdependent parameters is referred to as the extended Kalman filter (Gelb, 1975). Both the extended Kalman filter and the linearized Kalman filter in which $h_j (\hat{x}_j(-))$ is replaced by

$$h_j (\bar{x}_j) + \frac{\partial h_j (\bar{x}_j)}{\partial \bar{x}_j} (\hat{\underline{x}}_j(-) - \bar{\underline{x}}_j) \quad ,$$

where $\bar{x}_j$ is an a priori estimate of $\hat{\underline{x}}_j$, have been used in the retrievals. Between measurements the estimated parameters are assumed to remain unchanged except for the addition of white noise, i.e.

$$\frac{\partial \hat{\underline{x}}_j}{\partial t} = \underline{0} + \underline{\text{noise}} \quad \text{and} \quad \underline{x}_j{}^{\wedge}(-) = \underline{x}_{j-1}{}^{\wedge}(+) \tag{2.3}$$

where the noise vector elements are Gaussian random variables described by a noise covariance matrix $\underline{Q}$.

The measurements are processed through this algorithm both forwards and backwards in time, and the resulting two estimates are smoothed into a combined estimate based on the error covariance matrices for the backward and forward estimates (Gelb, 1975). A similar algorithm was applied by Kohri (1979) to the estimation of just the stationary wave parameters (A_i and ϕ_i).

3.   RESULTS

To illustrate the viability of the procedure, the algorithm has been applied to three sets of simulated measurements for a 10 day period. The first two simulations are for

$$\chi = 59 + 2 \sin \left[3\lambda + p(t-5.5)\right] + \text{noise} \tag{3.1}$$

where χ is the measurement, λ is longitude, t is time in days (between 0.5 and 10.5 days) and p is related to the phase speed of wavenumber 3. In the first example a constant value of p of 2.56 radians (of phase)/day (= 0.16 radians/90 minute timestep) is used.

The results for example 1 primarily demonstrate that the estimation procedure is stable and that when the state of the atmosphere is matched by the model (Eq. 2.1), the parameters may be retrieved in the presence of measurement noise. In particular, in contrast to the stationary wave estimation procedure, both the forward and backward phase estimates are correct and so is the combined estimate of wave amplitude. This example does, however, emphasize the need for a mathematical constraint on the solution: the solution possesses the ambiguity that wavenumber 3 moving at 2.56 radians/day is identical to wavenumber 4 moving at 3.72 radians/day in the opposite direction (i.e. "complementary" phase speeds). This ambiguity is resolved in the procedure by disallowing wave phase speeds greater than 3.14 radians/day.

Some existing atmospheric data sets (e.g. the LIMS) already provide estimates of wave amplitudes and phases based on a stationary wave model with time-dependent parameters (e.g. A_i and ϕ_i) in place of Eq. (2.1). For this reason and also because the stationary wave model is linear and results in a robust estimation of the parameters, stationary wave model estimates are used to provide a first guess to the moving wave model parameters. This is the basis ($\bar{x}$) for the linearized Kalman filter solution. To provide initial values of wave growth rates and phase speeds, linear fits to the stationary wave model results over 48-hour periods (this corresponds approximately to the sample period necessary to obtain independent estimates of the moving wave parameters) are employed. For the 48-hour periods at each end of the 10 day period, extrapolation is used to provide first guesses for the moving wave parameters.

The parameter estimates for this simulated data set have some sensitivity to the first guess — i.e. to the results inferred from the stationary wave model. Because the parameters being estimated have a nonlinear dependence (Eq. 2.1), it is possible for the solutions to converge to different answers based on different initial values. The stationary wave model estimation divides the energy between waves 3 and 4 moving at complementary phase speeds resulting in a substantial underestimation of the wavenumber 3 amplitude (see Fig. 1). Based on this initial guess the moving wave solution (2.1) yields a similar subdivision of energy between waves 3 and 4. If, however, the wave four phase speed is set to zero following the initial guess and maintained at zero in the moving wave solution and the amplitude of wavenumber 4 is zeroed out initially, the model obtains a wave amplitude of 2 for wavenumber 3 (see Fig. 1). These solutions indicate that excellent moving wave parameter estimates are obtained, despite initial guesses of wavenumber 3 amplitude which are 50% in error, provided that the mathematical ambiguity in phase speed is adequately addressed.

The second example uses data simulated via Eq. (3.1) but with

$$p = 0.48 + 0.0768(t-0.5) \text{ radians/day} \tag{3.2}$$

where t again runs from 0.5 to 10.5 days. The object of this exercise is, therefore, to determine how well the estimation procedure "keeps up" with a changing phase speed. Note, however, that in this example the phase speed is not p but $p' = 0.48 + 0.1536(t-5.5)$ radians/day.

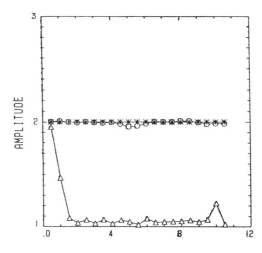

Figure 1.   Comparison of the amplitude estimates for a wavenumber 3 moving at 2.56 radians/day. The asterisks represent the correct result, the triangles indicate the stationary wave estimates and the circles show the moving wave results.

Figure 2 illustrates the forward and backward estimates of p. Note
that the estimates appear to lag the correct value of p by
approximately 2 days. This lag may be increased somewhat by reducing
the value of the diagonal element of the $\underline{Q}$ matrix corresponding to

this parameter. In all the examples in this paper the diagonal
elements of the $\underline{Q}$ matrix have been chosen to correspond to the minimum

averaging time necessary to obtain simultaneous estimates of all the
parameters (i.e. approximately 1 day for wave amplitudes and phases
and 2 days for rates of change in these parameters).

   The combination of the backward and forward estimates produces a
value of p different from the phase speed (Fig. 3). Coincidentally,
the estimate of p approximately coincides with the value of p given in
Eq. (3.2). An examination of the estimation procedure reveals that in
the forward retrieval (for example), the elements of the error
covariance matrix remain relatively large until t=5.5 days (i.e. the
center of the period of observations being analyzed). Prior to that
time the different terms of Eq. (2.1) are not orthogonal over the
interval (O,t). Specifically, the estimates of A and B in A+Bt and of
ϕ and p in ϕ+pt are correlated unless they are determined over an
interval centered on t=0. For this reason it is difficult to control
the averaging period for the backward and forward estimates of the
parameters after and prior to t=5.5 days respectively. After the
measurements at t=5.5 days have been incorporated into the backward

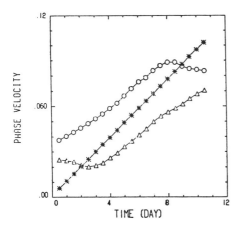
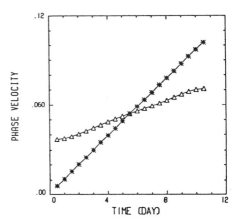

Figure 2. Forward (triangles) and
backward (circles)
estimates of phase
velocity for the moving
wave model (radians/90
minute timestep) for
example 2 (see Eq.
3.2). In this example
the correct result is a
linearly time dependent
phase velocity (the
asterisks).

Figure 3. The estimated phase
velocity (triangles)
obtained by combining
the forward and
backward estimates
shown in Fig. 2.

and forward estimates, the error covariance matrix elements decrease by approximately an order of magnitude. The averaging time then becomes more controllable. Of particular significance, when these backward and forward estimates are smoothed together, the combination is dominated by the estimate which contains the t=5.5 day measurement because of weighting by the inverse of the error covariance matrices. Thus the combined estimate of p is a "lagged" estimate at all times except near t=5.5 days. The error covariance matrices for the stationary wave estimation do not behave in this way because of better orthogonality. The combined estimate of any parameter in that case is found to be an approximately equally weighted combination of the forward and backward estimates. Nevertheless for both procedures further study of how best to combine the forward and backward parameter estimates would be useful.

Figure 4 shows the estimate of wave amplitude for this simulated data set. The stationary wave procedure results in a varying estimate of wave amplitude which worsens as the wave moves faster. In contrast the wave amplitude estimate (and the phase estimate which is not shown) for the moving wave model is excellent despite the occurrence of a lagged estimate of the phase speed as discussed in the previous paragraph.

The final tests reported on here utilized ozone mixing ratios obtained from the output of a three-dimensional numerical model of the atmosphere (Cunnold et al., 1975). Output from the model over a

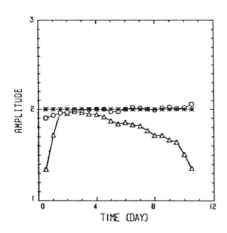

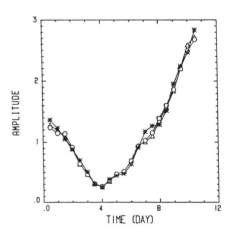

Figure 4.   The amplitude estimates for wavenumber 3 for example 2. The asterisks represent the correct solution, the triangles indicate the stationary wave results and the circles show the moving wave results.

Figure 5.   The amplitude estimates for wavenumber 1 for data simulated using a three-dimensional model (example 3). The notation is the same as for Figs. 1 and 4.

10-day period was sampled at the locations and times corresponding to a polar-orbiting satellite. This simulated data set contained all four wavenumbers with large amplitudes for wavenumbers 1 and 2 and small amplitudes for wavenumbers 3 and 4. All waves had phase speeds of less than 1 radian (phase)/day. All the wave parameters were estimated simultaneously; however the results for only wavenumber 1, which are the most interesting, will be discussed.

Figure 5 shows the temporal behavior of the amplitude of wavenumber 1 and its estimated value. Both the stationary and the moving wave models produce good estimates of wave amplitude because the wave is moving fairly slowly. Figure 6 shows the estimated growth in wave amplitude with time (i.e. the parameter B in Eq. 2.1) The backward and forward estimates of this parameter behave in a fashion analogous to the estimates of the parameter p in the previous example (Fig. 2). For similar reasons, because the growth rate is time-dependent, the estimated $\partial B/\partial t$ has a smaller magnitude than does the change of the wave growth rate with time. However, it is important to note that the simulation of the variation of both the wave amplitude (Fig. 5) and the phase (Fig. 7) is excellent. Thus the inclusion of the parameters B and p in the moving wave model produces non-lagged estimates of the amplitude and phase. If better estimates of B and p are desired, they may be derived a posteriori based on the temporal variations in the amplitude and phase estimates.

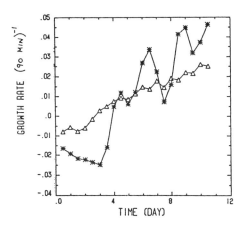

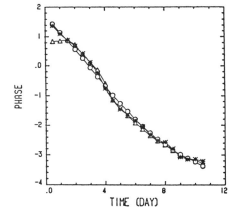

Figure 6. The growth rate estimates for wavenumber 1 in example 3 (units are per 90 minute timestep). The triangles represent the moving wave solution and the asterisks show the values calculated from linear approximations to the change of phase over 1 day periods.

Figure 7. The phase estimates for wavenumber 1 in example 3. The notation is the same as in Figs. 1 and 4.

4. CONCLUSIONS

Synoptic mapping of longitudinally-sparse measurements provided by a polar-orbiting satellite has been discussed. The algorithm which has been described accomplishes this objective by fitting a series of moving waves to the observations. The algorithm produces estimates of the amplitude, phase, amplitude growth rate and phase speed for each wave. Compared to the typical mapping procedure (here referred to as the stationary wave procedure) in which only amplitude and phase are estimated, the procedure described here allows both forward and backward estimates to keep up with typical temporal changes in wave amplitude and phase. This results in substantially better estimates of wave amplitude than are often produced by the stationary wave procedure.

The algorithm contains the mathematical ambiguity that two waves with adjacent wavenumbers and complementary phase speeds (phase speeds which differ by 2π radians/day) are indistinguishable. This ambiguity must be carefully accounted for in the algorithm; it is resolved by initializing to zero any wave which has a first guess phase speed exceeding π radians/day. Furthermore, over the time interval during which the first guess phase speed exceeds this value, the wave is constrained to be stationary (although with a temporally-varying amplitude and phase). The ambiguity is thus resolved in favor of the wave having a phase speed less than π radians/day.

Excellent retrieval of wave amplitude and phase has been demonstrated for several simulated data sets including a situation in which several waves were simultaneously active. It was noted, however, that the error covariance matrix is not diagonal in the moving wave estimation procedure (whereas it is in the stationary wave procedure). Because of this, it is difficult to control the averaging period until after the measurements in the middle of the data record have been incorporated into the forward or backward estimates. Moreover, the estimates calculated by combining the forward and backward retrievals are dominated by the retrieval which incorporates the middle measurements. Thus the combined estimates of phase speed and growth rate lag the correct values by approximately 2 days. If better estimates of these parameters are desired, they could be obtained a posteriori from the temporal variations in the amplitude and phase estimates.

REFERENCES

Cunnold, D., F. Alyea, N. Phillips and R. Prinn, 1975: A three-dimensional dynamical-chemical model of atmospheric ozone, _J. Atmos. Sci._, _32_, 170-194.

Gelb, A., (Ed.) 1974: _Applied Optimal Estimation_, MIT Press, Cambridge, MA, pp374.

Kohri, W.J., 1979: LRIR observations of the structure and propagation of the stationary planetary waves in the Northern Hemisphere during December 1976, Ph.D. Thesis, NTIS Springfield, VA 22161, PB-82-156639.

Salby, M.L., 1982a: Sampling theory for asynoptic satellite observations. Part I: Space-time spectra, resolution and aliasing, J. Atmos. Sci., 39, 2577-2600.
Salby, M.L., 1982b: Sampling theory for asynoptic satellite observations. Part II: Fast Fourier synoptic mapping, J. Atmos. Sci., 39, 2601-2614.

DISCUSSION

Gille: I notice the inferred waves didn't follow some of the rapid oscillations in example 3. Is that because you had a relatively long wave, or your averaging time was long compared to the period of those variations?

Cunnold: We typically used about a two-day averaging time. And to get enough information, you need a two-day averaging time. So I felt like we used the minimum that we could use consistent with the number of parameters that are being estimated.

Rodgers: Have you looked at more than one wave at a time in the non-linear case?

Cunnold: Yes I have.

Rodgers: Do you get mixing problems?

Cunnold: No. I haven't. The examples that I have looked at I haven't seen mixing problems.

OPEN DISCUSSION

SESSION E. - JOHN GILLE

Fleming: (Speaking for John Gille) John gave me some notes, so I will just read from them. There were three papers in his session and he said they were on very different but novel subjects. The subject of fractals looks promising and is of interest. More research related to using them will probably be fruitful. Further work on integration of satellite retrievals or radiances with forecasts is important and should be emphasized. Sequential estimation: there was little explicit mention of it at this time, other than a paper on mapping—still important. Multi-dimensional methods are a fruitful area also, but more work is needed.

I am now speaking for myself. In connection with using forecasts, let me just throw out this scenario. One of my initial concerns in using a forecast as the first guess in a temperature retrieval algorithm is the dependence of the solution on the first guess. Let's assume that the weather service can produce excellent forecasts 90 percent of the time using the radiosonde network and no satellite data. Our role in this case would be to let them know through our measurements that (in the remaining 10% of the time) a particular forecast in a particular area is moving in the wrong direction. Suppose their forecast is not intensifying a low-pressure system enough and that in that one area our satellite measurements are absolutely inconsistent with the way things are moving in the forecast. Now, if we use the forecast as the first guess, the retrieval may not respond to the degree it should to correct the forecast that is headed in the wrong direction because of the strong dependence on the forecast.

Wahba: If you tune your retrievals to the point where your retrieval is the first guess field and you put that into a numerical forecast model then it is as though you are simply integrating the model, forgetting about the satellite. And we know something about that problem. We know that weather forecast accuracy tends to degrade and so on. But if you set up the problem that way and then use your trade off parameters to add more and more satellite information into the retrieval and see the impact on the forecast, you would have a fairly clean way of looking at what the impact of this balance is.

Fleming: If you use the forecast as a first guess you are going to find that the retrievals are highly correlated with the forecast. Since the retrievals get weighted inversely with the correlation, that means that the satellite retrievals are going to get less weight in the model and, therefore, the retrievals will have little impact on the model. Most of the weight will come from the independent measurements, which are the radiosondes.

Wahba: If you take the satellite out of the picture and so for each integration step your first guess was the previous forecast, that is the way the dynamic models run. But now, you turn your dial and add satellite information into whatever mixture you want and ask the question: did I improve that integration or did I not, or in what way did it all change?

Fleming: This makes the forward problem very important. One way you can do the forward problem is to use the forecast. Then, if these calculated radiances agree with the measured ones, forget the retrieval. The forecast is as good as you are going to get it. Only when the calculated radiances and measured radiances disagree do you know that the measured radiances have information that the forecast doesn't have. In this case, you should bring that information in with full weight. But that does put the burden on being able to do the forward problem very well; otherwise, you don't know whether or not you are really doing better than the forecast.

Rodgers: To go back to your original scenario of the measurements disagreeing with the forecasts. This is really in the area of quality control. If you are using a forecast's first guess it must come with a covariance matrix and you should be able to tell whether or not there is something seriously wrong. Then you have to decide whether it is wrong with the forecast or wrong with the measurements. I suspect, in general, it is going to be something wrong with the forecast. When you come to combining retrievals with the forecast, when you have used the forecast as a first guess, you've got to provide the forecaster with the covariance matrix of your estimate. You should use the forecast really only as a linearization point.

Wahba: This problem has motivated the work that I have been trying to do and report on. So you find that you solve the forward problem with the forecast and you find that it doesn't match with the forecast and so you have a discrepancy. How do you know whether it is the forecast or the observations that are bad? I don't know whether you can do this in practice or not, but, in theory, if you know something about the nature of the errors in both cases and if the error correlation structure is quite different then maybe there is some hope of being able to find out which one is right and which one is wrong. We do know that forecast errors tend to be errors of phase sometimes. If the forecast errors are highly correlated and the measurement errors are "white," then there is hope of decomposing the difference of the two errors into a correlated and a white component.

OVERVIEW OF FRACTAL CLOUDS

Robert F. Cahalan
Laboratory for Atmospheres
NASA Goddard Space Flight Center
Greenbelt, Maryland 20771, USA

ABSTRACT

We summarize work being done at the Goddard Laboratory for Atmospheres on the fractal properties of clouds and their effects on the large-scale radiative properties of the atmosphere. This involves three stages:

1. analysis of high resolution cloud data to determine the scaling properties associated with various cloud types;
2. simulation of fractal clouds with realistic scaling properties;
3. computation of mean radiative properties of fractal clouds as a function of their scaling properties.

The approximation that clouds are scaling fractals with dimension 4/3 (Lovejoy, 1982) must be modified in three ways: First, a change in the fractal dimension is found at a characteristic size which depends upon cloud type, so that exact scale invariance does not hold. Secondly, the fractal dimensions are different for each cloud type in a given size range, so that exact universality does not hold. Finally, the more intense cloud areas are found to have a higher perimeter fractal dimension, perhaps indicative of the increased turbulence at cloud top.

Simulations of fair weather cumulus and stratocumulus clouds have been developed which take these properties into account. The simulations depend upon two scaling parameters which determine the distributions of cloud sizes and spacings, respectively, and also upon a maximum characteristic cell size.

The initial Monte Carlo radiative transfer computations have been carried out with a highly simplified model in which liquid water is redistributed in an initially plane-parallel cloud while cloud height and mean optical depth are held fixed. Redistribution decreases the mean albedo from the plane parallel case, since the albedo of optically thick regions saturates as optical depth is increased. The albedo of each homogeneous region may be computed from the thickness of each region independently only when the horizontal optical depth is large compared to the photon mean free path. The albedo of a region comparable in horizontal optical depth to the photon mean free path depends upon radiation from the sides. The mean albedo is insensitive to variations in optical depth on horizontal scales much smaller than the photon mean free path. These concepts have been illustrated with a simple one-parameter fractal model.

RSRM '87: ADVANCES IN
REMOTE SENSING RETRIEVAL METHODS
A. Deepak, H.E. Fleming, and J.S. Theon (Eds.)

371

1. INTRODUCTION

There is a growing body of observational evidence on inhomogeneous cloud structure. Derr and Gunter (1982) reported searching in vain for the sort of clouds assumed by cloud radiation modelers (whether plane-parallel, or a lattice of simple shapes). More recently, measurements of vertically averaged liquid water at one-minute intervals over a three-week period during the July 1987 FIRE experiment at San Nicolas Island in California stratus -- perhaps the closest thing to plane-parallel clouds -- show variations on all scales (J. B. Snider, private communication). Baker and Latham (1979) have suggested that such variations arise from inhomogeneous entrainment. Entrained dry air, rather than mixing uniformly with cloudy air, remains intact in blobs of all sizes, which decay only slowly by invasion of cloudy air.

A useful notion for discussing such cloud structure is that of a "scaling fractal" (Mandelbrot, 1983 and references therein), which is a set or object which is extremely irregular on all scales, while at the same time statistically invariant under certain transformations of scale. The simplest examples are the statistically "self-similar" fractals (op.cit. p. 350) -- geometrical patterns (other than Euclidean lines, planes and surfaces) with no intrinsic scale, so that no matter how closely you inspect them, they always look the same, at least in a statistical sense. While self-similar fractals are statistically isotropic, the so-called "self-affine" fractals scale differently in different directions (see e.g. Mandelbrot, 1986). The most general scaling fractals are the multidimensional fractals, or "multifractals" (Hentschel and Procaccia, 1983 and Frisch and Parisi, 1985) which exhibit a spectrum of fractal dimensions.

Lovejoy (1982) has presented empirical evidence that clouds are statistically self-similar in the horizontal plane from 1000 km down to 1 km, with a fractal dimension of approximately 4/3. If this were strictly true, than a satellite image of a cloud field, taken over a uniform dark ocean background with no evident geographical features, would be statistically indistinguishable from an image of any small portion of the scene if it had the same number of pixels as the original. Schertzer and Lovejoy (1986) have argued that clouds are self-affine in a vertical plane, and have considered multifractal cloud models (Schertzer and Lovejoy, 1988 and references therein). These idealizations provide important alternatives to the simple plane-parallel or lattice cloud fields, and it is important to consider their consequences and limitations.

The purpose of this paper is to give a brief summary of work being done at the Goddard Laboratory for Atmospheres to investigate the implications of the fractal properties of clouds for the large-scale radiative properties of the atmosphere. The empirical work is being done in collaboration with Mark Nestler of Science Applications Research, and Professor J. H. Joseph and his group at Tel-Aviv University. The radiative transfer computations are being done in collaboration with Warren Wiscombe of Goddard and William Ridgeway of Applied Research Corporation. Our approach involves three stages:

1. analysis of high-resolution cloud data to determine the spatial structure of various cloud types;
2. simulation of the spatial distribution of cloud liquid water and optical depth using a minimum number of parameters; and
3. computation of the mean radiative properties of the simulated clouds as a function of the cloud parameters.

The outline of the paper will follow these three stages. In the next section we begin by giving a brief introduction to fractals with a few simple examples. Section 3 summarizes the results of the analysis of LANDSAT data. Section 4 discusses a simple class of cloud simulations. Section 5 discusses the radiative properties of the simplest simulated cloud having a single fractal parameter. Finally, section 6 concludes with a brief discussion.

2. INTRODUCTION TO FRACTALS

A fractal is a mathematical set or object whose form is extremely irregular and/or fragmented on all scales. Clouds certainly qualify under this general but rather vague definition. To become more concrete and specific, we consider a self-similar scaling fractal, which is an object which remains similar (in the sense of statistical distributions) when subject to arbitrary magnification. (Fractals invariant under non-isotropic rescalings are termed self-affine.) Statistical properties such as the power spectrum, P, scale as

$$P(\lambda x) = \lambda^{f(d)} P(x),$$

(2.1)

where f is a scaling exponent which depends upon the fractal dimension d, which may be defined in terms of N_i, the number of objects with characteristic linear dimension r_i, as

$$N_i = C/r_i^d,$$

(2.2)

where C is a constant of proportionality. For a discussion of various types of dimensions in the context of chaotic attractors see Farmer et al., 1983.

In the following we give three simple examples of fractals, each introducing a new aspect needed in the fractal modelling of clouds. For a more systematic mathematical discussion of fractal sets, see Falconer (1985).

Perhaps the simplest example of a fractal is the von Koch snowflake. In order to construct this, we define a "kink operator" K, which puts a kink in every straight line segment of any figure by replacing its middle third by two equal line segments:

$$K(\text{———}) = \text{—}\wedge\text{—}$$

(2.3)

The snowflake is generated from an equilateral triangle by successive applications of K. The first application, $K(\Delta)$, produces a figure with $3*4 = 12$ sides, each 1/3 as long as the sides of the original triangle. After n applications, $K^n(\Delta)$ has $3*4^n$ sides, each $1/3^n$ as long as the original, so that the perimeter is $3*(4/3)^n$. In the limit $\to\infty$ of $K^n(\Delta)$, the fractal snowflake, $K^\infty(\Delta)$, has an infinite perimeter which encloses a finite area! Even in this simplest example, we see that fractals force us to modify our usual intuitions about geometrical objects.

The kink operator explicitly generates the multiple scales of a fractal, but the iteration of even the simplest operator may generate multiple scales, even when this is not explicit in the operator. Mandelbrot demonstrated this with the simple quadratic function. Consider the "Mandelbrot operator"

$$M(z) = z^2 + c,$$

(2.4)

where z and c are complex. For each value of c, we begin at $z_0=0$ and iterate, computing $z_1=M(0)$, then $z_2=M(M(0))$, and so on. The Mandelbrot Set consists of those values of c for which $z_\infty < \infty$. It is a compact set with a fractal boundary. Zooming in on the boundary, one sees miniature copies of the whole set which are roughly similar, just as in the snowflake. Unlike the snowflake, however, each copy has its own unique features, and increasingly complex structures appear as one zooms in. The iteration process is analogous to a dynamical process, with the operator M generating the time evolution from the initial state at z_0. The fact that such a complex attractor arises from a mapping as simple as (2.4) suggests the ubiquity of fractals in dynamical systems.

Neither of the above two examples contain any random element, which is necessary in the modelling of clouds. Thus as a final example in our list of "simplest" fractals, consider a random walk on a one-dimensional lattice with unit lattice spacing. In the usual non-fractal random walk, only nearest-neighbor jumps are allowed, and if both directions are equally probable, the probability of making a transition Δx is given by

$$p(\Delta x) = 1/2[\delta_{\Delta x,1} + \delta_{\Delta x,-1}],$$

(2.5)

and the corresponding structure function, the Fourier transform of (2.5), is

$$p(k) = \cos(k),$$

(2.6)

where k is the wavenumber. All other quantities of interest may be computed in terms of p(k). In particular, the single particle distribution becomes Gaussian at large times, with a variance increasing linearly with time, the well-known diffusion law.

To generate a fractal random walk, termed a "Levy walk" (Shlesinger, West and Klafter, 1987, and references therein) we allow not only nearest-neighbor jumps, but also jumps of length b with probability $1/a$, jumps of length b^2 with probability $1/a^2$, and so on, where a and b are integers greater than 1. The transition probability then becomes

$$p(\Delta x) = (a-1)/2a \sum_{n=0}^{\infty} 1/a^n[\delta_{\Delta x,b^n} + \delta_{\Delta x,-b^n}],$$

(2.7)

and the corresponding structure function is

$$p(k) = (a-1)/2a \sum_{n=0}^{\infty} 1/a^n \cos(b^n k).$$

(2.8)

This distribution is identical to Weierstrass' example of a function which is continuous, but nowhere differentiable. If the time required for each jump is assumed to be an increasing function of the jump distance, the variance of the single particle distribution at large times increases as a power greater than one, which Shlesinger et al. have termed "enhanced diffusion".

The record of a single realization of a Levy walk shows a property also seen in cloud fields : a heirarchy of clusters (Randall and Huffman, 1980 and Cahalan, 1986). A typical particle initially takes a number of nearest-neighbor jumps, then a larger jump of size b followed by more nearest-neighbor jumps, and so on, until eventually a jump of size b^2 occurs. After a number of b^2 jumps occur, there will eventually be a b^3 jump, and so on. Any given cluster of points, then, is part of a larger cluster created by the less frequent events.

3. EMPIRICAL RESULTS

The initial finding of Lovejoy (1982) that clouds are self-similar fractals was based upon a scatter plot of the area versus the perimeter of cloud areas determined by a threshold method from meteorological satellites and radar. This study was limited to clouds larger than a kilometer or so in horizontal dimension. Lovejoy and Mandelbrot (1985) studied the distribution of rain areas in GATE and compared them to a "fractal sum of pulses" (FSP) model. Lovejoy and Schertzer (1985) found somewhat better agreement with a "scaling cluster of pulses" (SCP) model, and Rhys and Waldvogel (1986) found violations of scaling in hail clouds. We have extended these results by employing the finer resolution available in LANDSAT data (see for example Cahalan, 1987, Cahalan and Joseph, 1988, and Wielicki and Welch, 1986.) We give a brief summary of the LANDSAT findings below.

In analysis of a large number of LANDSAT multispectral scanner (MSS) and thematic mapper (TM) images, over 200,000 cloud areas were identified. The procedure was to first determine the cloud fraction from the thermal channel by the spatial coherence method (Coakley and Bretherton, 1982). The brightness threshold was then adjusted to give the same cloud fraction in the reflected channels. This was termed the cloud "base" threshold. The method works well for the fair weather cumulus and stratocumulus clouds upon which the study focused, but fails for thin cirrus which show up only in the thermal channel. All contiguous areas above the threshold were identified for each scene. Finally, a second cloud "top" threshold was chosen in such a way that the brighter areas covered approximately 10% of the cloud fraction.

The distribution of cloud areas was first determined by binning in terms of the logarithm of the square root of the area, and plotting the logarithm of the number in each bin. If we let

$$r_i = (A_i)^{1/2},$$
(3.1)

where A_i is the cloud area, then if clouds are scaling fractals we expect a power-law $N_i = C/r_i^p$, which gives

$$\log(N_i) = \text{constant} - p \log(r_i),$$
(3.2)

so that the log-log plot should be linear with slope $-p$. In fact, we typically found two linear regions separated by a sharp break in slope at a characteristic size which is about 0.5 kilometers for fair weather cumulus. The smaller clouds follow the straight line quite closely, with a slope typically $p=0.6$.

Scatter plots of cloud perimeter versus the square root of the area on a log-log plot show a change in slope at the same characteristic scale. The smaller fair weather cumulus follow a slope close to Lovejoy's value of 4/3, but the larger fair weather cumulus have an average slope of 1.55.

The fact that the larger clouds are less probable and more irregular in shape suggests a random coincidence hypothesis. That is, the smaller clouds are generated by a scaling fractal process up to some maximum cell size of about 0.5 kilometers, and larger clouds occur only as accidental coincidences of the smaller ones. One test of this picture is to see if the smaller cloud areas have a simpler distribution of cloud brightnesses within each cloud area. Visual inspection of a few scenes seems to bear this out, since we observe that the smaller cloud areas have a single brightness maximum, while larger ones invariably have multiple brightness maxima.

Raising the threshold to a high level allows the determination of the fractal dimensions of the bright regions, those above the cloud "top" threshold. The brighter regions were found to have higher perimeter dimensions for both fair weather cumulus and stratocumulus. In the case of fair weather cumulus it may be that the thicker, and therefore brighter, cloud regions are more irregular because they arise from the random coincidence of the smaller cells. On the other hand, in the case of stratocumulus this may be associated with increased turbulence at the cloud top, where the convection is driven by radiative cooling (Caughey et al., 1982). Note that there is a limit to the increase of the perimeter dimension with threshold, since the brighter regions cover a successively smaller area as the threshold is raised.

Stratocumulus dimensions are intermediate between the smaller and larger fair weather cumulus, and the break in slope is less pronounced. Since stratocumulus convection is driven by cooling at the cloud top, rather than heating from below as in fair weather cumulus, the stratocumulus downdraft regions are of special interest. These take the form of long, irregular leads, not unlike those observed in sea ice.

The mean radiative properties of a cloud field depend not only on the shapes of individual cloud areas, but also upon the relative spatial arrangement of the clouds. Perhaps the simplest measure of this is the distance from the center of a given cloud to that of its nearest neighbor. One can show (Cahalan, 1986) that cloud centers distributed randomly and independently would produce a Weibull-type nearest-neighbor distribution, which is sharpley peaked near the mean spacing, so that large cloud clusters separated by large gaps have exponentially small probability. By contrast, the spacing of cumulus clouds in the LANDSAT images shows many large gaps and clusters. While a Weibull-type distribution is an adequate fit for small spacings, the larger spacings are better fit by a distribution generated by a Levy walk similar to that discussed above.

To summarize our empirical results:

1. the horizontal scaling properties of clouds depend upon characteristic sizes (e.g. larger fair weather cumulus are more irregular), cloud type (stratocumulus are more irregular than small fair weather cumulus), and brightness threshold (brighter regions being more irregular).
2. the greater irregularity of larger clouds and brighter regions of cloud suggest a hypothesis of random coincidence of smaller cloud elements.
3. cloud spacings are not independently distributed, but exhibit strong clustering similar to that of a Levy walk.

4. CLOUD SIMULATIONS

In the preceeding section we drew a number of conclusions about cloud structure based on observations of cloud reflectivity. What we really wish to know is how the cloud liquid water is distributed, since the reflectivity is computable from the distribution of liquid water, traditionally by specifying microscopic properties like drop sizes, and macroscopic properties like optical depth, etc.. As we shall see in the next section, the radiation field provides a kind of low-pass spatial filter, so that there may be small-scale variations of liquid water to which the LANDSAT data is completely insensitive. Analysis of data on cloud liquid water during the FIRE field program should help pin down the important relationship between reflectivity and liquid water.

For the sake of this discussion, we must assume a distribution of liquid water which depends upon a few basic input parameters, and then determine output parameters from the resulting reflectances which may be compared with the LANDSAT results. Let us consider the minimum set of parameters needed to describe the reflectivity of a fair weather cumulus cloud field :

$$
\begin{aligned}
&\alpha_r = 0.6, \text{ power of the } \sqrt{\text{area}} \text{ distribution,} \\
&A_r = (1/2 \text{ km})^2, \text{ maximum area,} \\
&\beta_r = 1.6, \text{ power of the spacing distribution,} \\
&x_r = 1 \text{ km, maximum spacing,}
\end{aligned}
\tag{4.1}
$$

where the subscript reminds us that these are parameters of the reflectivity field. We assume that the distribution of the square root of the area falls off with a power of -0.6 out to an area of 0.5 km, and that the nearest-neighbor spacing is determined by a Levy walk with a power of 1.6 out to a maximum spacing of 1 km. Larger cloud areas and larger spacings are assumed to occur by random coincidences.

One procedure for generating a field of liquid water which can reproduce the above parameters is as follows :

1. choose some simple initial distribution of liquid water, such as a homogeneous ellipsoid, which we shall term a "zero generation cloudlet", labelled c_0;
2. choose a power, α_{lw}, for the distribution of cloudlet horizontal size, so that the probability density of size goes as $(\text{size})^{-\alpha_{lw}}$ out to some specified maximum size, and let the orientations be uniformly distributed;
3. choose a power, β_{lw}, for the distribution of cloudlet horizontal spacings, so that the probability density of spacings goes as $(\text{spacing})^{-\beta_{lw}}$ out to some specified maximum spacing, and let the direction to the neighboring cloudlet be distributed isotropically;
4. generate a "first generation cloudlet" from the zero generation one, $c_1 = F(c_0)$, by superimposing c_0 cloudlets with sizes and spacings chosen from the above distributions, and successive locations being generated by a Levy walk; where cloudlets overlap, the liquid water is added up until a "saturation" value is reached, and the cloudlet in which saturation first occurs is defined to be c_1;
5. iterate the above, so that $c_2 = F(F(c_0))$, and so on.

As a result of the iteration, the liquid water field eventually becomes insensitive to details of the c_0 cloudlet, but depends upon the parameters of the size and spacing distributions. Parameters of the associated reflectivities, computed by Monte Carlo methods, also depend upon the liquid water parameters. They must be adjusted to reproduce (4.1), so that for example $\alpha_r(\alpha_{lw}, A_{lw}, \beta_{lw}, x_{lw}) = 0.6$.

Even this minimal attempt to reproduce the LANDSAT analysis involves a number of parameters, and the Monte Carlo computation introduces additional parameters, as we shall see in the next section. For the sake of theoretical simplicity in understanding the radiative results, it is useful to introduce a highly simplified fractal cloud model involving a single fractal parameter.

Our simplified fractal liquid water distribution has a constant mean vertical optical depth, $\tau_v(0)$, and a constant mean horizontal optical depth, $\tau_h(0)$. We begin with a plane parallel cloud which is infinite in one horizontal direction and has, for example, $\tau_h(0) = 4000$ in the other horizontal direction, and $\tau_v(0) = 10$ in the vertical. The cloud is then divided horizontally in half, and one of the halves is chosen at random in which τ_v is increased by a factor, say $f = 0.3$, while τ_v is decreased by the same factor in the other half. The total amount of liquid water and the mean optical depth remain unchanged. The cloud thickness is also kept fixed, so that the changes in τ_v are due to changes in the droplet density alone, and may be thought of as a transfer of droplets from one half to the other. Each half is then subdivided again in half, and the same process of droplet transfer carried out in each half, so that one has 2^2 regions, each with

$$\tau_h(2) = (1 \pm f)(1 \pm f)\tau_h(0)/2^2, \text{ and}$$

$$\tau_v(2) = (1 \pm f)(1 \pm f)\tau_v(0). \tag{4.2}$$

where the + and - signs occur with equal probability. The sum of all the values of $\tau_h(2)$ equals $\tau_h(0)$, so that the average horizontal optical depth is $\tau_h(0)/2^2$. The average vertical optical depth is of course $\tau_v(0)$. This subdivision process is then iterated down to a small "inner scale". In the above example the average τ_h is reduced to about one after twelve iterations, at which point τ_v varies between about 0.1 and 10 (see e.g. figure 1). The wavenumber spectrum of τ_v is approximately a power-law, with a power depending on f (see next section).

Figure 1 shows a horizontal cross-section of the vertical optical depth variations for one realization of the subdivision process described above. The optical depth is uniform in the perpendicular plane. The cloud height and the droplet size distribution are assumed constant, so that these variations may be thought of as variations in droplet density only. In the next section we discuss Monte Carlo radiative transfer computations for this highly simplified fractal cloud.

HORIZONTAL VARIATION OF OPTICAL DEPTH
AFTER N = 12 ITERATIONS

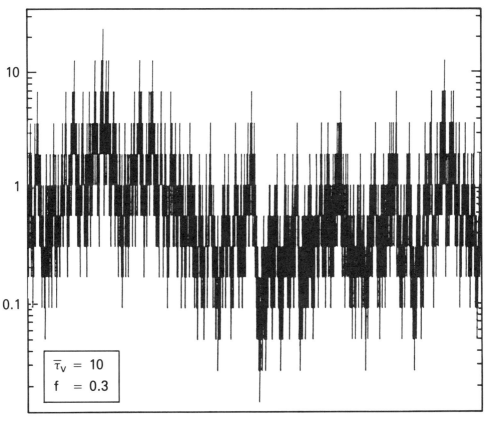

HORIZONTAL COORDINATE X

Figure 1. One realization of the horizontal variation of optical depth for the subdivision process described in the text after 12 iterations. (The optical depth is uniform in the other horizontal direction.) The initial horizontal optical depth of 4000 has been divided into 2^{12}=4096 segments with average horizontal optical depth of about 1 and vertical optical depth varying from about 0.1 to 10. The mean vertical optical depth is 10, and does not vary during the iteration.

5. MONTE CARLO COMPUTATIONS

The radiative transfer results described in this section are from Wiscombe et al. (1988). Although the results depend on only a single fractal parameter, the parameter f described in the preceeding section, there are additional parameters associated with the iteration process, and with the Monte Carlo computation itself. We first list the input parameters associated with the optical depth variations, along with our "typical values" :

1. $f = 0.3$:   fractal parameter;
2. $\tau_h(0) = 4000$:   initial horizontal optical depth;
3. $N = 12$:   number of iterations;
4. $NR = 10$:   number of realizations
5. $NP = 80,000$:   number of photons per realization;

and finally we list the input parameters in common with the usual one-dimensional radiative computations :

6. $\tau_v(0) = 10$:   mean vertical optical depth;
7. $\omega_0 = 1$:   single scattering albedo;
8. $\theta = 60°, \phi = 0°, 180°$:   sun angle;
9. $g = 0.85$:   asymmetry parameter
                             (Henyey-Greenstein phase function)

The power spectrum of the subdivision process may be determined from the covariance, which may be computed as follows: If, after N subdivisions, we let R be the ratio of vertical optical depth at some typical point to the mean vertical optical depth, $R=\tau_v(N)/\tau_v(0)$, then it is a product of N factors $1\pm f$, and the logarithm

$$\ln(R) = \sum_{k=1}^{N} x_k$$

(5.1)

is a sum of N independent identically distributed random variables, $x_k = \ln(1\pm f)$, and is therefore normally distributed for large N. If we consider two points separated by distance d, they will become separated into two segments after, say, n steps, where n is determined by

$$n = \{\text{smallest } k \ni d \geq L/2^k\}.$$

(5.2)

where L is the horizontal length corresponding to $\tau_h(0)$. The covariance may be determined using the two identities

$$R_1 R_2 = \exp[\ln R_1 + \ln R_2],$$

(5.3)

and

$$<\exp[y]>=\exp[<y>+\text{var}(y)/2],$$

(5.4)

where angular brackets indicate an ensemble average, and y is any normally distributed

OPTICAL DEPTH POWER SPECTRUM
10-REALIZATION AVERAGE

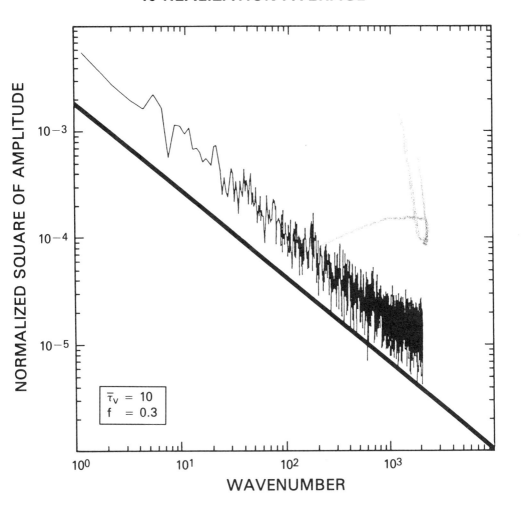

Figure 2. Average wavenumber spectrum of horizontal optical depth variations for 10 realizations similar to that shown in figure 1. The solid straight line corresponds to the theoretical form given by (wavenumber)$^{-0.86}$, where the exponent, determined by (5.8) in the text, is given by $1-[\ln((1+f)/(1-f))/2]^2/\ln2$ with f=0.3.

random variable with variance var(y). The first n terms in $\ln R_1$ and $\ln R_2$ (before the points are separated) will be identical, and the last N-n will be independent, so that

$$\ln R_1 + \ln R_2 = 2 \sum_{k=1}^{n} x_k + \sum_{k=1}^{2(N-n)} x_k$$

(5.5)

This is normally distributed for large N, so that the mean of (5.3) may be found from the identity (5.4), using

$$\text{var}(\ln R_1 + \ln R_2) = 4n\text{var}(x) + 2(N-n)\text{var}(x).$$

(5.6)

The n-dependent part behaves as

$$\langle R_1 R_2 \rangle \sim \exp[n\text{var}(x)] \sim (L/d)^{\text{var}(x)/\ln 2}.$$

(5.7)

The power spectrum corresponding to the covariance (5.7), given by the Fourier transform, behaves as

$$S \sim k^{-[1-\text{var}(x)/\ln 2]}.$$

(5.8)

where k is the wavenumber. This power-law satisfies the relation (2.1) expected of a self-similar scaling fractal. The average spectrum for 10 realizations of the subdivision process is shown on a log-log plot in figure 2, and decreases with a slope given by the exponent in (5.8). The random deviations from this power-law behavior are due to sampling error, and decrease with the square root of the number of realizations.

After carrying out the Monte Carlo computations with the parameters listed above, perhaps the simplest quantity to examine is the horizontally averaged albedo. Figure 3 shows a plot of the average albedo versus N, the number of iterations, for 10 realizations of the subdivision process. The error bars give the albedo range, and the solid line the mean albedo over the 10 realizations. As one subdivides to finer and finer scales, the albedo initially decreases linearly with N, and then approaches a constant after about N=8, where the horizontal optical depth of each segment becomes less than about 10, so that the aspect ratio becomes less than 1.

The linear regime, in which N is small and each segment is horizontally optically thick, can be approximated by computing the albedo of each segment independently as if it were plane-parallel. This approximation is shown as the line labelled "independent pixels" in figure 3. The redistribution of liquid water at each iteration decreases the mean albedo from the plane parallel case, since the albedo of optically thick regions saturates as their optical depth is increased, while the thin regions continue to darken as their optical depth is decreased.

The albedo of each homogeneous region may be computed from the thickness of each region independently only when the horizontal optical depth is large compared to the photon mean free path. As we approach N=8 in figure 3, each segment becomes comparable in horizontal optical depth to the photon mean free path, and the albedo depends upon radiation from the sides of each segment. As we approach the large N asymptotic regime, the mean albedo becomes insensitive to variations in optical depth occuring on horizontal scales much smaller than the photon mean free path, so that further subdivisions produce no further decrease.

MEAN ALBEDO
VS.
NUMBER OF ITERATIONS

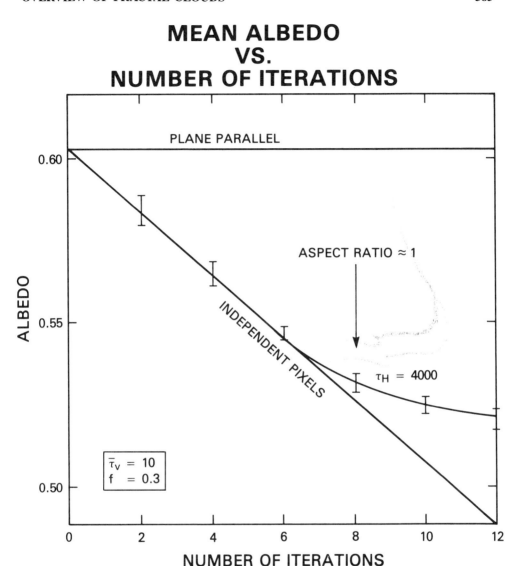

Figure 3. Mean albedo versus the number of iterations of the subdivision process. The error bars give the range of values for 10 realizations. The horizontal line gives the plane-parallel albedo for optical depth = 10. The line labelled "independent pixels" gives the average albedo obtained by setting the albedo of each segment equal to the plane-parallel albedo corresponding to its particular optical depth. As the number of iterations increases, the actual albedo determined by Monte Carlo follows the independent approximation until the horizontal optical depth becomes comparable to the vertical optical depth, which occurs near N=8 (since $4000/2^8 \approx 10$), beyond which the albedo becomes insensitive to further subdivision.

The effect of the photon mean free path on the horizontal variations in albedo may be seen in figure 4, which shows the average power spectrum of the albedo on a log-log plot for 10 realizations of the subdivision process. The large-scale (small k) behavior shows the same type of power-law decrease as the optical depth, while the small scale behavior is independent of k, indicating purely uncorrelated variations, or "white noise".

A feeling for the type of errors made by the linear "independent pixel" approximation can be gained by comparing its horizontal variation with that of the Monte Carlo result for a single realization. For each of the curves in figure 5 the white noise has been filtered out by applying a triangular running mean over the horizontal scale $\tau_h=\pm50$. The solid curve shows one realization of the Monte Carlo horizontal variation of albedo. The dotted curve shows the independent pixel approximation, in which the albedo is a monotonic function of the local optical depth. It clearly underestimates the albedo in each of the darker regions, near local minima of the optical depth, but provides a good approximation in the brighter regions. This difference is due to the radiation from the sides of the optically thinner regions, which considerably brightens them. Note that the solar radiation, which is incident from the left in the figure, tends to enhance the albedo on the left of each of the optical depth maxima. This solar zenith angle dependence also depends upon the fractal parameter f.

To improve the independent pixel approximation, we must estimate the effect of radiation transferred from the optically thick regions into the thin ones. One way is to employ an "effective" optical depth which depends on the environment of each segment. As a first guess, we simply apply the same ±50 triangular running mean to the optical depth to obtain an "effective" optical depth. This "filtered independent pixel" approximation is given by the dashed line in figure 5. It provides a clear improvement in the dark regions, and a better estimate of the mean albedo, but does not show the solar zenith angle dependence of the Monte Carlo result.

6. DISCUSSION

We have given a brief overview of work being done at Goddard Space Flight Center and Tel-Aviv University on the fractal properties of clouds and their implications for large-scale radiative properties of the atmosphere. Three examples of relatively simple fractals were described to illustrate concepts needed in modelling fractal cloud structure. The results of empirical studies of cloud structure determined from LANDSAT observations were then summarized, emphasizing the dependence of the fractal parameters on cloud type and threshold. A procedure for simulating the type of structure observed by LANDSAT was described, with a summary of the various parameters involved, and a one-parameter fractal model was introduced, which despite its lack of realism provides a useful tool for the investigation of radiative transfer through fractal clouds. The radiative properties of this "simplest" fractal cloud were summarized, with emphasis on the importance of the optically thin regions, and the role of the photon mean-free-path.

Work along these lines is continuing in our group and elsewhere (see e.g. references by Lovejoy and Schertzer, Rhys and Waldvogel, and Wielicki and Welch). Much remains to be done in terms of understanding the physical basis of fractal cloud structure and the process of radiative transfer through fractal clouds. This improved understanding may lead to improved remote sensing of cloud properties, and better parameterization of clouds in climate models.

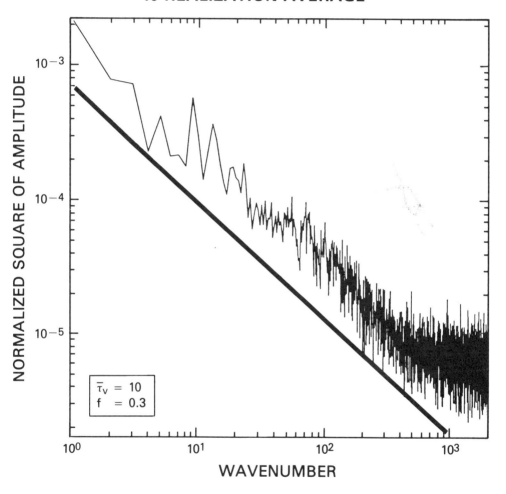

Figure 4. Average wavenumber spectrum of horizontal albedo variations for 10 realizations of albedos computed from optical depth variations like those shown in figure 1. The spectrum initially follows the power law indicated by the straight line, but becomes independent of wavenumber above about 400, which corresponds to horizontal scales smaller than about 10 optical depth units, the same scales to which the mean albedo in figure 3 is insensitive.

HORIZONTAL VARIATION OF ALBEDO

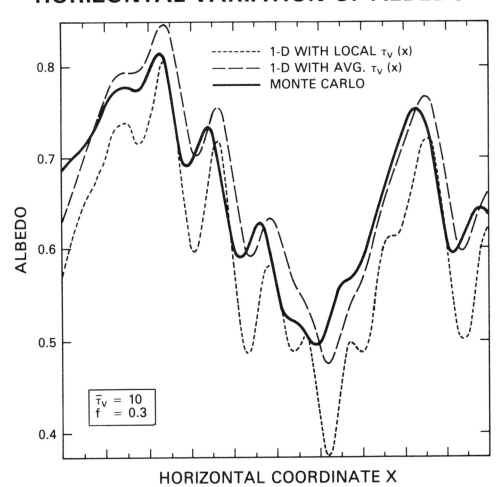

Figure 5. One realization of the horizontal variation of albedo, with the incident solar radiation from the left at a zenith angle of 60°. All three curves have been smoothed over horizontal optical depths of ±50 with a triangular filter. The dotted line is obtained by setting the albedo of each segment equal to the plane-parallel albedo corresponding to its particular optical depth. Thin regions appear very dark in this approximation. The solid line is the Monte Carlo result, which shows the brightening effect of neighboring segments on the thin regions. The dashed line is obtained by setting the albedo of each segment equal to the plane-parallel albedo associated with the <u>average</u> vertical optical depth over the region 50 units on either side. This produces a better approximation to the Monte Carlo result, especially in the thin regions, but tends to overestimate on the downstream side of the thick regions.

REFERENCES

Baker, M. and J. Latham, 1979: The evolution of droplet spectra and the rate of production of embryonic raindrops in small cumulus clouds, *J. Atmos. Sci.*, **36**, 1612-1615.

Cahalan, R. F., 1986: Nearest neighbor spacing distributions of cumulus clouds, Proceeding of the 2nd International Conference on Statistical Climatology, Vienna, Austria, June, 1986.

_____, 1987: LANDSAT observations of fractal cloud structure, in **Scaling, Fractals and Nonlinear Variability in Geophysics**, D. Schertzer and S. Lovejoy, ed., Reidel, in press.

_____, and J.H. Joseph, 1988: Fractal statistics of cloud fields, *Mon. Wea. Rev.*, to appear.

Caughey, S. J., B. A. Crease and W. T. Roach, 1982: A field study of nocturnal stratocumulus II. turbulence structure and entrainment. *Quart. J. R. Met. Soc.*, **108**, 125-144.

Coakley, J. A., Jr., and F. P. Bretherton, 1982: Cloud cover from high-resolution scanner data: detecting and allowing for partially filled fields of view. *J. Geophys. Res.*, **87**, 4917-4932.

Derr, V. and R. Gunter, 1982: EPOCS 1980: Summary Data Report-- Aircraft Measurements of Radiation, Turbulent Transport and Profiles in the Atmospheric and Oceanic Boundary Layers of the Tropical Eastern Pacific, NOAA Tech. Memo. ERL WPL-101, NOAA Wave Propagation Lab, Boulder, Colorado.

Falconer, K.J., 1985: **The Geometry of Fractal Sets**, Cambridge University Press.

Farmer, J. D., E. Ott and J. A. Yorke, 1983: The dimension of chaotic attractors, *Physica*, **7D**, 153-180.

Frisch, U. and O. Parisi, 1985: On the singularity structure of fully developed turbulence. in **Turbulence and Predictability in Geophysical Fluid Dynamics and Climate Dynamics**, M. Ghil, ed., Elsevier Science Publishers B.V., Amsterdam, The Netherlands, pp. 84-88.

Hentschel, H.G.E. and I.Procaccia, 1983: The infinite number of generalized dimensions of fractals and strange attractors. *Physica*, **8D**, 435-444.

_____and_____, 1984: Relative diffusion in turbulent media, the fractal dimension of clouds. *Phys. Rev.*, **A29**, 1461-1476.

Lovejoy, S., 1982: Area-perimeter relation for rain and cloud areas. *Science*, **216**, 185-187.

_____, and B. B. Mandelbrot, 1985 : Fractal properties of rain and a fractal model. *Tellus*, **37A**, 209-232.

_____, and D. Schertzer, 1985 : Generalized scale invariance in the atmosphere and fractal models of rain. *Water Resour. Res.*, **21**, 1233-1250.

Mandelbrot, B. B., 1983: **The Fractal Geometry of Nature**. W. H. Freeman and Co., San Francisco, 460pp..

_____, 1986: Self-affine fractal sets. Parts I., II. and III., in **Fractals in Physics**, L. Pietronello and E. Tosatti, ed., Elsevier Science Publishers B.V., Amsterdam, The Netherlands, pp. 3-28.

Randall, D. A. and G. J. Huffman, 1980: A stochastic model of cumulus clumping. *J. Atmos. Sci.*, **37**, 2068-2078.

Rhys, Franz S., and A. Waldvogel, 1986: Fractal shape of hail clouds. *Phys. Rev. Lett.*, **56**, 784-787.

Schertzer, D. and S. Lovejoy, 1986 : Generalized scale invariance and anisotropic inhomogeneous fractals in turbulence, in **Fractals in Physics**, L. Pietronello and E. Tosatti, ed., Elsevier Science Publishers B.V., Amsterdam, The Netherlands, pp. 457-460.

_____and_____, 1988 : Multifractal simulations and analysis of
      clouds by multiplicative processes. *Atmospheric Research*, **21**, in press.
Shlesinger, M. F., B. J. West and J. Klafter, 1987 : Levy dynamics of enhanced
      diffusion: application to turbulence, *Phys. Rev. Lett.*, **58**, 1100-1103.
Wielicki, B.A. and R.M. Welch, 1986 : Cumulus cloud properties derived using
      LANDSAT satellite data, *J. Clim. Appl. Meteor.*, **25**, 261-276.
Wiscombe, W. J., W. Ridgway and R. F. Cahalan, 1988: Radiative properties of quasi-
      fractal clouds, Proceedings of Lille Conference.

DISCUSSION

King: You know when you think of ice crystals they are hexagons because they represent a minimum energy configuration of hydrogen and oxygen atoms. My question is, is there anything in the physics of cloud formation, that is, nucleation, condensation, evaporation and so forth, that would lead one to a physical theory that they would obey fractal geometry, or in other words, is fractal geometry a phenomenological theory or is it a physical theory?

Cahalan: There is no physics in what I have talked about here. This is a physics-free description of the liquid water distribution in the cloud. If you try to put in some simple physics like you were talking about and have some sort of a minimum energy configuration, you tend to get clouds that are like soap bubbles. You start out with a bunch of little ones, little ones combine and beget bigger ones and pretty soon your whole region is covered with cloud. On the average, observed cloud fraction tends to be about 50%. The way I think of that is these cells that I was talking about that you are generating the larger ones from have a longer lifetime and you have a largest cell and that has the longest lifetime. If you start generating a bunch of these and let them die after a certain time, you are going to tend to get something like 50% cloud cover from that sort of model. And the physics is what determines the lifetime.

Neuendorffer: It would seem like you might be able to do another step improving it if you increased thickness as well as increase the amount of water. Is that going to be complicated?

Cahalan: So far we just took the simplest theoretical model because we have many parameters to vary and we've been burning CYBER time. But eventually the model will be motivated by the FIRE data analysis of both height and liquid water variations. The Monte Carlo calculations are set up to be able to deal with an arbitrary three-dimensional distribution.

Neuendorffer: It is very interesting work. Are you the only people doing this kind of thing?

Cahalan: Bruce Wielick at NASA/Langley is also doing analysis of LANDSAT data. Shawn Lovejoy at McGill in Montreal is working on radiative transfer theory but using a six flux type of model.

Speaker: I just wondered how confident are you in your Monte Carlo scheme?

Cahalan: We have error bars in this curve of albedo vs. number of cascades. Each of these points is a computation of 80,000 photons in a single realization and we show 10 realizations there. So, that is the spread that you would get and the solid line is just the mean of those.

Westwater: Many turbulence processes have a spectral decay which goes as -5/3rds in wave number. Is there any connection between the slope of that line and fractal dimension?

Cahalan: Originally, Lovejoy suggested that he was getting 4/3rds because if you consider a temperature field having independent Gaussian variations from point to point then areas above a threshold have a fractal dimension of 4/3. However, more recent data analysis finds that there is no single magic number. Different cloud types have different fractal dimensions. So, I think there is going to be more physics in it. The way the cloud forms is going to matter. For example, fair weather cumulus are driven by heating from below, whereas stratocumulus are driven by cooling from above, and have a strong inversion layer capping them. We find a difference in fractal properties between those two cloud types.

Kleespies: Once you have characterized a cloud field with fractal properties have you thought of any way of using the fractal numbers in paramaterizing the radiative transfer and, in particular, the radiation budget?

Cahalan: That is what we are working toward. These results that I showed you with this highly simplified model of the mean albedo depend on the fractal parameter F. If you believe that type of model (and actually it might apply in certain cirrus situations where you have long streaks of various optical depths) then, you can parameterize large-scale average albedo in terms of the single parameter F. But for other cloud types I think that the average albedo will depend upon additional parameters characterizing such things as cloud spacing and cloud height.

AUTOMATIC CLOUD FIELD ANALYSIS BASED ON SPECTRAL AND TEXTURAL SIGNATURES[1]

Jean Tournadre
IFREMER
Cedex, France

Catherine Gautier
California Space Institute
Scripps Institution of Oceanography
La Jolla, California 92093, USA

ABSTRACT

An automated method for analyzing cloud fields by means of cloud classification is presented. The method, an unsupervised three-step procedure, begins with a cloud field characterization based on a set of textural and spectral cloud parameters (e.g., minimum brightness) computed from pairs of visible and infrared (IR) satellite images. Second, a cloud feature extraction, accomplished by Typical Shape Function (TSF) analysis, is performed using the set of characterizing parameters. Lastly, a cloud field classification is made by regrouping the cloud features (represented by vectors in an n-dimensional space) into classes according to their similarity. The similarity criterion used is the correlation between the vectors describing the cloud features. An initial classification base is obtained from the independent classification of several visible and IR image pairs. The classes obtained for each image pair are then regrouped according to their similarity and a new global base of classes is formed and used to reclassify the image pairs. This approach provides an evolving classification base and thus permits continuous inclusion of new data sets. The method has been applied to a set of twenty five visible and IR image pairs, obtained from the GOES-6 satellite for a region over the Eastern Pacific where *in situ* measurements were collected during the Mixed Layer Dynamics Experiment, 1983. The image pairs have been classified and a base of eight cloud classes has been obtained. These classes correspond to clear conditions and characteristic cloud types such as: low-level broken stratus, low-level uniform stratus, vertically extending clouds (cumulus, nimbus), etc.

1. INTRODUCTION

Operational meteorological satellites have provided a wealth of new data from which only a small part of their potential information has been extracted. This is particularly the case for multispectral geostationary satellite data, obtained from the GOES satellite series' Visible and Infrared Spin Scan Radiometer (VISSR) instrument, because the data are voluminous (equivalent to 1 megabyte per minute per satellite) and require extensive data processing. Although individual satellite images can provide qualitative information such as

―――――――――――――
[1]This research has been supported through grant #N00014-86-K-0752 from the Office of Naval Research.

cloud and surface properties, in most cases satellite radiation measurements from a single image are not directly interpretable as meteorological parameters. Instead they require either interpretation models (statistical or physical based algorithms) to retrieve parameters of direct physical relevance or image combinations in time to provide a picture of atmospheric dynamics. Most available meteorological satellite data sets have been treated as "point" measurements, including the vertical dimension (obtained from the "inversion" of spectral information), but seldom has the data surrounding the point measurement (i.e., regional information) been used to estimate meteorological parameters (e.g., Debois et al., 1982; Garrand, 1986).

Clouds represent some of the most important information contained in visible, infrared, and microwave satellite measurements. Although it might appear simple to delineate cloud regions from visible or infrared satellite measurements, determining cloud boundaries is not trivial (e.g., Coakley and Bretherton, 1982). Extraction of cloud physical properties is even more complicated. The "inversion" of cloud radiances into cloud properties is usually an ill-conditioned problem, having many possible solutions. Nevertheless, using simple radiative models of clouds and the atmosphere, it is possible to extract some cloud properties (e.g., cloud top temperature, cloud optical thickness).

For many applications it is not necessary to know all cloud physical properties, and in many cases it is sufficient to classify clouds as a function of their type. This approach, nephanalysis, has been used for many years in meteorological and aeronautical applications. Such cloud mapping and classification, however, have not been straightforward to implement and, in particular, to automate. Many nephanalyses are still produced by hand. Although automated techniques have been developed (e.g., Parikh, 1977; Phulpin et al. 1983), they have not always made optimal use of data because of computing limitations. With increasing computing capabilities and renewed interest in cloud analysis, especially over large scales and using space observations (e.g., International Satellite Cloud Climatology Program--ISCCP), there is a means and a need to improve satellite based cloud type classification methods. Furthermore, cloud type classification methods can serve as a basis for more complex studies, such as those relating cloud types to other atmospheric or surface parameters (e.g., Garrand, 1986). Cloud type classification methods also compact data set size and can thus facilitate analyses.

This paper presents a new cloud type classification method that uses standard image parameters, such as textural and spectral information to characterize clouds. The proposed method simultaneously performs two important tasks: 1) cloud cover estimation, and, 2) cloud type classification. It is an unsupervised classification method and has the advantage over some other methods in that it uses most of the regional information contained in the satellite images.

2. METHOD DESCRIPTION

The classification method follows a three-step procedure: 1) a cloud field characterization based on a set of cloud parameters; 2) a cloud feature (type) extraction based on the sets of characterizing parameters; and 3) a cloud field classification based on characteristic features encountered in the data set. This implies that the choice of characterizing parameters is crucial, since they entirely define the cloud "properties" for the latter classification.

2.1 CLOUD SPECTRAL AND TEXTURAL FEATURES

Parikh (1977) demonstrated that spectral and textural features of visible and infrared satellite images could be used for cloud classification. The spectral features are functions of the density function $p(i)$ of the image gray level g (i.e., brightness or radiance); and the textural features are functions of the joint density functions, $d_{\theta\rho}(g_1,g_2)$, of gray level pairs (g_1, g_2) separated from each other in direction θ ($0^o,45^o,90^o,135^o$) by distance $\rho(1,2,4,8)$. The spectral features measure quantities such as reflectance (visible), cloud top temperature (infrared), or variations in reflectance and temperature (minimum, maximum, mean, and range). A total of 25 different spectral features can be computed; details can be found in Parikh, 1977.

The textural features are computed over np x np pixel cloud samples and provide a good estimate of factors such as overall homogeneity and the amount of local variation. If i represents the absolute value ($g_1 - g_2$) of the difference between gray levels g_1 and g_2 separated in direction θ by a distance ρ, then the frequencies $f(i)$ define the gray level difference histogram. The frequency, $f(i)$, is the number of times the difference i occurs between gray level pairs g_1 and g_2 (for specified ρ and θ) in the sample array. From the frequency histograms, $f(i)$, and assuming the gray levels range from 0 to 255, the following textural features can be defined as:

$$\text{Mean,} \quad M = \sum_{i=0}^{255} i \left(\frac{f(i)}{N}\right) \qquad (1)$$

$$\text{Contrast,} \quad C = \sum_{i=0}^{255} i^2 \left(\frac{f(i)}{N}\right) \qquad (2)$$

$$\text{Angular Second Momentum,} \quad ASM = \sum_{i=0}^{255} \left(\frac{f(i)}{N}\right)^2 \quad (3)$$

$$\text{Entropy,} \quad E = -\sum_{i=0}^{255} \left(\frac{f(i)}{N}\right) \log \left(\frac{f(i)}{N}\right) \quad (4)$$

where N is the total number of pair points in the sample array separated by distance ρ in direction θ.

Parikh estimated 143 different textural features, but showed, using a Fisher distance and a maximum likelihood operator for feature selection, that the most important features for cloud type characterization and classification are: minimum and maximum (or range) visible brightness and infrared temperature, visible textural, mean contrast, angular second momentum (ASM), entropy for distance 1, and infrared temperature entropy for distance 1.

2.2 CLASSIFICATION WITH TYPICAL SHAPE FUNCTIONS

Although classification can be performed in a number of ways, here we chose the classification scheme of Jakilee and Ropolevski (hereafter referred to as J & R). It is an unsupervised classification scheme that was primarily designed for temperature and humidity profile classification in order to organize profile sets as a function of the "kinks" or typical shapes in each profile. Its original goal was to regroup profiles that displayed similar characteristics at the tropopause (e.g., similar gradients at equivalent altitude). For the case of cloud field classification, the scheme characterizes the relationships that exist between the spectral and textural parameters, which form a vector in an n-dimensional space. For similarity with the classification's original use, such vectors will be called "cloud profile" in the remainder of the text.

Let us consider a set of p cloud profiles of m spectral and textural parameters computed over a $np \times np$ pixel cloud sample array. This set can be assembled as an ($m \times p$) matrix A (with $p > m$). The jth column vector a_j of A is then the jth "cloud vector". According to the singular value decomposition theorem, the matrix A can be written as follows:

$$A = X \Lambda Y^T \quad (5)$$

where: $\quad X X^T = X^T X = I_m \quad \text{and} \quad Y^T Y = I_m \quad \text{and} \quad Y Y^T = I_p$

The ($m \times m$) matrix X is the left singular vector matrix, Λ is the ($m \times m$) diagonal matrix of the singular values, and Y is the (p x m) matrix of the right singular vectors. It should be noted that the singular values, λ_i , are the non-negative square roots of the eigenvalues of the covariance matrix AA^T and are, by convention, ordered in such a way that $\lambda_i \geq \lambda_j$ (if i < j).

Following relation (5), a profile a_j can be expressed by the following expansion:

$$a_j = \sum_{k=1}^{m} x_k \, \lambda_k \, y_{jk} \qquad (6)$$

The portion of the overall variance V_j, explained by each left singular vector X_j, is proportional to the square of the associated singular value λ_j (i.e., the associated eigenvalue of the covariance matrix):

$$V_j = \frac{\lambda_j^2}{\sum_{k=1}^{m} \lambda_k^2} \qquad (7)$$

Since the data are highly correlated, the first few left singular vectors associated with the largest singular values explain nearly all the variance in the data. The remaining left singular vectors are associated with noise. The matrix A can then be approximated by a truncated expansion of m' singular left vectors, with $m' \leq m$; for a profile a_j we thus obtain:

$$a_j \approx a_j^{(m')} = \sum_{k=1}^{m'} x_k \, \lambda_k \, y_{jk} \qquad (8)$$

or in matrix form:

$$A \approx A^{(m')} = X^{(m')} \, \Lambda^{(m')} \, Y^{(m')T} \qquad (9)$$

where $X^{(m')}$ is the (m x m') matrix built from the m' first left singular vectors, $\Lambda^{(m')}$ is the (m' x m') diagonal matrix of the m' largest singular values, and $Y^{(m')}$ is the (p x m') matrix built from the m' first right singular vectors. It can be shown that $A^{(m')}$ is an optimum approximation of A in a least square sense (J & R).

As shown by J & R, the singular vectors that form the matrix X do not lend by themselves to an exclusive classification of the data. They only describe the main features. In order to classify the data, J &R demonstrated the need to rearrange the leading terms x_j in the approximation (i.e., find a linear combination of these terms) to form characteristic (or typical) shape functions representative of a unique class.

The matrix $A^{(m')}$ can be rewritten as:

$$A^{(m')} = \hat{A}^{(m')} \hat{Y}^{(m')T} \tag{10}$$

in which:

$$\hat{Y}^{(m')T} = \alpha Y^{(m')T} \tag{11}$$

and

$$\hat{A}^{(m')} = X^{(m')} \Lambda^{(m')} \beta^T \tag{12}$$

and where α and β are rotation matrices. Substituting equations (11) and (12) in (10), we see that, in order to preserve the approximation, the product $\beta^T \alpha$ must be equal to the identity matrix.

Following J & R, the matrix α is first computed using an iterative procedure so that the correlation between the rows of $Y^{(m')}$ is minimum, i.e.,

$$\sum_{i,j=1}^{m'} \sum_{k=1}^{p} \left(\hat{y}_{ki}\ \hat{y}_{kj}\right)^2 = \text{minimum} \tag{13}$$

This is achieved using Lagrangian coefficients. The minimization expressed in relation (13) ensures an exclusive classification, if at all possible (i.e., by the J & R method it means if the iterative computation of α converges). Once the matrix α is computed, the matrix β is determined and then the matrix $\hat{A}$ of TSFs is obtained using (12). Once the TSFs are obtained, they can be used to divide the data set into classes. For each profile a_j, the correlation (C_k) with each TSF $\hat{a}_k$ is computed:

$$C_k = \frac{\hat{a}_j\ \hat{a}_k}{\|\hat{a}_j\|\ \|\hat{a}_k\|} \tag{14}$$

The profile is then associated with the TSF class for which the correlation is the highest.

3. DATA SET, NORMALIZATION, AND DESCRIPTION

3.1 DATA SET DESCRIPTION

To test the cloud type classification method outlined above, the method was applied to a set of satellite images collected over a region of the Pacific Ocean where *in situ* oceanographic and meteorological measurements were made during the Mixed Layer Dynamics Experiment (MILDEX) from October 30 to November 18, 1983. The region analyzed covers 500 x 500 km^2 and is centered at 34° N and 125° W.

Both visible and infrared data were used from the GOES-6 satellite's VISSR/VAS instrument. VISSR's visible component is sensitive to upwelling radiant energy in the 0.5 to 0.8 μm part of the visible spectrum; its ground resolution at the latitude of the studied region is approximately 1.5 km. VISSR's infrared component is sensitive to upwelling radiant energy in the 10 to 12 μm atmospheric infrared window; its ground resolution is 8 times lower (i.e, about 12 km), but the data is recorded at a 4 times lower resolution.

The 50-day analysis period contains a variety of meteorological conditions and cloud types. To avoid the difficulty of a normalization procedure, which would be necessary to correct for sun angle variation effects on the visible brightness, we used visible and the infrared images acquired each day at the same hour, 18:45 GMT, near local noon.

3.2 NORMALIZATION

From the definition of the spectral parameters given in equation 2, it is obvious that these parameters range from 0 to 255. For the textural parameters, it can be shown that, using the two extreme possibilities for the image distribution (i.e., a "disorganized" and an homogeneous image), the parameter ranges are as follows:

$0 \leq$ mean ≤ 1

$0 \leq$ contrast $\leq (255^2)/3$

$1/255 \leq$ ASM≤ 1

$0 \leq$ entropy $\leq \log 255$

The spectral and textural parameters can therefore differ by up to two orders of magnitude. If the TSF analysis is performed on the unnormalized data matrix A, the differences between the spectral and textural parameters will overcome the differences between the individual profiles (which are highly correlated). This problem is eliminated by normalizing the parameters. The spectral and textural parameters are normalized by the mean and the variance of each parameter computed for the entire image data set. Thus each original profile a_j is transformed into a normalized profile a_{ij} through:

$$\bar{a}_{ij} = \frac{a_{ij} - \bar{a}_i}{\sigma_{ai}} \qquad i = 1, ..., P \qquad (15)$$

where a_i and σ_{ai} respectively represent the mean and standard deviation of parameter i over the entire data set P = 1, 2,...., m.

3.3 CLASSIFICATION OF THE DATA SET

For the cloud classification of the data set, 25 pairs of coincident visible and infrared images were selected. For each image pair, spectral and textural parameters are computed over a sample array of np x np pixels. Following Parikh, 1977, and after some tests for optimal cloud representation, seven visible parameters were computed: minimum, maximum textural mean, contrast, entropy, and ASM in direction 0 (zonal direction) and distance 1. Only three infrared parameters were computed: minimum and maximum temperature and entropy in direction 0 and distance 1. The textural parameters were computed only for direction 0 because differences in calibration and response function exist between the eight photodiodes that compose the visible component of the VISSR instrument and north-south differences in visible brightness could be misinterpreted for cloud texture in that direction.

The selected array size sample, np x np, covers 16 x 16 pixels (or approximately 24 x 24 km) for the visible data. This represents a compromise between a sample size allowing, on the one hand, the computation of statistically significant textural parameters and, on the other, one that covers an homogeneous cloud field. To keep the same area size for the infrared, the sample array selected was 4 x 4 pixels.

Because of computer limitations and our desire to build an evolving class set, we performed the TSF analysis on each individual image pair and then regrouped the obtained class for the entire set of image pairs analyzed. We were able to obtain P_j TSFs for each image pair or $P = \sum_{i=1}^{25} P_i$ TSFs for the entire set. The problem was then to determine among the set of TSFs $\{a_k\}$ k = 1, ..., P, those representing the same cloud type. This meant regrouping the TSFs according to some criteria. Our selected criteria, based on the correlation that exists between all the TSFs, are computed from:

$$C_k = \frac{\hat{a}_k \, \hat{a}_{k'}}{\| \hat{a}_k \| \, \| \hat{a}_{k'} \|} \qquad (16)$$

The TSFs that have correlations greater than 0.9 are assumed to represent the same cloud type. A TSF base of m_1 TSFs is then determined by averaging the TSFs representing the same cloud type. Each image pair is reclassified using the method described earlier, but now based on the new global TSF basis. Classes with less than 2% of the total number of profiles were temporarily discarded as not significant, but were kept for further comparison. The proposed regrouping approach also has the advantage of providing an evolving classification. For any new image pair analyzed, the correlation between its TSFs and the TSFs of the basis is computed. If correlations higher than 0.9 are found for any TSF, then the image is simply reclassified on the global TSF basis. However, if a correlation less than 0.9 is found for one of the TSFs, then a new class is added to the global set of TSFs.

4. RESULTS

4.1 CLASSIFICATION OF A PAIR OF VISIBLE AND INFRARED IMAGES

In our method an initial cloud field classification is performed on individual pairs of visible and infrared images using TSF analysis. As an illustration, we present in Fig. 1 on the next page the classification of the cloud fields observed on November 10, 1983. Figure 1 gives the visible (a) and infrared (b) GOES 6 images taken at 18:45 GMT on November 10, 1983 and used for the classification.

The spectral and textural parameters were computed on the visible image's 16 x 16 pixel array samples and the infrared image's 4 x 4 pixel array samples, and were then normalized by the mean and variance of the parameters computed on the whole data set (25 image pairs). The normalization coefficients are given in Table 1.

TABLE 1

Normalization Coefficients (mean and variance). Parameters: 1. minimum brightness; 2. maximum brightness; 3. mean brightness; 4. mean distribution (vis.); 5. contrast (vis.); 6. ASM (vis.); 7. entropy (vis.); 8. minimum radiance; 9. maximum radiance; 10. entropy (IR)

Parameter	1	2	3	4	5	6	7	8	9	10
	Min. Brght.	Max. Brght.	Mean Brght.	Mean Distr.	Contr.	ASM	Entro.	Min. Rad.	Max. Rad.	Entro. Rad.
Mean	72.	128.	96.	6.6	90.	0.25	1.56	105.	115.	0.94
Variance	28.	37.	32.	2.6	90.	0.07	0.29	27.	31.	0.51

After the singular value decomposition of the matrix built with the normalized spectral and textural parameters, we found that the first five left singular vectors explained 98.6% of the overall variance. Table 2 shows the first six singular values as well as the percentage of cumulative variance explained by each value.

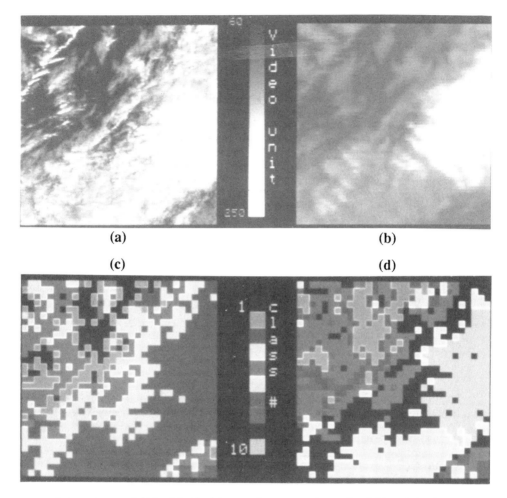

(a)   (b)

(c)   (d)

FIGURE 1: a) Visible GOES 6 image for November 10, 1983, 18:45 GMT. b)
Infrared GOES 6 image for November 10, 1983, 18:45 GMT. c) Classification of
16 x 16 pixel samples using the TSF's of Figure 2 (first step classification). d)
Classification of 16 x 16 pixel samples using the TSF base (final classification).

TABLE 2

Cumulative Variance of First Six Singular Vectors

Class	1	2	3	4	5	6
Singular values	90.4	61.2	21.7	16.2	13.7	10.3
Cumulative variance	62.7	91.5	95.1	97.2	98.6	99.4

The TSF analysis gave five TSFs, which are given below in Fig. 2. Figure 3 presents the mean normalized profiles of the samples associated with the five TSFs in Fig. 2. Note that Figs. 2 and 3 illustrate similar shapes for the TSFs and corresponding mean normalized profiles.

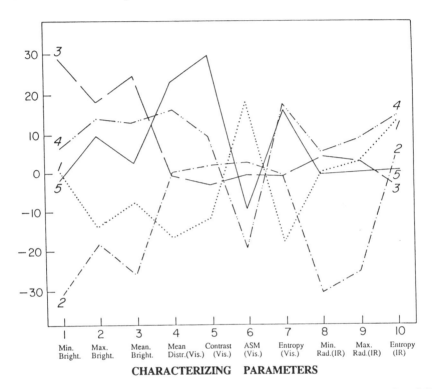

CHARACTERIZING PARAMETERS

FIGURE 2: Typical shape functions resulting from the classification of the visible and infrared GOES 6 images from November 10, 1983. Parameters: 1. minimum brightness; 2. maximum brightness; 3. mean brightness; 4. mean distribution (vis.); 5. contrast (vis.); 6. ASM (vis.);7. entropy (vis.); 8. minimum radiance; 9. maximum radiance; 10. entropy (IR).

Table 3 below shows the number of profiles found for each of the five classes, and presented back in Fig. 1c are the results of the first classification.

TABLE 3

Number of profiles per class for November 10, 1983.

Class number	1	2	3	4	5
Number of profiles	80	89	392	220	180

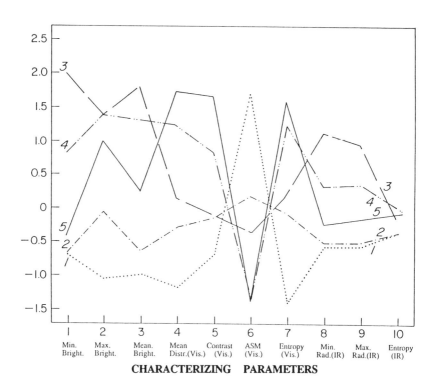

FIGURE 3: Mean normalized profiles for the five classes associated with the five TSF's in Figure 1. Parameters: same as Figure 1.

Class 1, as seen in Fig. 3, is characterized by low values of all parameters except the ASM. The low values of mean distribution, contrast, entropy (vis. and IR) and the high value of ASM indicate that the samples are homogeneous. The low values of the spectral parameters (vis. and IR) suggest that the samples are warm and dark. Class 1 (red in Fig. 1c) should be representative of clear sky conditions, which is confirmed by the classification results given in Fig. 1c. Such a clear sky class was found for each image pair classified.

The mean profile of class 2 in Fig. 3 shows the same low values of minimum brightness and infrared parameters as class 1, which suggest the presence of clear sky pixels within the samples. With the exception of ASM, all the others parameters, however, are larger. The samples should therefore be partly cloudy, and the clouds within the samples should be low and non-reflective with respect to the relatively low values of the maximum and mean brightness and infrared spectral parameters. Class 2 (green in Fig. 1c) represents samples of overlapping low clouds and clear sky and could be used to delineate low cloud layers.

Figure 3's Class 3 presents the high values of spectral parameters (vis. and IR) and mean values of textural parameters (vis. and IR). The samples should therefore be relatively homogeneous, high, and bright. Class 3 should be associated with a homogeneous cloud layer. Figure 1c partly confirms this association. Indeed, it can be seen in Fig. 1c that class 3 (dark blue) represents a layer of clouds that is homogeneous in the visible image (lower left corner), but the classification fails to separate this layer into two parts: one of lower clouds and another of higher clouds, as can be seen in the infrared image. This failure might result from the weak dynamic range of the infrared GOES radiances and from the fact that we use more visual parameters than infrared parameters, giving thus more importance to the visible data. However, it will be seen later that this problem can be overcome by a larger TSF base.

Class 4 (light blue) is mainly characterized by a medium range of visible brightness, a very low ASM value and high textural parameter values (vis. and IR). The samples are thus inhomogeneous and represent a relatively bright and cold, inhomogeneous cloud field. Class 4 appears on Fig. 1c as a field of mixed (low and high) clouds. Class 5 (pink) is associated with highly inhomogeneous samples because of the low ASM values and the high textural parameter values (vis. and IR). The contrast is especially high, the minimum brightness is very low and the maximum brightness is relatively high. This class could represent small cloud cells or cumuliform clouds.

At this stage of the classification, it is still difficult to exactly ascertain which type of cloud is associated with each class because of the great number of TSFs and the relatively small number of samples per class.

4.2 CLASSIFICATION OF THE ENTIRE DATA SET

To classify the entire data set, all twenty five visible and infrared image pairs were first independently classified using TSF analysis. Table 4 gives the number of TSFs (hereafter referenced as first step TSFs) found for each of the twenty five image pairs.

TABLE 4

Number of TSFs Per Image Pair (first step classification)

Image	1	2	3	4	5	6	7	8	9	10	11	12	13
Num. of TSFs	5	5	4	5	5	4	5	4	3	6	5	4	5

Image	14	15	16	17	18	19	20	21	22	23	24	25
Num. of TSFs	4	4	4	4	4	4	4	4	4	3	6	4

For the 25 pairs we classified, 109 different TSFs were found. The correlations ($C_{kk'}$) between all these first step TSFs were then computed and the TSFs for which the correlation was greater than 0.9 were assumed to represent the same cloud type. After this step, we found 14 different classes. The first line in Table 5 gives the number of first step TSFs attributed to each class.

TABLE 5

Number of First Step TSF's for Each TSF of the Base and Number of Profiles per Class Associated to Each TSF of the Base

Class	1	2	3	4	5	6	7
Number of TSF	23	27	19	7	15	8	2
Num. of prof.	2619	5478	3808	1961	3684	4414	56

Class	8	9	10	11	12	13	14
Number of TSF	1	1	1	1	1	1	1
Num. of prof.	610	755	129	14	161	169	167

The seven last classes contained only one first step TSF, which were not correlated to any other TSF. An intermediate base of the fourteen TSFs was computed by averaging the TSFs representing the same cloud type. Each image pair was then reclassified on this intermediate TSF base. The number of profiles associated with each TSF of the base is given in line two of Table Five. Classes 7, 10, 11, 12, 13, 14 were subsequently discarded as non-significant because less than 2% of the profiles (4805) were associated with them. We had a final base of 8 TSFs. Figures 4a and 4b show the TSFs of this final base.

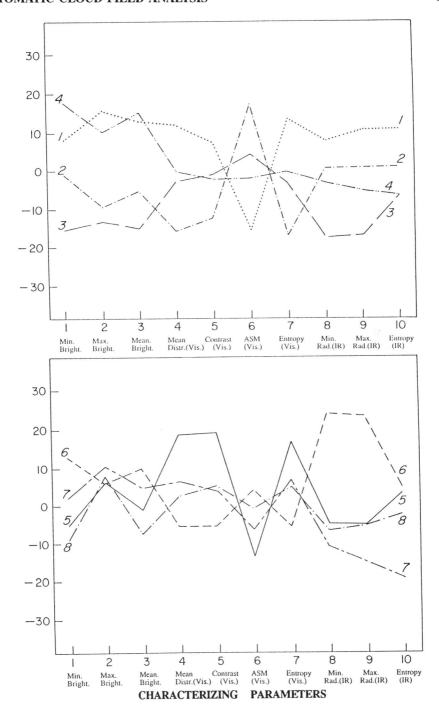

FIGURE 4a (top) and 4b (bottom): Typical shape functions of the TSF base used for the final classification. Parameters: same as Figure 2.

The image pairs were then reclassified on this final TSF base. Table 6 gives the number of profiles per class and corresponding percentage.

TABLE 6

Number and Percentage of Profiles Per Class for the Final Classification

Class	1	2	3	4	5	6	7	8
Num. of prof.	2135	4099	5137	2247	3877	4794	993	743
Percent of prof.	8.89	17.06	21.38	9.35	16.13	19.95	4.20	3.10

Before examining each of these classes in detail, let us come back to the first classification results for November 11, 1983. The five first step TSFs were associated, after computation of the correlations, to the TSFs of the final base as follows: 1 to 2, 2 to 3, 3 to 4, 4 to 1, 5 to 5. The data were then reclassified on the final base. The number of profiles associated with each class are given in Table 7 and the mean normalized profiles of each class are in Figs. 5a and 5b.

TABLE 7

Number of Profiles Per Class for November 10, 1983 for the Final Classification. (The results of the first step classification are given in parentheses.)

Class	1 (4)	2 (1)	3 (2)	4 (3)	5 (5)	6	7	8
Num. of Prof.	244 (220)	79 (80)	55 (89)	184 (392)	192 (180)	143	28	36

Figure 1d shows the resulting classification of the samples of the visible and infrared images of Figs. 1a and 1b. The number of profiles per class and the mean profiles are quite similar for both classifications. It is clear, from comparison of Figs. 1c and 1d, that the first step class 3 has been divided into two classes (classes 4 and 6) in the final classification. The mean normalized profiles of classes 4 and 6 are almost similar for the visual parameters (1 to 7) but class 6 presents very high values of infrared parameters as compared to class 4. Class 6 should thus be representative of higher (colder) clouds than class 4, which can be easily seen in Figs. 1a and 1b. The final classification overcomes the problem we pointed out in part 4.1. Class 4 is associated with homogeneous layer of low clouds and class 6 to homogeneous layer of higher clouds. In this case it is difficult to discuss the 7th and 8th classes because of the few number of samples associated with them.

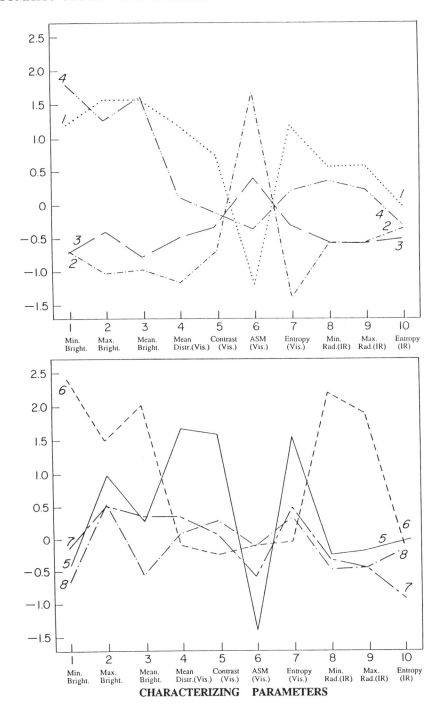

FIGURE 5a (top) and 5b (Bottom) : Mean normalized profiles of the samples associated with the eight final TSF's. Parameters: same as Figure 2.

5. DISCUSSION AND CONCLUSION

One of the problems with all classification techniques is the lack of validation and quantitative assessment. Results from automatic cloud classifications are often compared with classifications performed by expert nephanalysts. This was not the case here for we validated our method using an *a-posteri* evaluation of the cloud classes based on: 1)correlation between the visible and infrared image pair classified and the image of the classes obtained, and 2) quantitative analysis of the cloud feature profiles' physical significance. While this validation alone is limited and subjective, it provides us with a qualitative assessment of the method's classification skill. In the future, we expect to compare a subset of our results (e.g., cloud cover, number of cloud layers) with those obtained from other cloud analysis approaches, such as the spatial homogeneity method of Coakley and Bretherton, 1982.

The method proposed here has provided us with encouraging preliminary results but still requires testing, evaluation, and improvements. For instance, the comparison between the classification results of the first classification and those of the regrouped classification clearly indicates some deficiency in the selected parameters to characterize the cloud field. Specifically, too few infrared parameters were selected. A limited set of infrared parameters was chosen because of the VISSR infrared sensor's poor resolution, which limits the textural information that can be extracted from that data set. More infrared spectral parameters could, however, be used to give increased weight to the infrared data and thus better differentiate clouds using their cloud top information.

Other questions arise regarding the choice of characterizing parameters. First, the chosen set of parameters may not be optimal for a large ensemble of cloud conditions. We propose to develop analysis procedures, however, to derive the optimal set of parameters for all cloud types encountered. The analysis procedures we envision are the neural networks applied in recognition and classification modes. Second, the choice of only one direction (east-west) and one distance (one pixel) for computing the parameters limits the description of the cloud field, particularly its directionality and meso scale characteristics. Third, although the set of parameters chosen is standard in textural analysis, they are sometimes hard to interpret because of their lack of physical representativity.

Another issue of concern is the normalization of the characterizing parameters. In the present study, the normalization data set (i.e., mean and standard deviation) is computed for each individual image, which limits the direct comparison of classes over sequences of images, if the cloud fields are very different from one image to another. We will use, in the future, a universal normalization data set from a large ensemble of images selected to represent most cloud fields encountered in nature.

The classification chosen (TSF) differs from other classification methods (e.g., cluster classification) in that the classes are defined by the relationship between the parameters. This implicitly assumes that the distribution of the cloud profiles in a class is ellipsoidal, i.e., that the norm of the vectors is not important for its membership to a class. This assumption has been neither violated nor verified in this study. It validity will be verified in our future work.

ACKNOWLEDGMENTS

We wish to thank P. Collard and J. McPherson for their help in the cloud classification and B. Bloomfield for his assistance in preparing and editing this manuscript.

REFERENCES

Coakley, J.A., and F.P. Bretherton, 1982: Cloud Cover from High Resolution Scanner Data: Detecting and Allowing for Partially Filled Fields of View. J. Geophys. Res., 87, 4917-4932.

Debois, M., G. Seze, and G. Szejwach, 1982: Automatic Classification of clouds on METEOSAT Imagery: Application to high-level clouds. J. Appl. Meteor., 21, 401-412.

Garrand, L., 1986: Automated Recognition of Oceanic Cloud Patterns and Its Application to Remote Sensing of Meteorological Matters. Univ. of Wiscon., Madison, Thesis, 231 pp.

Jalickee, J.B., and C.F. Ropelewski, 1978: An Objective Analysis of the Boundary-Layer Thermodynamic Structure During GATE. Part I: Method. Mon. Wea. Rev., 107, 68-76.

Parikh, J., 1977: A Comparative Study of Cloud Classification Techniques. Rem. Sens. of the Environ., 6, 76-81.

Phulpin, T., M. Derrien, and A. Brard. 1983: A Two Dimensional Histogram Procedure to Analyze Cloud Cover from NOAA Satellite High-Resolution Imagery. J. Climate Appl. Meteor., 22, 1332-1345.

DISCUSSION

Gille: Within an image can you detect the boundary between two different types of clouds?

Gautier: Yes, in fact we have, which I have not discussed here. A lot of imagery classes are cloud edges. If you had clear and cloud classes you always will find a class for edges because the textural properties are different from the homogeneous conditions.

Johnson: Two quick questions. This was, of course, over oceans. Do you have any hopes of doing these things over land? Your clear cloud conditions will be much more heterogeneous than you see here.

Gautier: We are going to focus on the ocean for a while. I don't think we are far enough yet with the classification over the ocean, the easy case, before we can start the classification over land. We have some hope, but we are not there yet.

Johnson: Secondly, what will sub-pixel size clouds do to your classification scheme? Clouds that are not large enough to characterize by themselves an entire pixel but will, of course, change the values you see.

Gautier: As I said, there are several classes we find which are kind of combination of clear and cloudy on edges. So that would fit with the sub-pixel. The other answer I can give you to that question is it depends on the size. When we compute the parameters we compute on a super pixel, which is we take 16 by 16. We don't do it on a pixel by pixel basis. We take samples of pixels which are 16 by 16. We are not looking at a cloud individually but at a cloud field.

VLSI HARDWARE ACCELERATORS
FOR REMOTE SENSING

Gary K. Maki
Microelectronics Research Center
College of Engineering
University of Idaho
Moscow, Idaho 83843, USA

ABSTRACT

Standard stored program computers are unable to provide the computational power necessary to meet the requirements of next generation remote sensing. This paper discusses the limitations of present day computers, outlines requirements of future remote sensing systems and presents a technology to meet the computing needs for next generation systems that provide real time computation.

1 INTRODUCTION

The quality of science that can be advanced is directly related to the tools available to the scientist. One important tool is the digital computer that collects data and extracts necessary information. Until recently, the standard stored program computer has been adequate in meeting most scientific needs. However, new remote sensing techniques are requiring computational power of magnitudes that cannot be realized with stored program computers. Failure to provide solutions will greatly limit future advances that must be achieved in data retrieval and information extraction.

Many computational requirements associated with remote sensing cannot be done at all with today's fastest computers, much less be included as part of an instrument. This is true for the following reasons:

- Required computational power is too great; consider the following few examples:

  - Error Correction decoding utilizing a NASA-ESA standard requires 1600 Million Operations Per Second (MOPS) [5].

  - Synthetic Aperture Radar (SAR) requires 5312 MOPS [1] [2].

– Radiometric Correction for spectrometer instruments can require more than 50,000 MOPS [3].

• Physical size too large and weigh too much.

• Consume too much power.

It is clear that challenges like those listed above must be addressed to make major advances in remote sensing.

2   STORED PROGRAM COMPUTER LIMITATIONS

The problem of higher performance computing is not an electronic problem. Integrated Circuit fabrication and processing techniques have undergone major improvements in the last few years. Four years ago, 3 micron geometries were state of the art with typical gate delays of 3 nsec [7]. Today with 1.5 micron geometries, the typical gate delay is about 0.7 nsec. The speed and area metrics have each improved by a factor of 4.

The limitation is clearly one of architecture. Consider the following notion; let

• T_a = Time to perform an arithmetic/logic operation

• T_f = Time to fetch an instruction

• T_i = Time to perform an input/output operation

• T_r = Time to load an operand

• T_s = Time to perform a store operand operation

• K_a = Number of arithmetic/logic operations

• K_f = Number of fetch instructions

• K_i = Number of input/output operations

• K_r = Number of times to load an operand

• K_s = Number of times to store an operand

The total time to execute a given program is

$$T_e = K_a T_a + K_f T_f + K_i T_i + K_r T_r + K_s T_s$$

The only term in the above equation that produces useful "work" of transforming input data into some desired form is $K_a T_a$. All other terms specify the amount of time the machine is managing data, such as fetching instructions, storing temporary results and fetching data. If all the above terms were equally weighted, the stored program computer would be only 20% efficient. However, the largest term likely is $K_f T_f$ since the machine must fetch an instruction every time an instruction is executed. Moreover, in the execute phase of an instruction, the actual time an arithmetic/logic operation is performed is a fractional part of the phase.

The above does not take into account the impact of the software system. There is overhead due to the compiler, which translates a high level language into assembly language instructions. The compiler would keep track of temporary data storage and the like. The operating system and specific computer configuration (number of registers, memory etc.) also can have an impact on the efficiency of operation. Therefore, it is likely that the overall efficiency of the stored program computer is in the order of 1%.

The architecture is the limitation of the stored program computer and it is unlikely that this architecture can ever be expected to achieve the computational rates required by remote sensing of the future. However, there is an exciting new technology that is emerging that does provide solutions to these problems.

3 VLSI HARDWARE ACCELERATOR

A **Hardware accelerator** is a custom Very Large Scale Intergrated (VLSI) system that affords a unique high computational performance solution to a specific problem. Attractive features of a hardware accelerator are

- Direct implementation of equations or algorithms into hardware

- Generally no software, programs or complex operating systems

- Highly parallel and efficient execution

- Very low space, weight and power requirements (often less than 10 VLSI chips)

- Fast - 10 to 100 times the speed of the best supercomputer

- Greater reliability and incorporation of fault tolerance is easier

Hardware accelerators are not only possible, but practical using today's technology.

- Computer Aided Design (CAD) tools exist that perform the operations necessary for complex VLSI design

- Sophisticated integrated circuit processing is common

- Design time has been reduced by a factor of 100 in the past 10 years

- There are centers of excellence, such as the Microelectronics Research Center (MRC), that have produced and are producing accelerators now

- Low cost when compared to a supercomputer

Critical applications for hardware accelerators within the remote sensing scientific community include:

- Data Communications

- Navigation

- Data Compression

- Digital Control

- Robotics

- Image Processing

- Pattern Recognition

- Real Time Information Processing

- Computer Vision

3.1   DESIGN PROCEDURE OVERVIEW

The design steps to producing a custom VLSI hardware accelerator are illustrated next:

- The first step is to translate the fundamental mathematical equations into an architecture. Depending on the complexity of the equations and the interconnect (communication between modules on the chip), this step can range from simple and straight forward to very complex. Often times, major amounts of effort are required by both the scientist and the VLSI system architect. The result of this effort is to specify needed functional units, such as multipliers and adders, and the interconnect requirements.

  General mathematical systems of equations that have been implemented in a hardware accelerator are solving sets of ordinary differential equations and polynomials.

- After the basic architecture is defined, the functional units (multipliers, adders, etc.) are designed and the interconnection strategy specified. A logic simulator is the key design tool for the logic design phase.

- After the logic design, it is necessary to specify the transistor sizes (and any other parameters if a technology like GaAs is used) in the electronic design phase. The correct transistor sizes will guarantee proper speed performance. Programs such as SPICE are used in this phase.

- Cell layout is the last design phase where the electronic cell specification is translated into a set of polygons that represent the various transistor and interconnect layers. Each electronic fabrication process, such as CMOS, has a set of design rules that must be met; these rules specify minimum geometries and feature sizes. A Design Rule Checker (DRC) program is used to aid the layout technician.

- Verification is the last step in the design process. From the logic design phase, a logic simulation file is generated that specifies each component (transistor) and the interconnection matrix. Every node is specified and its relationship with other elements are known. The verification phase takes the data generated by cell layout and compares it with the logic simulation; they must agree. Verification also extracts actual capacitance values from the layout and provides a feedback loop with the electronic design phase such that speed performance is guaranteed.

3.2   DESIGN EXAMPLE

The following example is used to illustrate the power of a hardware accelerator to a communication problem. NASA and the European Space Agency have specified a high performance Reed Solomon Error Correcting Code [4]. This code is a 16 symbol error correcting code, where each symbol is an 8-bit byte and there are a total of 255 symbols per message. The complexity of the decoding algorithm can be summarized as follows [5]:

- The syndrome generator evaluates 32 polynomials of order 254.

- The Euclid divide module applies Euclid's algorithm recursively to a pair of polynomials of degree 32 and 31.

- A Polynomial module evaluates three polynomials 255 times of the following degree.

  - Error Location polynomial of degree 16.

  - Error Magnitude polynomial of degree 15.

  - First derivative of the Error Location polynomial with is degree 15.

- The Correction module determines 255 quotients and adds result to uncorrected message polynomial.

Table 1 shows the number of calculations per message and the number of operations per second that need to be realized for the specified 80 Mbit/ second data rate.

Table 1. Summary of Computation Power of NASA Chip Set

Module	Number of Operations per message	Millions of Operations/sec
Encoder	16,320	640 MOPS
Syndrome	16,320	640 MOPS
Euclid	3,110	122 MOPS
Polynomial	20,910	820 MOPS
Correct	1,020	40 MOPS

Clearly the decoding algorithm could not have been accomplished on an existing stored program computer. The above VLSI processor was implemented in 4 custom VLSI circuits in 3 micron CMOS. In 1.5 micron CMOS, it is thought that this processor could be implemented on a single chip.

4   FUTURE DIRECTIONS

Research and industrial centers need to begin to address the means by which future remote sensing are going to be implemented such that real time computing can be achieved within the instrumentation. Indeed, future engineering systems will consist of sophisticated high speed data acquisition, data processing and control systems. Space borne computer systems must achieve higher performance and require less power and weight. Moreover, real time information extraction will have to be incorporated to reduce the total amount of raw data that must be transmitted, stored and processed. In some cases, obtaining real time information is vital as demonstrated by the medical example where a computer that was used to monitor and compute a patient's condition performed at such a slow rate that the doctor would be informed that the patient had died 10 minutes after the fact.

Clearly, stored program (traditional) computers do not have the required performance capabilities and struggle with a software burden. Custom VLSI hardware accelerators meet performance criteria and contain no software in the traditional sense. High performance is achieved through parallel/pipelined architectures customized to solve a particular problem.

The design philosophy associated with a custom hardware accelerator is radically different than the design of traditional computer systems. In the traditional computer system design, someone conceives an architecture and all users attempt to map a given problem into that architecture; some problems fit well, many do not. Efficiency in a stored program machine can be less than 10%. In a custom VLSI design, the architecture is designed such that the problem to be solved maps directly onto the hardware resulting in very efficient operation.

Engineers and scientists need to be exposed to the potential afforded by hardware accelerators such that the advantage of hardware accelerators can be exploited in future remote sensing systems. A new generation of engineers and scientists need to be trained to design supercomputing hardware or are able to utilize the power of this hardware to advance science.

The research and development to achieve hardware accelerators must be cross disciplinary involving scientists who understand the problems of remote sensing and VLSI engineers who are able to produce custom VLSI architectures specifically oriented to solving particular problems. The best approach is to involve elements from the national research laboratories, industrial experts and university researchers in solving this problem.

References

H.M. Assal and J.F. Vesecky, 1986: *Spaceborne SAR Azimuth Processor: VLSI Implementation Assessment*, Stanford Center for Radar Astronomy Special Report D909-86-1 for JPL.

B. Bowen, W. Brown, 1985: *Systems Design*, Prentice-Hall, Inc.

M. Herring, 1987: *High Resolution Imaging Spectrometer: Instrument Description*.

G. Maki, P. Owsley, K. Cameron, and J. Shovic, 1986: *VLSI Reed Solomon Encoder: An Engineering Approach*, Custom Integrated Circuits Conference, pp 177-181.

G. Maki, P.Owsley, K. Cameron, and J. Venbrux, 1986: *VLSI Reed Solomon Decoder*, IEEE Military Communications Conference, 46.1-46.6.

J. G. Nash, 1986: *Concurrent VLSI Architectures for Image and Signal Processing*, IEEE Potentials, pp 12-14.

Weste, N., Eshraghian, K. 1985: *Principles of CMOS VLSI Design: A Systems Perspective*, Addison Wesley Publishing Company, pp. 163-172.

DISCUSSION

Kleespies: Once you have the custom chip designed, and I presume that your ultimate goal is to put one into manufacturing, what kind of rough dollar amounts are we talking about?

Maki: The prices I have been hearing from some of my friends in industry to do the fabrication, generation of the masks, and packaging may be about 60- to 100-thousand dollars. To set up the test system with automatic IC testers is going to be a function of chip complexity, which might add another 100-thousand dollars. But once all of that thing is set up, then the prices to fabricate more chips come down quite a bit. It is hard to answer the question.

THE USE OF HUFFMAN CODING
TO COMPRESS SATELLITE DATA

Larry M. McMillin
National Oceanic and Atmospheric Administration
National Environmental Satellite, Data, and Information Service
Satellite Research Laboratory
Washington, D.C. 20233, USA

ABSTRACT

When satellite data are used to obtain profiles of temperature, measurements at a number of frequencies are obtained in order to produce the entire temperature profile. Because the measurements are limited in vertical resolution, the high order bits of the various measurements are highly redundant. This makes the satellite data a prime candidate for compression. However, weather is really the departure from the extreme case. When enough bits are saved to produce the extremes accurately, the compression is limited and most measurements are still sent with redundant information.

Huffman coding is gaining wide use as a data compression technique. In Huffman coding, the length of a bit pattern representing a number is inversely proportional to the frequency of the number in the data sample. Since the bit length is variable, the code has the property that no number starts with a bit pattern that represents a lower number. Thus each bit pattern is unique.

The combination of standard statistical techniques with Huffman coding is an ideal solution of the dilemma of the amount of accuracy to maintain. It is demonstrated that the data can be compressed to less than half of its length without losing any information. Using this technique, the compressed data can be recovered to the original bit level. In the terms of information theory, the procedure seeks to increase the entropy of the code, first by taking advantage of what is already known to code only the difference between a given number and its estimation based on recovered data, and second by using Huffman code to maximize the entropy of the coding procedure.

1. INTRODUCTION

Satellite measurements used for temperature sounding are obtained from three instruments, the 20 channel High resolution Infrared Radiation Sounder (HIRS), the 4 channel Microwave Sounding Unit (MSU), and the 3 channel Stratospheric Sounding Unit (SSU). These three instruments are referred to as the TIROS Operational Vertical Sounder (TOVS). Data from these instruments are coded as 11 bits and sent to the ground. They are also archived as 11 bit data. The high order bits of these data are highly redundant. However, when the

RSRM '87: ADVANCES IN
REMOTE SENSING RETRIEVAL METHODS
A. Deepak, H.E. Fleming, and J.S. Theon (Eds.)   419

data are transmitted to the ground and archived, bandwidth and media
storage are sized to accommodate both the independent and redundant
information. One of the consequences of this requirement is that
accuracy is traded against practical considerations of transmission
and storage costs. Although these problems are bad enough for the 27
channel TOVS, it has recently (Smith et al. 1987) been determined
that an instrument containing on the order of 1000 to 2000 channels can
significantly increase the accuracy of satellite retrievals. With
2000 channels, the redundancy problem is multiplied many times.

One property of geophysical data is that the multiple channels
contain redundant information. Conventional coding techniques fail
to account for the redundancy. Once the value for one or more
channels is known, the uncertainty of all the other channels is
greatly reduced. Knowledge of some of the channels can be used to
reduce the number of bits required to determine the values of the
remaining channels. For example, data are frequently represented by
a selection of eigen vectors. When this is done, there is still the
question of the accuracy level to preserve. This is of particular
concern for data transmission from a satellite where anomalous
signals may indicate a pending problem. If a compression method does
not represent these cases accurately, the spacecraft can be
endangered.

This paper estimates the compression possible with satellite data
to determine if the effort to do a compression is justified.
Radiances, both clear and cloudy, are sent as radiances
from the satellite with accuracies of about 0.3 $\mu W/(\text{sec } m^2 \text{ sr } cm^{-2})$
in the 15 μm region. For this study, a set of clear column
brightness temperatures was used. To compensate for the differences
in the data sets, the accuracy was set to 0.1 K rather than the
higher number indicated by the radiance accuracy.

Barath (1987) discusses some of the principles of information
theory and Amsterdam (1986) discusses data compression with Huffman
coding. When dealing with satellite data, the code is generated by
taking all the possible numbers generated by the instrument and
determining the frequency of occurrence of each bit pattern. Then
the two patterns with the lowest frequency are determined. The
numbers corresponding to the patterns become a node on a decision
tree. The frequencies of the two patterns are summed and used to
replace frequency of one of the two patterns. The bin associated
with the other pattern is deleted, reducing the number of bins by
one. The search for the two bins with the lowest frequency is
repeated. The procedure is iterated until only two bins remain. As
the procedure is iterated, a tree is formed in which the leaves are
the numbers to be encoded and each node has two branches. By
assigning zeros to left hand nodes and ones to right hand nodes, a
code is created that represents the number at the leaf of the tree.
The numbers with the lowest frequencies are at the leaves of the
shortest paths and the numbers with the highest frequencies are at
the leaves with the longest paths. Amsterdam (1986) gives an
excellent description of the Huffman procedure and shows that the

procedure results in the shortest bit pattern that can represent the numbers.

2. APPROACH

Much of the approach has been outlined in the introduction. A set of satellite data was selected for the evaluation. These data consist of brightness temperatures for a set of about 4000 profiles taken in May 1987 that were selected to equally cover all latitudes and seasons. It was decided that the decoding would proceed as a boot strap. One channel would be selected to be decoded first, then it would be used to predict a second channel. The number sent for the second channel is the difference between the measured value for the second channel and the value predicted from the first channel. It is added to the predicted value for the second to obtain the measured value. Then the values for the two channels are used to predict a third channel. The number sent for the third channel represents the difference between the measured value for the third channel and the value predicted from the first two channels. Since some channels can be predicted very accurately from a linear combination of other channels, few bits are required to represent them.

This procedure is dependent on the order of the channels. The first channel was selected as the one with the lowest standard deviation. The second one is the channel most accurately predicted from the first, the third from the first two, ... , and the last from all but the last. When the order of the channels was selected, the standard deviations of the selected alternatives were noted.

Since the bits representing numbers in Huffman coding are dependent on the frequency, the bit patterns for the data in question must be known along with the code. For small amounts of data, this limits the compression possible, since the bit patterns are an additional overhead that adds to the length of the file. For sounding data this is not a serious limitation since the frequencies of the numbers can be kept constant with little loss in compression. However, it is convenient for many purposes to recognize that the deviations of the numbers being coded are nearly Gaussian. Then, given a single number, a tree structure and bit patterns for the numbers can be generated. The number required is the ratio of the standard deviation to the accuracy denoted by a single bit. It will be convenient to refer to this number as the normalized standard deviation. It is then necessary to send only the normalized standard deviation as it uniquely defines the tree structure and the bit patterns. In fact, it would be possible to define bit patterns for a fixed set of Gaussian distributions that span the range of normalized standard deviations. Since the storage penalty for using a slightly inaccurate normalized standard deviation is small, it would be adequate to simply pick the nearest one. Figure 1 shows the average number of bits per number required to store values as a function of the normalized standard deviation.

CODE LENGTH (BITS)

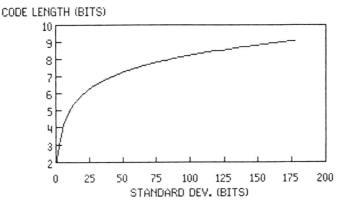

FIGURE 1.   The average number of bits required for a Huffman code as
a function of standard deviation in bits.

It was mentioned earlier that there are cases where it may be
important to recover every bit that is sent. The Huffman code
strings for practical values of the normalized standard deviation are
in the range of several hundred patterns or less. This is much
smaller than the 2048 bit patterns required to represent every
possible 11 bit number. However, the solution is simple. Simply
reserve one bit pattern as an escape flag. When this flag is
encountered, the next 11 bits are a normal 11 bit number. This
increases the number of bits required to send these cases, but the
number of occurrences is so small that the effect on the length of
the file is negligible. In fact, even this procedure is nearly
optimal. A major cause of these cases is transmission errors. These
errors have the property that the probability of the error affecting
any given bit is equal to the probability that it would affect any
other bit.   Amsterdam (1986) demonstrated that the Huffman procedure
for equally probable data results in the bit pattern used for normal
integers. The optimal code for a background of equally probable
numbers superimposed on a Gaussian distribution is simply the code
for a normal integer proceeded by a flag with a length such that the
combined length of the flag and the number fits in right place in the
sequence of Huffman codes. However, the penalty in file length for
getting a single number out of sequence is not very great, so it is
practical just to pick a number for the escape flag and not worry
about its appropriate length. For this study, it was decided to have
a single bit represent zero. Thus in the coding, one pattern was
assigned to the interval 0 to 1/2 the bit accuracy, one to 1/2 to
1+1/2, 1+1/2 to 2+1/2 etc. When the sign was considered, this
resulted in one pattern for 0 to +1/2 and one for 0 to -1/2. A
single value for -1/2 to +1/2 is required so one of the two patterns
was selected and the other was used as the escape flag.

3. APPLICATION

The file length was estimated for a sample of about 4000 radiance
profiles that were part of a sample of clear column radiance
measurements collocated with radiosondes. The sample was selected to

represent all latitude zones equally and was collected in May of
1987. During the testing period, some channels of the SSU had failed
and were being filled in from another instrument, so they were left
out of the comparison, leaving 24 channels. The visible channel was
also left out because it is not well predicted from other channels
but is related to other information. For example, the half of the
orbit that the satellite is viewing night, the visible channel
measures zero. Leaving it in its present form would have over
estimated the length of the code, while defining a means to predict
it would have greatly increased the complexity of a simple
evaluation. It was left out on the assumption that it could be sent
in a code of the average length of the other 23 channels. Channel
21, the MSU window channel, is strongly affected by the surface
emissivity which is near unity over land and near 50% over water. It
was included in its original form. It is noted that the uncertainty
of this channel could be reduced by utilizing information about its
location, but it wasn't.

A single mean was generated for the sample and departures from
the mean were calculated using

$$\overline{bt} = (\sum_n bt)/n \tag{1}$$

$$\sigma_{bt} = (\sum_n (bt - \overline{bt})^2)/n \tag{2}$$

where bt is the brightness temperature, n is the sample size, the bar
denotes the average, and σ denotes the standard deviation. The
channel with the smallest standard deviation about the mean was
selected. This channel was used to predict the other channels using
an equation of the form

$$\hat{bt}_j = a_{0,j} + \sum_{\substack{i=1 \\ i \neq j}}^{m} (a_{i,j} * bt_i) \tag{3}$$

where the subscripts i and j denote channels and the a's are the
regression coefficients. For each remaining channel, the standard
deviation of the difference between the measured and predicted values
was obtained from

$$dbt_j = bt_j - \hat{bt}_j \tag{4}$$

where dbt_j denotes the difference between the observed and predicted
brightness temperature for channel j. The standard deviations of the
difference were searched for the smallest value to determine the next
channel. It was added to the recovered channels, increasing the
number by one, and the recovered channels were used to predict the
remaining channels. The process was iterated until the order of all
23 channels was determined. It should be noted that the resulting
order is not necessarily optimal. For example, it could be possible

that sending a single channel with a large standard deviation first
would greatly reduce the standard deviations of the remaining
channels. Probably the only way to know is to try all possible
orders. On the other hand, the difference in code length between the
optimal order and several other orders is likely to be small. For
this initial evaluation, the search of all possible combinations to
get one that might be only marginally better was not justified.

Once the standard deviations were determined, then the standard
deviations in physical units were divided by the accuracy represented
by a single bit to convert the standard deviations to bits. Then the
frequency of occurrence of each number represented by the original 11
bit coded was determined using an approximation for the Gaussian (see
Hastings 1955) given by

$$P(x) = 1. - 1./[(((a4*x+a3)*x+a2)*x+a1)*x+1] \qquad (5)$$

where P(x) is the probability that a number is greater than x, and a1
is .278393, a2 is .230389, a3 is .000972, and a4 is .078108.

TABLE 1. PROBABILITIES (FREQUENCIES) OF OCCURRENCE OF
BIT VALUES FOR A STANDARD DEVIATION OF 50 BITS.

X	PROB	X	PROB	X	PROB	X	PROB	X	PROB	X	PROB	X	PROB
0	.0223	10	.0218	20	.0192	30	.0157	40	.0119	50	.0084	60	.0054
1	.0224	11	.0216	21	.0189	31	.0153	41	.0115	51	.0080	61	.0049
2	.0224	12	.0214	22	.0185	32	.0149	42	.0112	52	.0077	62	.0046
3	.0224	13	.0212	23	.0182	33	.0146	43	.0108	53	.0074	63	.0044
4	.0224	14	.0209	24	.0179	34	.0141	44	.0104	54	.0071	64	.0042
5	.0224	15	.0207	25	.0175	35	.0138	45	.0101	55	.0068	65	.0039
6	.0223	16	.0204	26	.0171	36	.0134	46	.0097	56	.0065	66	.0037
7	.0222	17	.0201	27	.0168	37	.0130	47	.0094	57	.0062	67	.0035
8	.0221	18	.0198	28	.0164	38	.0127	48	.0090	58	.0059	68	.0034
9	.0220	19	.0195	29	.0161	39	.0123	49	.0087	59	.0056	69	.0032

Table 1 shows the frequencies for a standard deviation of 50. Given
the frequencies, the Huffman procedure was applied to generate
Huffman code for the Gaussian distribution. Table 2 shows the code
for selected numbers. Notice that the shortest codes are assigned to
the numbers with the highest frequencies. In generating the code,
codes were generated for one side of the distribution, and a sign bit
was added to cover the other side. In addition, the frequency of the
first interval was taken as the frequency of 0 to 1/2 of a bit times
two as

$$p(0) = P(0<x<1/2)*2 \qquad (6)$$

where p(0) is the probability that the value of a number is zero and
P(0<x<1/2) is the probability that x is between 0 and 1/2 the
accuracy represented by a single bit. This results in one code for

+0 and one code for -0. One was arbitrarily selected to represent zero, and the other became the flag for the escape code to indicate the start of the 11 bit code to represent the rare numbers.

TABLE 2.   HUFFMAN AND 11 BIT CODES FOR A
STANDARD DEVIATION OF 50 COUNTS.

#	HUFFMAN CODE	11 BIT CODE
0	000000	00000000000
1	000010	00000000001
2	000100	00000000010
3	000110	00000000011
4	000101	00000000100
5	000011	00000000101
6	000001	00000000110
7	0111111	00000000111
8	0111110	00000001000
9	0111101	00000001001
10	0111011	00000001010
20	0101111	00000010100
30	0100000	00000011110
40	0010001	00000101000
50	01001110	00000110010
60	011110010	00000111100
70	001010100	00001000110
80	0010101011	00001010000
90	00110111000	00001011010
100	001001001101	00001100100
110	0010000011001	00001101110
120	00100000110000	00001111000
130	0101011110100011	00010000010

Once the codes were generated, the length of a file was calculated by summing the product of the number of bits for a particular number by its frequency, i.e.

$$L = \sum_{i=0}^{K} p(i)*CL(i) \tag{7}$$

where L is the total length, p(i) is the probability for the given number, CL(i) is the length of the Huffman code for the given number, and K is the number of numbers coded. This was repeated for all 23 channels. Standard deviations and standard errors are given in Table 3 and the corresponding lengths in Fig. 1. The standard deviations of the 23 channels were supplemented to fill in gaps and the results were used to produce the curve shown in Fig. 1. Since many physical measurements have distributions that are nearly Gaussian, the curve can be used to estimate the code length for many of these measurements.

Finally, the code lengths for all 23 channels were summed to calculate the length of the entire profile as well as the average length per channel. The average length per channel is slightly greater than 5 for a total length 116 bits for 23 channels. Given the data compression due to the boot strap regression, the code length corresponding to the maximum entropy was calculated from the relationship

$$L = \Sigma\ p(i)\ \ln\ p(i) \qquad\qquad\qquad (8)$$

TABLE 3. STANDARD DEVIATION ABOUT THE MEAN AND STANDARD ERROR OF THE PREDICTED VALUES FOR THE TOVS CHANNELS.

CHANNEL NUMBER	STANDARD DEVIATION	STANDARD ERROR	CHANNEL NUMBER	STANDARD DEVIATION	STANDARD ERROR
23	2.34	2.34	6	8.85	0.34
4	3.11	1.79	11	8.57	1.16
16	2.65	1.43	7	11.85	0.56
5	6.32	1.34	14	11.80	0.45
15	7.67	0.83	9	14.94	1.66
3	2.68	1.00	13	13.90	0.29
2	3.12	0.54	10	15.21	0.46
24	5.73	1.00	8	17.74	0.91
1	3.83	1.09	18	17.61	1.66
17	5.10	0.91	19	16.70	1.04
12	6.30	3.51	21	16.34	12.37
22	8.66	0.53	20	147.85	68.08

This gave a total length of 115 bits for 23 channels, so the Huffman code produced results very close to the theoretical limit. Another item of interest was the compression that resulted from the regression procedure. To evaluate this, the original data were compressed using the Huffman code. This produced an average length of 7.82 bits per channel for a total length of 180 bits, 64 bits more than 116. Thus the regression reduced the code length by 64 bits and the Huffman code reduced the code length by 73 bits. Both steps make significant contributions to the reduction in length. Thus for 23 channels, the 11 bit radiance data takes a total of 253 bits. Regression reduces this by 73 bits to 180, and the Huffman procedure reduces it by an additional 73 bits to 115. The total reduction is about half of the original length.

It is also possible to consider the original length and accuracy of the temperature data. It was coded as temperatures in 16 bit integers to an accuracy of one part in 64. To estimate the code length, we refer to Fig. 1 and note that the average length of bits corresponds to a standard deviation of about 10 bits. Multiplying this by 6.4 gives a standard deviation of 64 bits. From the curve, this corresponds to a code length of 7.5 bits. This assumes the increase in length of all codes is proportional to the increase in

length of the average code.   In practice the increase in length of
the shortest codes is greatest, so the increase of the average length
under estimates the average increase in length.   To allow for this
factor, the length can be increased from 7.5 to 8.

The net result of this study is that the combination of
regression and Huffman coding can reduce the code length required for
TOVS data to 1/2 its original length.   This is based on data from a
single spot.   Additional compression could be obtained by classifying
data and taking departures from a class mean by using an additional
parameter to identify the class.   Data from a single scan have
spatial consistency that could also be used to reduce the code
length.   These procedures increase the complexity of coding
procedure, and the gain in length must be balanced against the
increased complexity.   This study demonstrates that a relatively
simple coding procedure can compress the code to half its original
length without losing any information.   The data can be recovered
bit for bit.

The compression due to the method is a function of the redundancy
in the data.   Probable future instruments (Smith et al. 1987) with
large numbers of channels will have a much greater redundancy than
the TOVS.   The procedure should produce code lengths of 2 to 4 bits
per channel for this type of data, and should be considered.

It should be mentioned that the procedure examined in this paper
has an advantage over conventional Huffman coding.   In Huffman
coding, it is necessary to send the code and this becomes an overhead
that reduces the compression possible.   By coding Gaussian
distributions, it is necessary only to send the standard deviation
plus the mean of the data vector.   This procedure is general enough
that it is possible to think in terms of a hardware implementation of
the Huffman code for a Gaussian distribution.   Such a chip could be
used to store many types of geophysical data, and would decrease the
coding and decoding time, although both the regression and the
Huffman procedures are relatively fast, especially on a machine that
is designed to handle bits efficiently.

REFERENCES

Amsterdam, J., 1986: Data compression with Huffman coding.   *Byte*, 99–
     108.
Bharath, R., 1987: Information theory.   *Byte*, 291-298.
Hastings, C.H. Jr., 1955: *Approximations for digital computers*.
     Princeton University Press, Princeton, New Jersey. 200 pp.
Reingold, E.M. and W.J. Hansen, 1986: *Data structures in Pascal*.
     Little, Brown and Company, Boston.
Smith, W.L., H.E. Revercomb, H.M. Woolf, H.B. Howell, D. LaPorte, and
     K. Kageyama, 1987: Improved geostationary satellite soundings
     for the mesoscale weather analysis/forecast operation. Proc.
     Symp Mesoscale Analysis & Forecasting, Vancouver, Canada, 17-19
     August 1987, ESA SP-282 (August 1987). 79-83.

DISCUSSION

Neuendorffer: It looked like that data you were working with was predominantly cloud-free infrared data. Is that true?

McMillin: Yes.

Neuendorffer: But the satellite is basically going to see cloud-contaminated infrared data most of the time. Are you going to onboard process to just send down the cloud-free stuff or how do you plan to do that?

McMillin: No, but in the cloud data a number of the channels are even closer together than they are in the cloud-free data. So, there is nothing in the method that you can't use it on cloudy data, in fact, it would probably be a little bit better.

Westwater: Have you examined any retrievals using compressed data versus the original data to see if there is possibly a degradation in the retrieval accuracies due to the compression of the data?

McMillin: I don't know how to answer that because I can recover every bit that was in the original data. I am just eliminating the bits that are used for redundant information.

ADVANCES IN DATA EXCHANGE METHODOLOGY

Mary G. Reph and Lola M. Olsen
National Space Science Data Center
NASA Goddard Space Flight Center
Greenbelt, Maryland 20771, USA

William B. Rossow
NASA Goddard Space Flight Center
Institute for Space Studies
New York, New York 10025, USA

ABSTRACT

Solving problems caused by the wide variety of data formats can be very time-consuming and costly both for data archives and for projects that require the exchange of data among investigators. Cooperative efforts between the NASA Climate Data System (NCDS) and the First International Satellite Cloud Climatology Project (ISCCP) Regional Experiment (FIRE) have yielded a family of formats which is flexible enough to meet the needs of individual investigators without necessitating the overhead of major reformatting programs. This set of formats is designated the FIRE Standard Data Format (SDF).

The FIRE SDF is defined in a way which minimizes the amount of human intervention (programming, software maintenance, and data entry) needed to catalog, inventory, read, or copy the data sets submitted for archival. However, the format definition is flexible enough to allow the investigators to incorporate all needed data in a way that makes sense to them and to make modifications in the future, without imposing a great deal of overhead in the format. Tapes submitted in the FIRE SDF consist of a header file, a volume table of contents file, a test file, optional ancillary files, and data files. The header file documents each tape's contents and internal structure, giving the basic information needed to catalog the data and defining the basic logical structure of the data set in a way that requires minimal recourse to external documentation. The volume table of contents file documents the contents of each data file on the tape, giving the information needed to inventory the files for later retrieval.

The NCDS hopes to transfer the knowledge gained by working on the FIRE SDF to other projects which the NCDS may support in the future, including the Tropical Rainfall Measurement Mission to be launched in the 1990s.

1. INTRODUCTION

The National Aeronautics and Space Administration collects large amounts of data of many different types. These data consist of a variety of instrument measurements and derived parameters – maintained at various levels of data reduction (at different spectral, spatial, and temporal resolutions). Typically, the producers of data products specify a format for data exchange which is tailored to the mission (such as, one satellite, one instrument, or one experiment) and specific application functions. This practice has resulted in a proliferation of data formats and structures. The variability between these formats causes problems both for researchers, especially if they are involved in projects requiring exchange of data between investigators, and for archives or other groups responsible for making data available to the research community in a readily usable form.

RSRM '87: ADVANCES IN
REMOTE SENSING RETRIEVAL METHODS
A. Deepak, H.E. Fleming, and J.S. Theon (Eds.)   429

This paper describes the *format* which has resulted from a cooperative effort between one mission – the First International Satellite Cloud Climatology Project (ISCCP) Regional Experiment (FIRE) – and the NASA Climate Data System (NCDS). The FIRE is a U.S. cloud climatology research program to validate and improve cloud/radiation parameterizations used in general circulation models (Cox, 1987). It spans a number of years and includes an extended time observation period, as well as intensive data collection periods. The success of the FIRE is dependent upon over fifty principal investigators being able to perform multi-data studies in a timely and cost-effective manner (Starr, 1987). The NCDS is an interactive scientific information management system which enables scientific users to locate data sets of interest, preview the data sets using graphical and statistical methods, and extract interesting subsets for further analysis at their own sites. It manages a large collection of data of interest to NASA's research community and is designated as the central archive for the FIRE data. The success of the NCDS as an information management system is dependent upon the development of a methodology and techniques which are readily applicable to all data sets (GSFC, 1977). Otherwise, the system would be too costly to maintain and expand for providing support of additional data sets or for providing additional functionality. The specification of an exchange format assists the FIRE and the NCDS in working together for mutual benefit, allowing both to forego many of the usual problems with data formats (FIRE, 1986).

2. JUSTIFICATION

Researchers and archivers confront a number of problems with data sets (OAO, 1979). One such problem is knowing which data sets to access. Knowledge of data sets appropriate for meeting specific research objectives is difficult to obtain, considering the unique characteristics of data from the many missions (or satellites) and experiments (or sensors). An important function of the NCDS is to help a researcher determine what data exist (or are planned) for supporting his or her research efforts and the characteristics and condition (quality, usability, etc.) of these data. The NCDS must provide the researcher with the information which he needs to make informed decisions about obtaining particular data sets. This information must be up-to-date and accurate and must be presented in a form which will allow the researcher to readily compare the characteristics of two or more data sets. This information, which comprises the NCDS's catalog (somewhat like a mail-order catalog in function), must be succinct but still sufficient for enabling the researcher to determine whether to retrieve and use data from the data sets. However, it is costly for the NCDS to obtain and maintain this information. With each new data set, manpower has to be expended in examining the literature and interviewing the data producers to obtain complete information. This information then has to be stored in a form which will allow the NCDS to meet the researchers' needs. Once the information is available to users, it has to be maintained so that it continues to be accurate and up-to-date, requiring continual review of the research literature and contact with data producers and archivers.

The next problem encountered by a researcher is actually obtaining the data in a ready-to-use form. Usually a researcher has to accept straight copies of the data in the same computer representation as the original data. This means that he might obtain several tapes of data although only a small proportion of the data meet the temporal and spatial characteristics needed for his research. Because of costs, the data archive is generally unable to maintain an inventory (like a department store's inventory showing which shelf has the item the customer is requesting) of its data holdings which is both accurate and detailed enough to allow the archive to retrieve only the data meeting the user's specifications.

Once a researcher obtains the data, he often has problems with interpreting and preparing them for further analysis by applications software (CCSDS, 1987). A researcher is typically dependent upon external documentation to interpret the data format. Testing of this external documentation is often accomplished only via manual means, with a programmer concerned about the particular computer representation of the values in both the researcher's machine and that of the data producer. To resolve these problems, the programmer usually resorts to contacting the data producer (which may be impossible if the production team has been disbanded). Even with well-written documentation, there are problems with verifying that the version of the document corresponds to the version of the data. This leads to a duplication of effort, which may become very costly.

This multitude of problems exists when the exchange media is primarily magnetic tape. However, more and more researchers are requesting data sets on other media, such as diskettes for their particular personal computers, optical disk, or compact disks. And as communications capabilities improve, more and more users will want to transfer data directly from the archive (or the data source) to their computer facilities. These factors will compound the already critical data format problems, which can only be magnified unless standards are defined and followed.

3. APPROACH

In order to circumvent some of the usual problems with data set formats, a set of rules was defined for the construction of data sets to be submitted to the FIRE Central Archive. These rules are referred to as the FIRE Standard Data Format (SDF) and actually refer to a family of formats, not just a single, fixed format.

The FIRE exchange format is written in a way which will allow the FIRE Central Archive to minimize the amount of human intervention (programming, software maintenance, and data entry) needed to catalog, inventory, read, or copy the data sets submitted to the archive, while still allowing the investigators the flexibility to incorporate all needed data and to make modifications in the future. The SDF should greatly facilitate the exchange of data among investigators and the use of the data by other scientists after the data are released to the public. This standard should also allow the archive to provide more sophisticated access services than otherwise possible. Investigators may obtain subsets of data sets by specifying the type of records and the values of many data elements, such as the latitude, the longitude, and the time of the data – and not just by the physical storage medium, such as the specific tapes. Investigators may also take advantage of general-purpose capabilities already developed at the archive facility. This will allow them to perform initial analyses of the data with the archive's data manipulation and display capabilities.

Additionally, the archive can deal with some of the differences between computer representations of numerical values, instead of putting the burden on the investigator. Under the current agreement, the archive's programs will read data sets using either IBM or VAX integer representations and provide investigators with data subsets in their choice of either of these representations (although the format is defined in a way which would allow other computer representations without redesigning the format). The generic software routines needed to perform these functions can also be made available to the investigators. Even if these routines are not directly usable, they should provide some guidance in developing software for reading other data in the standard data format. Because the staff of the archive facility will be familiar with the standard data format, they will also be able to provide limited assistance with problems encountered.

The holdings of the FIRE Central Archive will consist of certain subsets of the total data set selected by the FIRE Science Evaluation Team (FSET) for special study or identified by the FSET as especially important. These data will be submitted to the central archive in the SDF. The SDF will facilitate team interaction and distribution of data. Investigators have also been encouraged to adopt the standard format for individual holdings of data collected for the FIRE. This will allow them to take advantage of any formatting software developed for the SDF by other investigators or by the central archive.

A FIRE data set begins its life cycle with the definition of the logical and physical structure of the data it is to contain. This format definition is expressed in explicitly specified fields of a tape header file which is written in ASCII and can easily be read by standard computer utilities and output for human interpretation. Then source data, and sometimes ancillary data from an archive, are processed to generate data files. These files conform to the structure specified in the header file. While processing the data, the producer also generates a table of contents file for each tape and a test file which can be used in validating the decoding of the records. All of these files are sent to the archive, where appropriate information is extracted from them for retrieval from the catalog, inventory, and data dictionary. When a researcher needs data, he checks a directory (like telephone "yellow pages", pointing users to different stores they might want to contact) to find out which archive to contact. Then the catalog of that archive directs him or her to the data set meeting his or her needs. When the researcher specifies the data set selection criteria, the system locates the data using the inventory, and then retrieves the data. As it retrieves the data, the systems uses the data dictionary to interpret the data, extract appropriate portions, and output the data in a useful form. This cycle is shown in a simplified form in Figure 1.

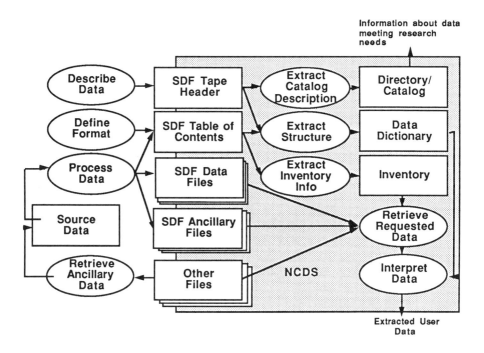

FIGURE 1. Life-cycle of a Data Set in Standard Data Format

4. IMPLEMENTATION

The FIRE data tape format is standardized only in terms of certain tape characteristics, file arrangements on the tape, a few file structure and data characteristics, and a few data organization guidelines. In other words, the actual data arrangement within a data file is not standardized except for certain imposed simplifications. This approach compromises between the minimum of reprocessing by the investigators and the maximum convenience of a single data format for data comparison and analysis. The required characteristics simplify the data structures sufficiently to eliminate most format problems.

A FIRE Standard Data Format tape will contain a tape header file (volume ID), a volume table of contents (TOC) file, a test data file, possibly one or more ancillary data files, and one or more data files. Each file will contain one or more physical records (blocks), consisting of a fixed number of bytes. The general layout of the tape and further organizational breakdowns described below are shown in Figure 2.

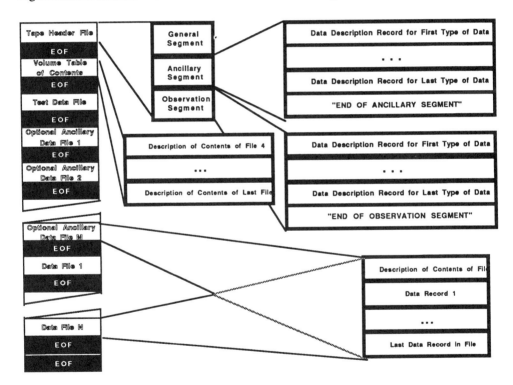

FIGURE 2. Organization of FIRE Tape in Standard Data Format

4.1 TAPE HEADER FILE (VOLUME ID)

The tape header file is always the first file on a data tape and is composed entirely of ASCII character data, thereby making it easy for a person to read. It contains general information describing the data contained on the tape and documentation of the tape structure. This file consists of one (or more if needed) physical record (block). NCDS programmers developed a program for creating this header interactively. This program was distributed on a sample tape in FIRE SDF which the archive sent to all interested investigators. The tape header file is made up of three segments (see Figure 2): the general segment which describes the overall content of the tape, the ancillary segment which describes the ancillary data and their structure, and the observation segment which describes the observed data and their structure. A sample of the general segment is provided in Figure 3. Information contained in the general segment includes the following: a unique sequence number for the tape, the time range of the data on the tape, the producer's name and address, the date the tape was created, the name of the computer and operating system used to create the tape, the computer representation used for the numeric values, and a plain language description of the contents of the tape.

```
FIRE. MSIFO.PSUS.000001.0.19870702.19870718
MR. WILLIAM J. SYRETT
DEPARTMENT OF METEOROLOGY
503 WALKER BUILDING
THE PENNSYLVANIA STATE UNIVERSITY
UNIVERSITY PARK, PA  16802
USA
19870831


19860901
6250
DEC VAX 8200 UNDER VMS 4.5
VAX
0123456789=:>   /STUVWXYZ,(-JKLMNOPQR*];÷ABCDEFGHI.)[<
THIS TAPE CONTAINS WIND SPEED AND DIRECTION DATA GATHERED BY THE PENN STATE
DOPPLER SODAR DURING PROJECT FIRE ON SAN NICOLAS ISLAND. THE DATA SET CONSISTS
OF HOURLY AVERAGED WIND SPEEDS AND DIRECTIONS AT 50-METER VERTICAL RESOLUTION
FROM 800 TO 100 METERS ABOVE SEA LEVEL.  A DATA QUALITY INDICATOR IS INCLUDED;
QUALITY RANGES FROM 0 TO 5, HIGHER NUMBER INDICATES GREATER CONFIDENCE IN DATA.

END OF GENERAL SEGMENT
```

FIGURE 3.   Sample General Segment of Tape Header File

The ancillary and observation segments are structured the same and are made up of data description records describing each type of data record contained on the tape. The Data Description Record (DDR) describes each variable in the data record, as well as general information about the record (such as size) and the specific source of the data (mission/satellite and instrument/sensor). A sample data description record is shown in Figure 4. For each variable in the record, the DDR gives its name, field type (such as integer stored in 1, 2, or 4 bytes), measurement units, display format, precision, resolution, minimum and maximum acceptable values, and any functional dependencies on cyclic (repeating) values.

```
RECORD NAME: PSUSODAR - HOURLY WIND SPEED, DIRECTION, 15 LEVELS      200    4800
MR. WILLIAM J. SYRETT
DEPARTMENT OF METEOROLOGY
503 WALKER BUILDING
THE PENNSYLVANIA STATE UNIVERSITY
UNIVERSITY PARK, PA  16802
USA
PSUSODAR SNI   :PENN STATE DOPPLER SODAR AT SAN NICOLAS ISLAND, PROJECT FIRE
DOPSODAR    :DOPPLER ACOUSTIC SOUNDER
YEAR         YEAR  I*4      I2      31       0      1      1       87      87
MONTH        MONTH I*4         I2      31      0      1      1      6       7
DAY          DAY   I*4      I2      31       0      1      1       1      31
HOUR         HOUR  I*4      I2      31       0      1      1       0      23
HEIGHT       HEIGHT 00 M         F4.0    31       0      5      50      100     800
WIND DIRECTION WDR   I*4 DEGREES    I3      9      -99     5      1       -99     360HEIGHT:  800
                           . . .
WIND DIRECTION WDR   I*4 DEGREES    I3      9      -99     5      1       -99     360HEIGHT:  100
INSTRUMENT DESCRIPTION:
DOPPLER SODAR, TIME RESOLUTION=1 HOUR FOR THIS DATA SET, VERTICAL RESOLUTION=
50 METERS.  SODAR LOCATED AT 33.27N, 119.58W, ROTATION ANGLE=72 DEGREES
TEMPORAL CHARACTERISTICS:
WIND TABLES PRODUCED EVERY 2 MIN, 3 MIN, 6 MIN, 10 MIN, 12 MIN OR 15 MIN
ARE USED TO COMPUTE THE HOURLY DATA RECORDS
SPATIAL CHARACTERISTICS:
NORTHWEST TIP OF SAN NICOLAS ISLAND, 33.27N, 119.58W
CALIBRATION INFO:
ASSUMED 1-KM MEAN-LAYER TEMPERATURE OF 288K.  CREATES +/- 5M HEIGHT
UNCERTAINTY AT ALTITUDE OF 800M.
REFERENCES:
CONTACT BILL SYRETT AT ABOVE ADDRESS FOR INFORMATION ON DATA PROCESSING
 SOFTWARE AND SODAR REFERENCE MANUAL.
SOFTWARE:
```

FIGURE 4. Sample Data Description Record

4.2 VOLUME TABLE OF CONTENTS FILE

The Volume Table of Contents File (TOC) is always the second file on a data tape and is composed entirely of ASCII character data. It contains text describing the tape contents. This textual information is arranged in the form of a table showing a file by file listing of contents. The following information is provided for each data file on the tape:

Data file sequence number on tape
Data sequence number within data set
Record type in the file (short name as defined in the header file)
Temporal range of data in file
Spatial range of data in the file, in terms of ranges of latitude and longitude
Where appropriate, description of viewing geometry in file

The data file sequence number indicates the file number containing the listed data. This number allows identification of any data file and any data record on the tape by comparison to the sequence numbers given in each record. The data sequence number refers to a numbering of observations within an observation set (e.g., image number, orbit number, flight number) that is used by the investigator to relate the observations on this data tape to other observations on other data tapes.

4.3 TEST DATA FILE

The Test Data file is written entirely in ASCII and represents the contents of the first observation data file of the tape. This file is written using the display format specified in the DDR. This file should be easily read and can be used to validate the contents of the first observation data file.

4.4   ANCILLARY AND OBSERVATION DATA FILES

The data records of any one file contain only data for one defined record type. The record type is documented in the first physical record of the file (so interpretation of data is not dependent upon the sequence of the data files, since sequence errors can easily occur and later expansions will not be limited to sequential access media). This physical record contains the same descriptive information about the data in the file as the corresponding record in the TOC file. This descriptive information is composed entirely of ASCII character data.

It is recommended that certain variables be included in all records of a data set. Though an investigator is not required to follow these guidelines, use of them should ease data set comparison. These variables are described in a DDR in the same way as the other variables. Some of these variables are for preventing loss of synchronization by input/output errors, while others are for ease of selection of data meeting the experiment needs. The recommended variables are listed below with a short description:

FNUM	Data file number on tape
RNUM	Data record sequence number within the file
TSEQ	Data record sequence number within tape
DSEQ	Data record sequence number within data set
YEAR	Year of the data observation (for example 1986)
MONTH	Month of the data observation (1 through 12)
DAY	Day of the data observation
HOUR	Hour of the data observation
MIN	Minutes of the data observation
SEC	Seconds of the data observation
MSEC	Milliseconds of the data observation
LAT	Latitude of the observation
LONG	Longitude of the observation

Investigators are also encouraged to include other appropriate location fields, such as altitude, or fields indicating viewing geometry (solar zenith angle, relative azimuth, viewing zenith angle, or other equivalents).

The actual data records should not differ greatly from what would normally be produced. The values of the variables are written in the record as described in the DDR.

5.   CONCLUSION

The ultimate utility of any standard is in its use. It is hoped that the techniques described in this paper can be transferred to other projects. In fact, they are being proposed for another project currently in the planning stages – the Tropical Rainfall Measurement Mission to be launched in the 1990s. Some would argue that it is just too much trouble to conform to the standard. If this standard just adds to the proliferation of formats, this argument is justifiable. However, both programmers and investigators are greatly benefited if the standard reduces duplication of effort – if they can use the same analysis tools to explore multiple data sets or easily obtain tools from others.

The FIRE Standard Data Format allows descriptive information on the contents of a data set to be recorded with a data set (i.e., on the data set tape), ensuring that the documentation stays with the data, thereby increasing their useful lifetime. It also allows the data to be easily decoded using simple computer programs. Using the specifications laid out for the FIRE, the formatting aspects of data exchange have been brought under control.

The NCDS has already shown the potential of these techniques by writing software programs for extracting subsets of data in SDF and outputting them in a form readily usable by investigators and ready for preview with the NCDS tools. Sample data sets submitted by several investigators have been examined for conformance to the format rules, and graphical displays such as that in shown in Figure 5 can now be produced for any data in FIRE SDF using the tools of NCDS.

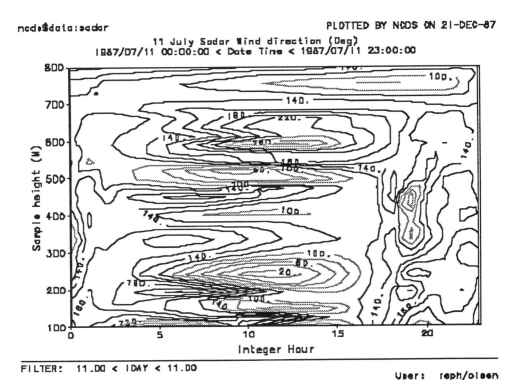

FIGURE 5. Sample of NCDS Graphics Output with FIRE Data

The distribution by the archive of a sample tape containing software for building the tape header, where the structure of the tape is defined, assisted investigators in conforming to standards – and is one of the first areas where a duplication of effort was hoped to be avoided.

Conformance to the standard format has been fairly easy to validate using automated methods. Eight tapes have been examined so far. One of these was in total conformance to the standard; four contained small errors that were easily corrected; two needed some reworking but the problems could be clearly identified.

The ultimate success of the Standard Data Format remains unknown at this time. Cooperative efforts are being made to achieve success, and several FIRE participants have expressed their support. Some have found the format to be very appropriate for their needs and have encouraged its use with data other than FIRE.

ACKNOWLEDGMENTS

The authors wish to acknowledge the contributions of the NCDS programming staff. In particular, Nick Iascone contributed the graphics used to illustrate this paper and set up programs for reading FIRE SDF tapes and determining whether they met specifications. We would also like to thank William J. Syrett, a FIRE Investigator at the Pennsylvania State University, for allowing us to use his data for illustrating the format concepts.

REFERENCES

Consultative Committee for Space Data Systems, February 1987. *Space Data Systems Operations with Standard Formatted Data Units: System and Implementation Aspects.* CCSDS 610.0-G-5. Code-TS. NASA. Washington, DC.

Cox, Stephen C., David S. McDougal, David A. Randell, and Robert A Schiffer, February 1987. "FIRE - The FIRST ISCCP Regional Experiment." *Bull. Amer. Meteor. Soc.* 68:2. pp 114-118.

FIRE, December 1986. *FIRE Marine Stratocumulus Intensive Field Observations – June 29 - July 19, 1987. 1987 Operations Plan.* Langley Research Center. Hampton, VA

Goddard Space Flight Center, November 1977. *Proposed NASA Contribution to the Climate Program.* NASA/GSFC. Greenbelt, MD.

OAO Corporation, October 1979. *User Requirements for a NASA Climate Data Base Management System.* Beltsville, MD.

Starr, David O'C, February 1987. "A Cirrus-Cloud Experiment: Intensive Field Observations Planned for FIRE." *Bull. Amer. Meteor. Soc.* 687:2. pp 119 - 124.

DISCUSSION

Gautier: You have told us that you had some feedback from the investigator. I suppose that one of the difficulties is the time it takes for a PI to put his or her data in the right format. Do you have any idea of the time they have spent, those who have really followed your format, on doing this work?

Reph: I haven't gotten specific feedback about the time. As I said, we've got our programs that help them out a little bit. Their programmers usually contact our programmers and we've been able to help them out, but I have no specific idea as to the time.

CHRP:
CONFIGURABLE HIGH RATE PROCESSOR

John E. Dorband, Dan A. Dalton, James R. Fischer,
Jose M. Florez, Warner H. Miller, H.K. Ramapriyan,
and John V. Rende
NASA Goddard Space Flight Center
Greenbelt, Maryland 20771, USA

ABSTRACT

The variety, resolution, and anticipated data collection rates of remote sensing instruments expected to be deployed in the last decade of this century and early 21st century imply requirements for on-board reduction of data volumes which maximize the scientific return from space in the face of limited transmission bandwidth. Such data reductions can be achieved either through lossless or lossy data compression or through on-board analysis and information extraction prior to transmission of results. Recent advances in computer science and hardware technology make it feasible to consider the development of on-board computer systems with sufficient capability to accomplish the above tasks. The Configurable High Rate Processor (CHRP) project was initiated in FY88 as part of NASA's Civil Space Technology Initiative in data systems to address these issues by developing the concept of an onboard, high rate, high capacity data system for the 1990's. CHRP will be an aggregate of high speed data processors, storage devices, and I/O devices which are interconnected on a very high speed data network. The data network will support point to point communications between up to 16 nodes at over 1 gigabyte per second. Mass storage nodes will have a capacity of up to a terabit. Several processors will achieve gigaflop computation rates. The network will support high speed input from multiple flight sensors as well as high speed output to telemetry. This flight processing system will be programmed and controlled from work stations on the ground and will allow test and debug while actual production work is being performed.

1. INTRODUCTION

Advances in sensor technology are providing *individual* instruments capable of producing data at rates far in excess of total downlink communications bandwidths. Several current and proposed high rate instruments are listed below along with their data rates in megabits/second.

High Resolution Imaging Spectrometer (HIRIS)	512
Synthetic Aperture Radar (SAR)	300
Thematic Mapper (TM)	85
Moderate Resolution Imaging Spectrometer (MODIS)	20
High Resolution Solar Observatory (HRSO)	16

The evolution of these instruments implies that significant onboard processing will be needed on future space and Earth science missions to perform either lossless or lossy data compression or on-board analysis and information extraction prior to transmission of results. Such onboard processing will require very high speed processors and associated peripheral equipment such as large capacity onboard data buffers and high rate data buses. This paper describes a five year strategy for identifying the on-board processing requirements, expressing them in terms of candidate computer architectures, selecting an architecture for development in an evolving ground based test bed and demonstrating actual end-to-end processing on the testbed.

2. REQUIREMENTS

The CHRP project is driven by the need for a flexible high rate/high capacity on-board processing capability for NASA missions in the 1995-2005 time frame. The objective is to develop and demonstrate a system approach for providing high rate/high capacity data processing of space and Earth sciences sensor data onboard spacecraft. Oversight by a steering committee of high rate instrument investigators who provide regular reviews and guidance is essential to its success.

The system capabilities must be tailorable to meet widely diverse and evolving mission requirements from the simplest one-instrument/one-processor mission to the most complex missions in a cost effective manner. The high degree of commonality and interchangability that is needed for hardware and system software elements will necessitate standard interfaces and modular interconnect designs.

The system must be able to provide the total onboard data management support required for scientific operations from instrument interface to communication link transmitters and receivers. This includes all formatting, coding, buffering, processing, editing, storage, and multiplexing required by complex heterogeneous payloads operating at hundreds of megabits per second.

The system must be configurable in real time to adapt to changes in the operating environment. Support of multi-processing, including multiple processor types performing simultaneous independent or correlated tasks, is expected to be required. Interface compatibility to the Jet Propulsion Laboratory (JPL) SAR and HIRIS processors as well as the NASA/Ames symbolic processor will be actively sought. The CHRP will build upon the technology base established by NASA's Office of Aeronautics and Space Technology investments in the Massively Parallel Processor (MPP), the terabit buffer, Gallium Arsenide (GaAs), Very High Speed Integrated Circuit (VHSIC), and the Star*Bus.

Space hardening, fault survival, and minimization of size, weight, and power are expected to be challenging. The development of all elements will focus on the flight adaptation of existing systems with minimum new designs. The insertion of VHSIC and/or GaAs technologies is of primary importance and will be critical performance drivers in all hardware elements.

3. SPECIFIC TASKS

3.1 ARCHITECTURE DEFINITION PHASE

The development strategy begins with an Architecture Definition Phase during which alternative computational scenarios will be created and utilized to evaluate the efficiency and flexibility of various architectural options. The Earth Observing System (EOS) Project is expected to provide the baseline inputs to this evaluation process and participate in all CHRP reviews.

During the first eighteen months, an On-*board Computational Requirements Study* will be conducted to identify and cultivate those candidate scientific investigations that would be enabled by the on-board CHRP, and to analyze the candidates and determine their system requirements. Generation of CHRP applications requires cultivation of visionary thinking between the CHRP design team and members of the space and Earth science communities who may need its capabilities. Identification of the interested scientists to serve on a CHRP Scientific Advisory Board will take place through direct contact with the EOS Project and through co-sponsorship of a Data Compression Workshop being set up by the OAST Data Systems Working Group. We will analyze the applications identified by the workshop to determine their feasibility, range of plausible execution speeds, and critical CHRP resource requirements. The intermediate results of our analyses will be presented to the Scientific Advisory Board and at subsequent Data Compression Workshops to subject them to iterative refinement. As the onboard computational requirements are firmed up the *CHRP Architecture Requirements* will be determined and documented, followed by the *Specification of the CHRP System Elements*.

3.2 CHRP TESTBED

The operational CHRP Testbed is an essential aspect of the five year development. It will be located at Goddard in the Space Data and Computing Division/Code 630 and will support the evolutionary development of the CHRP hardware, system software and application demonstrations. Nationwide access to it will be via networks including the NASA Science Network (NSN) and ARPANET. Two generations of testbed are envisioned:

The initial testbed will become operational during FY89 and will be based on commercially available hardware and software performing functions similar to those required, but likely at lower rates, with lower reliability, and using hardware that is physically too large and that consumes too much power.

The enhanced technology testbed will become operational during FY91 and will evolve from the first testbed through changeouts of hardware and system software. This testbed will focus on demonstrating selected components operating to critical specifications. These components are expected to be: the very high speed data network, the processor nodes, and the mass storage nodes. Incorporation of simulated or actual instruments into the testbed to generate data is essential. Incorporation of the JPL SAR and HIRIS processors as well as the Ames symbolic processor is desired.

Emphasis in the testbed, as in the CHRP, is on flexibility such that changeout of hardware components is accomplished quickly through standardized hardware and software connections with the system software configuring itself automatically to the resulting set of nodes.

The system software for the CHRP testbed will be based on the concept of network computing which enables the distribution of computation as well as data among multiple networked storage devices and heterogeneous CPUs. This approach will enable the science users of CHRP as well as the software developers to use the software rich environment of workstations located anywhere in the world where network access is available. Through the workstations, software development and CHRP program execution control will be performed as though the workstation and the CHRP hardware were one unified system. The workstation will also provide end-to-end simulation capability in software of the CHRP hardware to enhance the software development process. The user interface, based on windows and icons, will allow a scientific user to rapidly learn to use the on-board CHRP computational capabilities. The CHRP will be programmed through a single high level language to allow all heterogeneous CPUs to be viewed by the software developer as differing only with respect to speed of execution of certain instructions, i.e. vector and parallel CPUs will execute operations on vectorizable and parallel data structures much faster than scalar CPUs will.

3.3 CHRP APPLICATION DEMONSTRATIONS

Algorithms, which will reduce the effective bandwidth required to transmit useful information to scientists, will be developed and demonstrated on the testbed. These algorithms are generally categorized as data compression and data analysis. Both lossy and lossless data compression will be considered. Analysis algorithms will be implemented in order to demonstrate the CHRP concept that the scientist can, from his laboratory, program the CHRP in space, by uploading the program, executing it, and receiving meaningful results at his workstation. Data compression algorithms will be identified, implemented and tested on classes of data such as those from the Advanced Very High Resolution Radiometer (AVHRR), the Airborne Visible and Infrared Imaging Spectrometer (AVIRIS), the Coastal Zone Color Scanner (CZCS), the Scanning Multichannel Microwave Radiometer (SMMR), and the Solar Optical Ultraviolet Polarimeter (SOUP). These sensors are representative of those to be flown during the Space Station era. In designing processing and analysis algorithms, emphasis will be placed on implementation of currently proposed techniques for HIRIS, MODIS and SAR processing. HIRIS and MODIS data will not be available until 1995. Therefore, substitute data such as those from AVIRIS and CZCS, which are now available, will be used. Prior to availability of the CHRP testbed, implementation of existing algorithms on compatible processors is an intermediate goal.

3.4 CHRP TECHNOLOGY DEVELOPMENT

The CHRP hardware repertoire (Fig. 1) will contain network, processor, mass storage and interface elements. Three of the components that are being targeted for development are: a data network with gigabyte/second capacity; a processor array with gigaflop speeds when performing data compression; and a terabit mass storage device. The preliminary specifications described below will be influenced by the results of the architecture study.

Data network: A cross-bar switching system using fiber optic technology will be developed so that up to 16 point to point data communications paths can be established, each passing data at greater than 1 gigabyte/second.

Processor array: The processor array compute server will be a very high performance number cruncher, able to sustain at least one gigaflop per second when performing the data compression algorithms developed under this program. Its range of applications will be much broader than signal and image processing, including high speed simulation and graphics generation. It will be able to sustain a data transfer rate at the full bandwidth of the CHRP's central data network. The array will be a 2-dimensional grid containing 65,536 processors arranged as a 256 x 256 mesh. Its control processor will be able to submit 20 million instructions per second to the array.

Mass storage: A mass storage technology element with a capacity of 10^{12} bits and with a data transfer rate greater than or equal to 100 megabytes/second and access times less than 100 milliseconds will be developed. In addition, a solid state

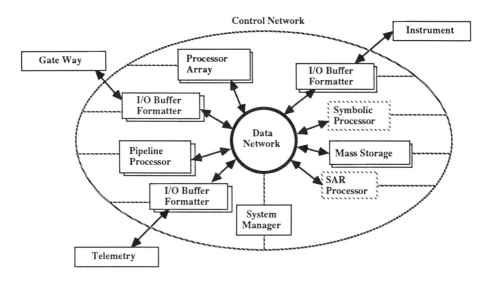

FIGURE 1. The Configurable High Rate Processor.

buffer with capacities in the 10^8 bit range will be developed to sustain a data transfer rate equal to the full bandwidth of the central data network.

The specifications stated above are intended to press the state-of-the-art and may have to be selectively lowered if our technology evaluation shows they will be infeasible. We believe, however, that a machine meant to be used in a production mode 10 years from now must be forward looking, thus visionary, due to the long lead time to put it in space and the rapid pace of technological evolution.

4. CONCLUSION

Because current and proposed imaging sensors will collect more data in space than can be sent to the ground, discrimination through computation must be exploited. An aggressive yet feasible architecture such as CHRP will provide such a computing capability. CHRP is a system approach including high speed computation, high speed interprocessor communication, and integrated multitasking on heterogeneous processors, serviced by an evolving remote host development environment. It can just as easily be used by an astronaut located at a workstation in space, as a scientist located in a ground based laboratory. Its features will form a versatile and user fiendly high performance remote computing facility to meet the in-space supercomputing requirements through the end of this century.

DISCUSSION

Johnson: Just to put this in perspective, I am not familiar with what existing computational capabilities have flown yet. What sort of capabilities have been used in space?

Dorband: At the moment all I'm familiar with in space flight is an old version of a serial processor that was built some time in the early 70's, that is being used for most of those space flight projects or uses that I know of. So, super-computing has not been addressed as yet for flight. Not even much in the middle levels as far as I know.

Treinish: You propose a highly flexible hardware system and then show that it would be addressable by work stations presumably on a scientist or operators desk. Have you begun to address the software architecture necessary to allow someone at a workstation to easily take advantage of that type of hardware flexibility?

Dorband: Right now we are looking into SUN and MACINTOSH workstations and they will both be running UNIX. We are planning on looking to UNIX as a baseline under networking with TCPIP, so that is where we are looking right now. We are primarily looking at commercial operating systems and software and hardware. In fact, our initial test bed for this will be made primarily of commercial hardware and a second test bed will actually be integrating in components that are lightweight low power and high speed. There is some work being done, even at JPL, the max system, which is a homogeneous network of nodes, that is a message-passing-type system. And what we would like to do is extend that, or talk to them about extending that, into a heterogeneous realm where the processes will have to be tied to specific processors but may be one of several specific processors and we may have multiple processors. So that is the sort of approach on the network itself.

Speaker: Who funds this operation? Is it NASA?

Dorband: Yes, this is NASA-funded under the CSTI plan, that is Civil Space Technology Initiative.

OPEN DISCUSSION

SESSION G. - H. K. RAMAPRIYAN

Deepak: Dr. Ramapriyan mentioned that he liked this idea of including the session on data compaction in this retrieval methods workshop. His initial impression was that this topic wouldn't go too well, but later on he heard some very good comments coming from the workshop participants, so he feels very encouraged that we should continue to include this topic in future workshops on retrieval methods.

AN INTERACTIVE, DISCIPLINE-INDEPENDENT DATA VISUALIZATION SYSTEM

Lloyd A. Treinish
National Space Science Data Center
NASA Goddard Space Flight Center
Greenbelt, Maryland 20771, USA

Abstract

Critical to the understanding of data is the ability to provide pictorial or visual representations of that data, particularly in support of correlative data analysis. To implement this concept for the space and earth science research community, the National Space Science Data Center (NSSDC) has an on-going program to develop new, generic (*i.e.*, data-independent) techniques for the display of multidimensional data as well as associated information about these data or metadata. These techniques utilize the latest methods in computer graphics and imaging, and *state-of-the-art* hardware. As part of these efforts, the NSSDC has developed the NSSDC Graphics System (NGS), an interactive discipline-independent toolbox to support the visualization of data. In order to utilize the NGS, data of interest must be stored in terms of the NSSDC Common Data Format (CDF), a data-independent abstraction for multidimensional data structures. The CDF has been used to develop a number of generic data management, display and analysis tools for a wide variety of disciplines at the NSSDC. The CDF development efforts are evolving into a *standard* method for storing space and earth science data for a variety of applications. (Refer to L. Treinish and M. Gough, *A Software Package for the Data-Independent Management of Multidimensional Data*, EOS, **68**, pp. 633-635, 1987 for a description of CDF.) The NGS supports the ability to display or visualize any arbitrary multidimensional subset or slice of any data set by providing a large variety of different representation schemes, all of which are supported by implicit animation. In addition, the design of the NGS provides an open-ended framework for discipline-independent data visualization, so that new capabilities can be added. New tools are being implemented as a result of NSSDC's research in several areas of computer science.

1. INTRODUCTION

In July 1987 the Fourteenth Annual Conference of the Association for Computing Machinery (ACM) Special Interest Group on Computer Graphics (SIGGRAPH), the premiere international scientific meeting on computer graphics, was convened. This conference covered all aspects of academic and industrial graphics user/computer communications and manipulation - hardware, software, languages, data structures, methodology and applications. A group convened at this conference to discuss the issues and problems in dealing with the understanding of very large and complex (*e.g.*, multidimensional) data streams that are available today from such sources as spacecraft instruments and supercomputer-based models and simulations. This problem is generally not being addressed and the problem will be compounded by the next generation of data sources, which will literally bury the scientific community in bits. Despite advancement in data generation and computer technology over the

RSRM '87: ADVANCES IN
REMOTE SENSING RETRIEVAL METHODS
A. Deepak, H.E. Fleming, and J.S. Theon (Eds.)   449

last few decades, methods of analyzing large and complex data streams basically have not changed, leaving significant fractions of data not fully understood or scientific information ungleaned. The National Science Foundation (NSF) recognized this problem, and commissioned a panel to study the situation, and the impact of one area of technology, computer graphics and imaging (NSF, 1987). The panel that produced the NSF's report, *Visualization in Scientific Computing,*, and the SIGGRAPH group included representatives from most major scientific disciplines and non-classified supercomputer facilities (public, private, industrial and academic) used for scientific applications in the United States.

2. THE PROBLEM - GENERIC VISUALIZATION

Critical to the understanding of data is the ability to provide pictorial or visual representations of that data, particularly in support of correlative data investigations. This is a concept that the National Space Science Data Center (NSSDC) at NASA's Goddard Space Flight Center (GSFC) has recognized for several years and as cited above has begun to grow in importance among the scientific community at large only recently. To implement this concept for the space and earth science research community, the NSSDC has embarked on an applied computer science research and development effort in the following areas:

- Generic data and metadata modelling and representation

- Advanced data structures supporting graphics as well as data analysis and management applications

- Correlative visualization and analysis techniques for multiple parameter/ dimensional data sets

- Parallel rendering algorithms

- Portable, operational visualization environments

This work has developed new, generic (*i.e.*, data-independent) techniques for the display of multidimensional data as well as associated information about these data or metadata. These techniques utilize the latest methods in computer graphics and imaging, and state-of-the-art hardware and are embodied as the NSSDC Graphics System (NGS), an interactive discipline-independent toolbox to support the visualization of data, on the NSSDC Computer Facility (NCF) DEC VAX 8650 and VAX-11/780.

3. COMMON DATA FORMAT (CDF)

The NSSDC has developed the *first*, self-describing data abstraction for the storage and manipulation of multidimensional data to support discipline-independent scientific applications. This abstraction, which consists of a software package and a self-describing data structure, is called the Common Data Format (CDF) (Treinish and Gough, 1987). CDF provides true data independence for applications software that has been developed at the NSSDC. Scientific software systems at the NSSDC use this construct so that they do not need specific knowledge of the data with which they are working. This permits users of such systems to apply the same functions to different sets of data. In addition, the CDF provides a simple means for the transport of data among different research groups in a format-independent fashion. The users of data-independent NSSDC systems rely on their own knowledge of different sets of data to interpret the results, a critical feature for the multidisciplinary studies inherent in the earth and space sciences. Such CDF-based software can use the information available through the CDF software package to inform a user about contents, history, and structure of data supported in a given CDF, and allow such a user to concentrate on the scientific nature of the data of interest rather than its format. CDF has been used to develop a number of generic data management, display and analysis tools for a wide variety of disciplines at the NSSDC, including the NGS. CDF, through its software package, provides to the applications programmer a mechanism for uniformly viewing data of interest via a data structure oriented to the user of the data (*i.e.*, a scientist). CDF is a mechanism for the flexible organization of interdisciplinary data into generic, multidimensional structures consistent with potential scientific interpretation provides a simple abstract conceptual environment for the scientific applications programmer who works with data, but also encourages the decoupling of data analysis considerations from those of data storage.

The CDF software package is a toolbox of programming primitives for managing multidimensional data ensembles; it provides a simple abstract view for random access of arbitrary blocks of data. Applications such as the NGS must be built into higher-level software that employs CDF. The programmer that utilizes the CDF data abstraction views the CDF software package as consisting of thirteen operations. These abstract routines are designed to make it easy for a programmer to utilize data in terms of CDF, independently of the complexity of the data. FORTRAN language bindings for CDF are now operational on DEC VAX/VMS computer systems (Gough, 1987). C language bindings have just been developed and are being tested for VAX/VMS, IBM MVS and VM, and UNIX (*e.g.*, Sun, A. T. & T.), etc. environments, coupled with conversion utilities to transparently move the physical files composing a CDF from one computer system to another. These developments will be enhanced for distributed access over local area networks.

CDF is designed to be portable so that copies can be made available on computer systems outside of the NSSDC to promote the exchange of both software and data. In fact, over *80* organizations outside of the NSSDC representing various NASA laboratories, research groups, current and future flight projects, etc. as well as other government agencies, universities, corporations and foreign institutions are now becoming ß-test sites for the CDF software package for their respective development of software to archive, manage, manipulate, display or analyze data in a variety of disciplines. Such ß-testing will permit these organizations to evaluate CDF for potential use in future flight projects as well as to support specific scientific investigations. As a result the CDF development efforts have become a standard method for storing space and earth science data for a variety of applications. In addition, the CDF has been critical to the success of the NSSDC's current Coordinated Data Analysis Workshop (CDAW)-8 activity in magnetospheric physics, and the ability of the participants to produce scientific results via data interpretation and analysis at an unprecedented rate. CDF was used to manage a diverse CDAW-8 data base (more than 50 different data sets) that was served by various generic analysis tools, especially the NGS (Manka, 1987; and Manka *et al.* 1987).

4. THE SOLUTION - THE NSSDC GRAPHICS SYSTEM (NGS)

In order to develop the required graphical capabilities in as timely and cost-effective manner as possible, NSSDC has employed several off-the-shelf capabilities. The NGS employs NASA's Transportable Applications Executive (TAE) as an easy-to-use, consistent, uniform user interface (TAE Support Office, 1986). As TAE evolves into supporting an object-oriented, dynamic, window-based interface, the NGS will adopt these improvements. To support two- and three-dimensional interactive graphics on any type of graphics hardware, the NGS employs the Template package, developed by Template Graphics Software (Template Graphics Software, 1986). In addition to providing the environment and the tools to generate graphics, Template provides the NGS with sophisticated graphics device-independence for both direct as well as post-processing of graphical objects. As this package evolves and expands in capability, the NGS will take advantage of these enhancements.

The NGS supports the ability to display or visualize any arbitrary multidimensional subset or slice of any data set by providing a large variety of different representation schemes, all of which are supported by implicit animation. In other words, any field within one or more CDF-based data sets can be used for any axis, including sequencing for animation (Gough, 1986). In addition, the NGS places a strong emphasis on complete annotation of its graphical products, and extensive use of color. The NGS supports the following ways of displaying data:

- Two-dimensional histograms
- X-Y plots, including optional multiple axes, pseudo-color and polar coordinates
- Multiple panel displays
- Location maps
- Contour plots with and without maps
- Surface diagrams with and without maps, including optional pseudo-color-
- Pseudo-color images with and without maps
- Solid modelling without maps

In addition, the NGS will support these visualization techniques in the future:

- Solid modelling with maps
- X-Y-Z plots with and without maps
- Two-dimensional vector field plots with and without maps
- Three-dimensional vector field plots with and without maps
- Three-dimensional histograms with and without maps
- Scatter diagram matrix

The NGS also provides a number of options for each representation scheme, such as curve fitting, gridding, scaling, filtering, font selection, statistics, graphics metafile generation, etc. Specific options are associated with all representation schemes (*e.g.*, controlling the range and increment of isolines on contour maps).

The NSSDC strongly emphasizes that the NGS develop very accurate, high-performance graphics tools for data visualization. For example, to support the visualization of large, geographic data sets the NGS employs very flexible world mapping capabilities that are not only quick but also very precise to eliminate any distortion in mapped displays (Ni and Gough, 1987). Currently, 22 general projections are supported, while more can be easily added by user request. The user has complete control over the specification of the pole point and viewing window for any of the display techniques available with world maps, to support arbitrary reprojection of any data set. Both low- and medium- resolution world coastline data bases are supported, the latter with political boundaries. In the future, a very high-resolution world coastline data base will be added as well as a world topographic data base.

5. APPLIED COMPUTER SCIENCE RESEARCH

The design of the NGS provides an open-ended framework for discipline-independent data visualization, so that new capabilities can be added. New tools are being implemented as a result of NSSDC's research in several areas, which is an outgrowth of previous efforts in interactive solid-modelling techniques employing NASA's Massively Parallel Processor (MPP) (Treinish *et al.* 1986). For example, the following specific new techniques have been recently developed:

- Advanced data structures supporting graphics as well as data analysis and management applications to assure rapid display and manipulation of large, complex data sets:

- storage and sampling of three-dimensional data via generic oct-trees

- polygon expansion via quadtree-based rectangular subdivision for pseudo-color imagery

- Rendering and manipulation algorithms with serial (*e.g.*, VAX) and parallel (*e.g.*, MPP) implementations that can operate on *any* data object or geometry:

- n-dimensional gridding

- ray-tracing via recursive spherical triangle subdivision

Like CDF, the NGS is designed to be portable so that copies eventually can be made available on computer systems outside of the NCF to promote the exchange of both software and data. The NGS is already being ß-tested at several sites on the Space Physics Analysis Network (SPAN) to evaluate it for potential use in future flight projects as well as to support specific scientific investigations. In addition, the NGS is currently available operationally to the users on SPAN of NSSDC's Network Assisted Coordinated Science (NACS) system in support of the CDAW activities. It will be available to users of NSSDC's NASA Climate Data System (NCDS nee Pilot Climate Data System, PCDS) in the future (Treinish and Ray, 1985, Reph *et al.* 1986). To properly support the diverse use of the NGS, the software is maintained with strict configuration control under NSSDC management.

6. USAGE OF THE NGS AND EXAMPLES

The NSSDC is currently helping a number of scientists in a variety of disciplines solve their problems in scientific data visualization through the tools available in the NGS in addition to the aforementioned ß-testing and evaluation of the NGS for future flight missions, etc. Unfortunately, the limited forum of this paper prevents the power of visualization techniques that utilize color, three-dimensions or animation to be shown. In addition, the breadth and depth of these applications cannot be illustrated effectively. Therefore, the following examples are offered as basic examples of the capabilities of the NGS. Keep in mind that each of these visualization techniques, as well as those not shown in this paper, can be applied to *any* data.

Figure 1 is a simple histogram derived from an all-sky infrared image prepared from the zodiacal history file of data from NASA's Infrared Astronomy Satellite (IRAS). The data were provided to the NSSDC as a a collection of pixels with associated intensities laid out in galactic coordinates. The NGS allows such three-dimensional data to be treated as one-dimensional for the preparation of statistical summaries like histograms. The number of times a brightness value occurs within the illustrated collection of brightness bins is shown along with the mean and standard deviations as well as percentile levels. The brightness data shown have been scaled to exclude values above 100 MJanskys/Steradian. NASA's COsmic Background Explorer (COBE) Project, which will map the remnants of the Big Bang, is currently evaluating the NGS for its utility in data analysis after the spacecraft is launched. The IRAS data provide an excellent testbed for such an evaluation.

Figure 2 illustrates data about the Earth's magnetic field derived from the International Sun-Earth Explorer (ISEE)-1 spacecraft. These data were among the ensemble of data sets that formed the aforementioned CDAW-8 data base. This plot is divided into three panels, one each for the x, y and z component of the magnetic field. Each of the traces are for some five hours of *in situ* observations by the ISEE-1 magnetometer on January 28, 1983. The right-hand series of three traces is for the magnetic field components in GSE coordinates in nanoTeslas, which are marked by a circle. The left-hand traces are for the variance of the components in nanoTeslas^2.

Figures 3 and 4 illustrate data from the Total Ozone Mapping Spectrometer (TOMS) on board NASA's Nimbus-7 spacecraft. The data are available as daily world grids (37440 cells per grid) from late 1978 through the present at the NSSDC. These data have become increasingly valuable as they indicate the presence of the so-called ozone hole over the south pole. The total ozone content in all of these displays is in terms of Dobson Units. The ozone minimum of last year reached its lowest point on October 13, 1986. Both of these displays show various aspects of these data during this period. The NGS has the ability to arbitrarily reproject any geographic data, whether continuous grids or images, or scattered points. These gridded TOMS data are provided in their own unique grid. In order to preserve equal area or equal distance in a map, focus on a specific feature of the data or compare to another data set, it is necessary to support arbitrary geographic mapping of data. These figures show what can be achieved even with simple contour displays.

Figure 3 is a world map of the total ozone for October 13, 1986 in a Mollweide projection. The contour lines are incremented every 20 Dobson Units. The statistics are based upon all of the data for that day. The ozone minimum appears as the region at the bottom of the map below the gradient near the Antarctic coast.

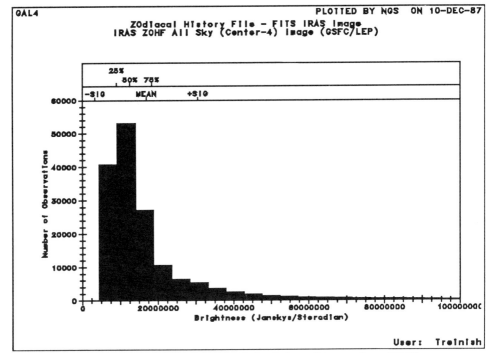

FIGURE 1. Histogram of an all-sky infrared image from IRAS.

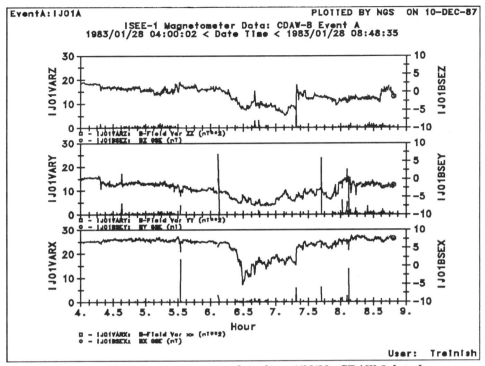

FIGURE 2. ISEE-1 magnetometer data from 1/28/83: CDAW-8 data base.

Figure 4 is a zoomed in display of the data, showing the internal structure of the ozone hole projected in an azimuthal equidistant map centered over the south pole. The statistics in this example are based upon the southern hemisphere data only. In this example, the map has been rotated from a center at the equator and prime meridian in Figure 3 to the south pole. The magnification of the viewing window has been expanded until Antarctica fills the window.

Figure 5 illustrates data derived from the International Satellite Cloud Climatology Project (ISCCP), which is supported by the NSSDC's NCDS effort. ISCCP is generating world grids of cloud parameters several times each day for five years (1983-1988), which are derived from observations taken from a suite of international spacecraft. Traditionally such global data sets are displayed in large-scale geographic displays like those that are supported by the NGS, which include contour maps like Figure 3 or pseudo-color images. Although such techniques are important for seeing the spatial distribution of data across some geographic field, they may not be sufficiently quantitative at a detailed level for some applications. In other words, it is often useful to treat continuous or gridded data as scattered points and vice versa. Figure 5 shows how the NGS can arbitrary slice data for display purposes. In this example, ISSCP world grids at 00:00 GMT on July 4, 1983 are selected, and sliced into 10-degree latitude bands as an animation sequence of 18 frames. Cloud top data from the sixth frame of this sequence (40 to 30 degrees south latitude) are plotted as a function of longitude. Specifically, the individual pressure and temperature cell values from that latitude band are extracted from the world grid and shown in this scatter diagram as circles and squares, respectively. In addition, the two collections of points are fitted with separate cubic splines. Such an analysis shows the correlation of the pressure and temperature values at the cloud tops at a microscopic level.

Figure 6 visualizes the mean sea surface over the Aleutian Trench, which reflects changes in the local bathymetry. In this picture an estimate of the mean sea surface derived from altimeter data from NASA's SEASAT and the GOES-3 satellites has been viewed from the southwest. The data are prepared as a world grid at one-eighth degree resolution. This three-dimensional image has been tilted in the horizontal by 73° and it is viewed at an inclination of 36°. The image has been illuminated from the southwest with an artificial light source that has been placed at the same location as the eyepoint of the viewer. Over this small region, the mean sea surface reflects the shape of the ocean bottom rather than subsurface changes in the composition of the earth. The large depression of the sea surface over the trench amounts to 20 meters of height change, whereas the depth of the sea floor changes by more than 3000 meters across the trench. The data analyses to generate this picture were performed at NASA/GSFC by James Marsh, Chet Koblinsky, and Dave Wildenhain. Conventional visualization methods cannot show all of the subtle structure apparent in high-resolution data such as these sea surface heights derived from altimeter data. A generic ray-tracing algorithm available in the NGS permits a solid, light-shaded object to be generated from any data, as in this picture, to bring out such details.

In addition, the NGS has been one critical tool in the analysis of daily rainfall data from northwestern Peru as part of an overall effort to understand the meteorology in Peru during the great El Niño episode of 1982-1983 (Goldberg *et al.* 1987).

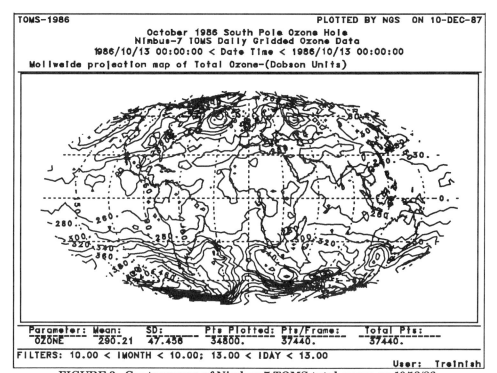

FIGURE 3. Contour map of Nimbus-7 TOMS total ozone on 10/13/86.

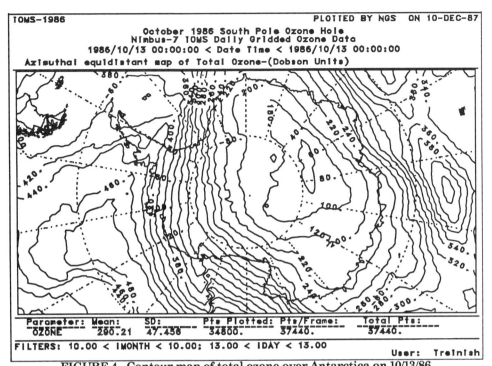

FIGURE 4. Contour map of total ozone over Antarctica on 10/13/86.

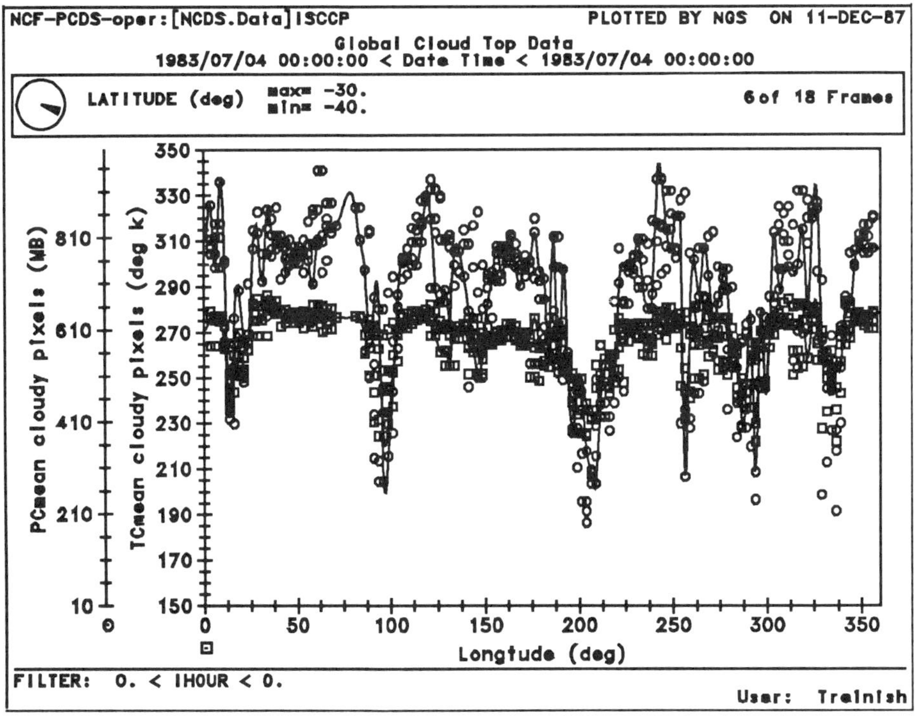

FIGURE 5. Cloud top temperatures and pressures from ISCCP grid on 7/4/83.

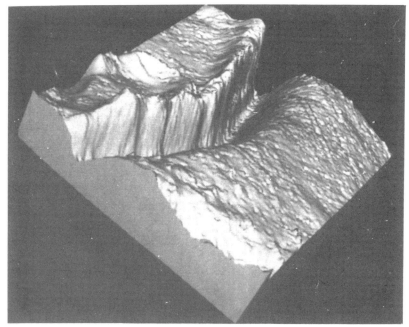

FIGURE 6. Mean sea surface of the Bering Sea showing the Aleutian Trench.

7. CONCLUSION

In the future it is hoped that this work will be expanded to support the implementation of the NGS into low-cost graphics workstations (*e.g.*, Apple Macintosh II) as well as high-performance graphics workstations (*e.g.*, Silicon Graphics IRIS), and the support of production of presentation-quality animated visualizations. In such an operational environment an analyst would use the various conventional NGS tools to examine data of interest at different levels of detail, and then interactively view a solid geometry model of the data in three dimensions on a graphics workstation, including the examination of sequences of such models. A user could optionally invoke the MPP or the NASA Space and Earth Sciences Computing Center CDC Cyber 205 to prepare the geometries or otherwise manipulate the data in a timely and straightforward manner. After such an examination, a user could visually choose an appropriate orientation for the geometries, and then request a ray-traced image(s) or animated sequence to be generated on the MPP or Cyber 205. The results of such a request would be a sequence of high-resolution photographs or slides, or animation on film or video tape.

ACKNOWLEDGMENTS

The research reported in this paper has been supported mainly by NASA's Office of Space Science and Applications (OSSA). The author wishes to thank James L. Green, Director of the NSSDC, for his continued support of the concept of discipline-independent software systems to support scientific research, and the excellent staff of the NSSDC, including Michael Gough, Craig Goettsche, David Wildenhain and Kuo Ni, whose talent and creativity enabled the author's dream of generic visualization become a reality.

REFERENCES

Goldberg, R. A., G. Tisnado M. and R. A. Scofield, *Characteristics of Extreme Rainfall Events in Northwestern Peru during the 1982-1983 El Niño Period* , **Journal of Geophysical Research, 92**, n. c13, pp. 14225-14241, December 1987.

Gough, Michael, *A Generic Tool for the Generation and Display of Animated Sequences* , **Proceedings of the Third Annual Conference of the TEMPLATE User Network,** Megatek Corporation, February 1986.

Gough, Michael, **NSSDC Common Data Format Implementer's Guide**, NSSDC, NASA/ GSFC - SAR 87-01, February 1987.

Manka, Robert H., *CDAW Space Data Analysis Progresses* , **Eos Transactions American Geophysical Union, 67**, n. 52, December 30, 1986.

Manka, R. H., D. N. Baker, J. L. Green, R. E. McGuire and L. A. Treinish, *Coordinated Data Analysis Workshops (CDAW): Applications in Geophysics* , presented at the **Fall 1987 Conference of the American Geophysical Union**, December 1987.

Ni, Kuo P., Michael Gough and Lloyd Treinish, **A Flexible, Template-Based Software Package for Generating Maps**, NSSDC 88-05, NASA/GSFC, February 1988.

National Science Foundation, **Visualization in Scientific Computing**, July 1987.

Reph, Mary G., Lloyd A. Treinish, Carey E. Noll, Thomas D. Hunt and Shue-Wen Chen, **Pilot Climate Data System Users' Guide**, NASA/GSFC TM 86084, Revised, January 1986.

Template Graphics Software, **Template Reference Manual Version 5.5**, TGS, Inc., San Diego, 1986.

TAE Support Office, **TAE User's Guide Version 1.0**, TAESO, NASA/GSFC, 1986.

Treinish, Lloyd A. and Surendra N. Ray, *An Interactive Information System to Support Climate Research* , presented at the **First International Conference on Interactive Information and Processing Systems for Meteorology, Oceanography and Hydrology**, January 1985, American Meteorology Society, pp 72-79.

Treinish, Lloyd A., Michael L. Gough and David Wildenhain. *Animated Computer Graphics Models of Space and Earth Sciences Data Generated via the Massively Parallel Processor* , presented at **Frontiers '86: The First Symposium on the Frontiers of Massively Parallel Scientific Computation**, September 1986.

Treinish, Lloyd A. and Michael L. Gough, *A Software Package for the Data-Independent Storage of Multi-Dimensional Data* , **Eos Transactions American Geophysical Union, 68**, pp. 633-635, 1987.

DISCUSSION

Gautier: I have two questions, first you said you were supporting several devices. What are they?

Treinish: The underlying structure of this system is a proprietary package that supports a couple of hundred different devices. Hence, it supports most of the standard terminals, plotters, and other equipment of that sort.

Gautier: What is the availability of the package, if it is a proprietary one?

Treinish: We have used a proprietary package underneath our software, because the cost of that package is minuscule compared to developing its capabilities ourselves. It is a package like other proprietary packages that typically had dozens of man-years worth of effort that you can buy for a fraction of what it costs to have someone program that kind of capability. We are beginning to make the system available to individual groups both at Goddard and outside. I suggest we discuss that separately as to how we could do that.

Gautier: The second question concerns color which is quite important in display. How are you choosing your color mapping? Do you have any standards, have you developed special criteria for mapping procedures?

Treinish: We are looking at a couple of things. Since most of our users have very inexpensive or very simple devices, we use colors that are relatively fixed. That could change in specific environments. I derived these colors after doing some specific investigations into the ideas behind phsychology of color and examined the mapping of data to the CIE diagram, etc. I tried to come up with some simple ideas that were relatively straightforward to view so you would not have to be an expert to read the colors. That limits you in some flexibility but it provides a common basis for essentially arbitrary data since the software really doesn't know the specific nature of the data. In that way, we can build a common framework. Hence, the idea of mapping data to a simple visible spectrum, for example, doesn't require a lot of education to learn to use. There are other techniques that use repeatable color scales, brightness rather than RGB, etc, that can be used and potentially can convey more information, but then there is a problem of the education and the flexibility for the devices. That is our basic approach. Our framework is such that we can move in other directions as requirements dictate. So we try not to lock ourselves out of other methods.

MODERN WEATHER DATA ANALYSIS AND DISPLAY ON A PERSONAL COMPUTER

Owen E. Thompson
University of Maryland
College Park, Maryland 20742, USA

1. INTRODUCTION

The basic goals of this project were: (A) to develop a relatively inexpensive, image-processing workstation for displaying high technology weather satellite, weather radar, and conventional weather data for purposes of research and education ; (B) to adapt this station for the production of videotape and film materials of weather and climate phenomena for study and presentation ; (C) to adapt this technology to UNIX-based Apollo workstations connected in a local area network to provide both MS-DOS and UNIX workstations accessing shared image and alphameric data files ; (D) to promote arrangements between the private and public sector to economically transfer this technology into the educational arena. This report summarizes the progress toward these goals.

2. BACKGROUND

In cooperation with Environmental Satellite Data, Inc., and with assistance from the National Center for Atmospheric Research, a micro-computer weather display system suitable for research and education was assembled under this project. This development was intended to serve the needs of universities, colleges, and high schools for real-time weather information processing and display, and to set the stage for educational spin-offs of Project UNIDATA. The term "UNIDATA"[*] refers to a project sponsored by the National Science Foundation, ATM Division, one goal of which was to design an efficient, cost-effective system for delivering weather data to universities.

[*]Project UNIDATA is managed by the University Corporation for Atmospheric Research (UCAR) in Boulder, Colorado to which the Principal Investigator took sabbatical leave in 1985-86 as Director of Educational Affairs.

RSRM '87: ADVANCES IN
REMOTE SENSING RETRIEVAL METHODS
A. Deepak, H.E. Fleming, and J.S. Theon (Eds.)   461

The UNIDATA project had developed a data broadcast concept for efficiently delivering weather data to educational institutions. This concept was instituted by contract with the Zephyr Weather Service, Inc. whereby data collected by the NOAA National Weather Service (NWS) was communicated by land lines between the World Weather Building in Suitland, Maryland and a Zephyr satellite uplink in Chicago. The data is uplinked to the WESTAR V communications satellite from which it is rebroadcast on a video carrier, receivable by ground station antennae generally anywhere in the country. The signal is demodulated at the ground station and accessible, thereby, as a digital data stream.

The UNIDATA Project contract with Zephyr Weather Service focused primarily on the communication of conventional weather observations, (E.g. surface and upper air measurements of pressure, temperature, humidity, winds, etc.). However, ESD, Inc. also provided to UNIDATA-Zephyr users a digital data stream of Geosynchronous Orbiting Environmental Satellite (GOES) visible, infrared cloud, and infrared water vapor images, and an additional image service of NWS Manually Digitized Radar (MDR), current and forecast weather conditions, surface temperature and dewpoint temperature analyses, and other weather analyses. The availability to UNIDATA users of this high tech, digital satellite, radar, and analysis imagery was by separate business agreement between ESD and Zephyr, and not generally highlighted in the UNIDATA Project.

ESD, Inc. had already developed an IBM-PC/AT based workstation which would ingest, manage, and display the GOES imagery, MDR radar imagery, additional weather analysis imagery, as well as test, ingest, manage, plot, and display conventional weather data. In addition, ESD had developed hardware/software procedures to ingest NWS Remote Radar Weather Display System radar sweeps (RRWDS), Lightning Position And Tracking System (LPATS) lightning stroke data (by telephone connection), Collins Doppler Radar images (by direct connection), and to manipulate all data on a single IBM PC/AT based workstation. This capability to provide a complete, weather data workstation on a relatively inexpensive PC/AT chassis opened the door to enhancing weather study at both university and pre-college levels.

3. THE ESD-UNIVERSITY OF MARYLAND UVAS - UNIDATA VIDEOGRAPHICS ANIMATION SYSTEM

In this section, a brief summary is given of the technical capabilities of the ESD-Maryland UVAS. In this case, "technical capabilities" do not mean bits and bytes but, rather, the capabilities for locally receiving, managing, and displaying modern weather data.

Figure 1 shows a map of the basic computer graphics core of the UVAS. The UVAS receives conventional weather data and *digital* GOES satellite data via a Zephyr Weather Service satellite downlink. The incoming signal is demodulated, and the NWS Domestic Data Service and the ESD 1 Digital Satellite Service are stripped off and presented to the IBM PC/AT workstation. The ESD Weathergraphix (TM) software then samples each data stream and compares "headers" in each stream with user-prepared acquisition lists to determine whether or not a particular data file is to be ingested or ignored. The desired data files are ingested and stored on the PC/AT hard disk (or floppy disk). The software also utilizes user-declared limits to automatically maintain a continuously moving local archive of data of a given time length (E.g. a local hard disk archive of images and conventional weather data for the 24 hours prior to the current time).

Additional weather data, such as RRWDS radar or LPATS lightning data, can be ingested through a telephone modem in a more traditional demand-download mode. Such data are ingested and stored on the hard disk as well. All data, once stored, are available for processing, plotting, superposition, and display by other modules of the ESD software. One important feature of the software package is that all processes operate compatibly (using appropriate interrupts) in such a way that the broadcast data streams are continuously monitored no matter what other processes are in progress. The addition of an electronic digital drawing tablet allows the user to perform a wide variety of art and graphics tasks for annotation of images in the system.

In the University of Maryland UVAS setup, the basic ESD computing system was interfaced with a variety of video and film units to demonstrate the feasibility of preparing take-away educational materials from this system. The RGB graphics output from the ESD system was input to a National Television Standards

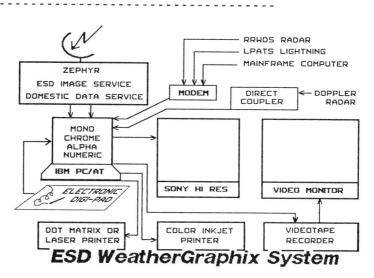

FIGURE 1.   Diagrammatic representation of the ESD-University of
Maryland UNIDATA Videographics Animation System computer
components.

Committee (NTSC) video encoder to produce a standard video output compatible with ordinary television monitors and tape recorders. (A integrated circuit board version of an NTSC encoder is piggy-backed to the internal graphics board on the IBM PC/AT bus so that the simultaneous output of RGB and encoded video is available at the back panel of the computer. An external RF modulator could be added to provide video input to the least expensive television tuners without composite video input.) The addition of this standard video feature is important for educational applications. Low-cost television monitors and tuners can be used (although at reduced screen resolution compared with the high resolution RGB monitor) throughout a classroom or laboratory, and the graphics output can be immediately videotaped for later user with an ordinary video cassette recorder. The digital art tablet allows the user to annotate any image produced on the system, and to create separate graphics slides and video animation loops for classroom presentation. With videotape materials in mind, several additional functions were added to the system. An optical bench for transferring 16mm film and 35mm slide material into video signals (via a television camera) was added to the system so that material in these media could be edited into videotape educational modules. In order to provide hard copy take away materials, a color inkjet printer and a black/white dot matrix printer were added so that maps and color images could be produced for student use. Finally, to allow for hard copy educational materials production at the higher resolution of the RGB graphics output, an additional high resolution monitor is to be interfaced with 35mm-still and 16mm-animation cameras to provide both slide and movie film animation capabilities. The complete UVAS system is illustrated in Figure 2.

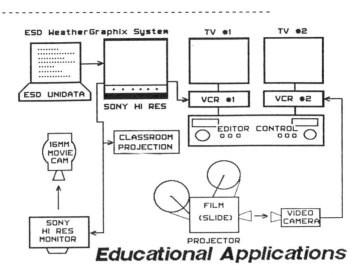

FIGURE 2. Diagrammatic representation of a complete UVAS central educational system. The ESD Weathergraphix system is interfaced with video projection, videotape, film, slide, and hardcopy printers for broad capability for educational use.

4. WEATHER DATA CAPABILITIES OF THE ESD-UNIVERSITY OF MARYLAND UVAS

In this section, an overview of the types of weather data display and user manipulation of that data is given.

1. GOES Satellite Images
Digital visible and infrared images of cloud cover from GOES EAST and GOES WEST are available every hour. Screen resolution is either 768x480 or 384x240 pixels, depending on the particular image, and digital brightness levels are up to 256 levels for GOES Infrared, and up to 64 levels for GOES Visible. GOES infrared water vapor channel images are also available every 6 hours. Digital image data can be color enhanced by temperature-profile calibrated color tables, or dynamically from the keyboard. Eleven fixed, and three floating sector views give broad coverage and close-up views of North America. Full disk and half disk views also give coverage of Central and South America, and extensive portions of the Pacific and Atlantic Oceans. Images can be zoomed,

panned or remapped to custom projections specified by user. Data are included in the ESD-1 Digital Satellite Service and are available through the Zephyr satellite downlinks employed in the UNIDATA Project, or by telephone dial-up. Image data are automatically ingested, stored, and managed by the UVAS and made available for display or print at user's command.

2. GOES Satellite Composite Images
Composite images of satellite and radar, satellite and surface pressure, satellite and precipitation are broadcast regularly by the ESD-1 Digital Satellite Service to the Zephyr downlink, or are available by telephone dial-up.

3. National Radar Summary
Color images of the National Weather Service (NWS) Manually Digitized Radar (MDR) network summary are broadcast hourly by the ESD-1 Digital Satellite Service to the Zephyr downlink, or are available by telephone dial-up.

4. Surface Weather Maps
Color maps of surface temperature, dewpoint, windchill, humidity, surface analysis, precipitation summary, jet stream analysis, and forecasts of high/low temperatures, fronts, and pressure centers are broadcast by the ESD-1 Digital Satellite Service to the Zephyr downlink, or are available by telephone dial-up.

5. NWS Domestic Data Service
The National Weather Service Domestic Data Service data stream, available through the UNIDATA Zephyr downlink, can be automatically ingested, stored and managed by the UVAS. The UVAS software allows the user to scan, print, or plot data on screen, or plot data on hardcopy. Registration software allows data to be accurately plotted onto any screen image (e.g. satellite images or map backgrounds) which have been defined by the user. NWS severe weather watch boxes may also be plotted and displayed on the field. The Domestic Data Service data are also available by telephone dial-up from ESD.

6. Arithmetic and Contouring Analyses

Domestic Data Service variables can be plotted in the form of
isopleths registered to a map or image projection specified by the
user.   UVAS also allows the user to perform arithmetic operations
on   variables   in   the   database   to   produce,   plot,   and   contour
user-defined quantities on map backgrounds or images as desired.

7. Remote Radar Weather Display System (RRWDS)
Radar echo images from National Weather Service RRWDS weather
radar can be ingested, stored and managed by the UVAS.   Data
ingest is by telephone dial-up to NWS RRWDS radar sites.   Once
formed into color images, the frames can be manipulated in the
same way as any digital image in the UVAS system.   (E.g. See
sections   10-15)

8. Doppler Radar
Radar images from Collins Doppler Radar can be ingested, stored,
managed and displayed by the UVAS at user's command.   Data ingest
is by direct connection to the Doppler receiver/transmitter unit
using an ADC interface device.

9. Lightning Position And Tracking System (LPATS)
LPATS data can be ingested, stored, managed and displayed by the
UVAS.   Data ingest is by telephone dial-up to the LPATS Central
Analyzer located in the area of interest.   LPATS strikes can be
animated on-screen using color table rotation animation routines
built into the UVAS graphics software.

10. Art and Graphics Capabilities
The UVAS art/graphics software includes facilities for drawing
and plotting in several line and shape modes; line color and width
selection; color fills and floods "on top" or "underneath" drawn
lines; titling with a variety of fonts; user definition of custom
fonts and symbols; zooming, panning and cut-and-paste.   Art work
may be done from the keyboard or from electronic digipad.   Screen
resolution is up to 768x480 with a palette of up to 256
simultaneous colors selected from over 16 million possible colors.
Cut-and-paste facility allows user to paste a selected subset of
foreground colors of one image onto a second image providing

professional-grade production of graphic images suitable for slides, films, videotapes, or as part of an automatic computer weather display. (See section 13)

11. Map Building and Image Remapping

Maps can be built by the UVAS in any of a number of map projections at any scale. The World Data Base II map point sets are available to a high level of detail world-wide. Numerical data can then be accurately and automatically plotted to custom built maps. Images with known projection parameters can be remapped to such custom built maps by the PC/AT resident ESD software.

12. Image Animation

All images, whether ingested from external sources or created through the art/graphics software, can be animated on screen using color table rotation techniques, or by "looping" up to 20 separate color images. The ESD software includes macros for easily producing animation of stored images. The local data management software automatically updates each image version, and deletes the oldest of a pre-specified number of versions of that image specification. Thus, for example, user could configure the system to save the most recent 4 hours of radar network data, the most recent 18 hours of US satellite views, the most recent 7 hours of northeast sector views, the GOES water vapor image for one time each day, and so on. Fast Frame Loop programs then automatically animate the most recent versions of each image type over the preselected time interval.

13. Sequencing Programs

Image display, animation loops, color table rotations, electronic slide shows, data plotting, map building, radar sweeps, and virtually all other system software functions can be integrated into *sequencing programs* for weather briefings, seminar presentations, classroom use, and student self-study with the UVAS. To simplify creation of sequencing programs, the UVAS has a special "record-mode" which records the manual keystrokes of a user while he or she operates the computer, and automatically provides the record as a callable sequencing program for later use.

14. Standard Video Output

In addition to the 768x480 pixel graphics display, the UVAS can be fitted with an internal standard television encoder and GENLOCK module to produce a composite color video output signal for video taping or TV monitor/display. This display mode is simultaneous with the higher resolution RGB graphics display. The addition of an external RF modulator produces a signal tunable by the lowest cost television receivers if desired.

15. Hardcopy Options

Color image hardcopy can be done using a color inkjet printer while alphameric data and plots can be printed on a standard dot-matrix printer or on a laser printer. Film copy in 35mm slide, or 16 mm movie format, can be made from the high resolution graphics monitor. With the video encoding option, videotape hardcopy of animations can be easily made. All files can, of course, be copied to floppy disk and archived or shared with other ESD system users throughout the U.S.

16. Multiple UVAS Workstations

For teaching laboratories, the UVAS can be configured as a non-display FRONT-END for data ingest, storage and management. The FRONT-END can then be used to provide data to other UVAS display workstations via a local area network, as a file server accessible by telephone or hardwire, or can be set-up as a local data *re-broadcaster* to a chain of UVAS workstations.

17. Under Development

Development is underway to provide NCARPLOT, GEMPLOT, and other mainframe plotting and graphics applications for UVAS so that those applications can be easily utilized for educational applications and materials development. Software is available to provide wide capability of upper air and thermodynamic analyses, and international data analysis. The Maryland-ESD team is currently working on a migration of the UVAS application to Apollo brand micro-computers to provide local data management, display, plotting, and animation in either MS-DOS or UNIX operating systems utilizing common data files. The Apollo-based experiments are directed toward high-end use of the applications in a high speed local area network.

5. INITIATIVE TO PRODUCE AN EDUCATIONAL
PRICING SCHEDULE FOR UVAS

While the Principal Author was on sabbatical leave to UCAR as Director of Educational Affairs, an effort was initiated to negotiate a special pricing schedule with ESD, Inc. to provide their weather data management, analysis, and display technology to educational institutions to stimulate the use of weather information in education.   An agreement was negotiated to substantially discount the ESD Weathergraphix(TM) system, and to substantially discount the price of the full ESD 1 Digital Satellite Image Service.   This agreement was completed and disclosed to the UCAR in late 1986, and to the NSF and NASA in 1987.

6. CONCLUSIONS

In this report, a powerful weathergraphics workstation developed by the private sector has been adapted to the modern weather research and educational needs of universities, colleges, and high schools.   The PC-based system has a wide variety of capabilities suggestive of significantly more expensive research workstations under development elsewhere.   Moreover, through negotiations by the Principal Author with ESD, Inc officials, this system has been vastly discounted for bonafide educational institutions to stimulate the use of modern weather science and technology in classrooms and teaching laboratories.   Thus, the products of private sector development costing in excess of $2 million has been brought to the doorsteps of universities and schools at a quite reasonable unit cost.   Although the ESD-University of Maryland UVAS was *not* developed using NSF UNIDATA Project funds, the characteristics of the system, and the details of the special pricing agreement, have been provided to UCAR and NSF in support of educational community goals. Details of this agreement may be obtained from the author at: Department of Meteorology, University of Maryland, College Park, MD   20742.

This development project was funded by the National Science Foundation, Science and Engineering Education Directorate; the National Aeronautics and Space Administration; the University of Maryland; and Environmental Satellite Data, Inc.   Additional thanks are owed to Randy Eastin, and other staff members of the National Center for Atmospheric Research, Scientific Computing Division.

DISCUSSION

Kleespies: I note in the representation of the imagery used, at least what you presented here, you are using the brightness temperatures to represent the land as being green and brown, the ocean as being blue, and that is fine for the media but I don't think it belongs in the classroom. It is one of my pet peeves and you lose so much information from the clouds when you do that. You just have the really gross cloud features and the smaller features are completely wiped out.

Thompson: We can take the color out if you like. We can control the color tables. I think I miss your main point then. What does that have to do with the color of the background?

Kleespies: When you play that trick and do color the background, you lose the cloud features.

Thompson: The cloud features that you are worried about were taken out by the slicing method. It is sort of independent of what colors you use. The way you posed the issue concerning color is not an issue. I can make the backgrounds black, but I think your real issue is that there has been a slicing that is a suppression of low-level cloudiness. And that is correct. In those 16 color-images there has been suppression of low level cloudiness by processing the data. However, I showed you the full visible and the full infrared. One need not do that.

Kleespies: The problem is the examples that you showed for the 8-bit IR and the 6-bit visible demonstrated the same effect. Just let the background be what it is. I think the students learn more about what the data really looks like if you do that.

Thompson: In the infrared, what it really is, is invisible to your eye and mine. We must do some color coding.

Smith: At Wisconsin we are trying to do similar things, use PCs in the classrooms. And we find it takes enormous investment to support such a system. We have a hard time getting the funding needed to give us a full-time programmer and the teaching assistants that are needed to really make a thing like this viable. I was just wondering if you could share some of your experience at Maryland with me on that.

Thompson: The software I use is a fully-supported commercial package. The maintenance management problem of the PC is that I called the commercial vendor. That is part of the contract of purchasing their software and subscribing to the data base. So, to some extent, by that technique I have offloaded a part of the problem, because at Maryland we clearly cannot compete with Wisconsin in this area. We do not have the expertise to do that. So, avoiding that problem represented one of the strategic decisions leading to a commercial package. The overall cost is another issue. The way that I would represent the cost of this to me, besides the hardware, was approximately the cost of a third of a graduate student for one year. That is what I paid for the software.

OPEN DISCUSSION

SESSION H. - OWEN THOMPSON

Thompson: I would like to make a couple of comments about the data compression session and then I will try to put it all together. I was particularly interested in seeing that we are rapidly getting to the point of integrating things that we know about algorithms, things that we know about microprocessing and display systems, integrating these things into some sort of a giant configuration of sensors, data processors, algorithms, and analysis systems. Yet, we still talk about individual pieces of this. I have some fear about this which resulted in a very short story that I wrote about 6 minutes ago. In the year 2005 a particularly tough inverse algorithm invented by Henry Fleming overheats a microprocessor chip on board a spacecraft which, in turn, emits an additional burst of thermal energy, contaminates the next scan of 6 instruments on an EOS platform and that error propagates very rapidly through the assimilation cycle and destroys the validity of the 14-day weather forecast. Well, so much for the joke. The part that is not so funny, though, is that in such a scenario the optimization problem goes well beyond any particular algorithm to do one piece of that problem. On another subject, I have a fear that we are never going to be able to agree on exactly how to compress data. After 25 years of systems and methods development, it is still not very easy for scientist A to get a hold of data set B and go about his work. I think that we will get more people involved in this field and solve the scientific problems as soon as there is an easy way of getting data shipped back and forth. Finally, based on some of the discussions on the EOS program it is quite clear that remote sensing in the future is going to be able to provide huge amounts of data to every citizen of the world. I am also concerned about who is going to be around in the year 2005 to study that data. The students working on their Ph.Ds in the year 2005 using EOS data are now at about the elementary school level and that is rarely a proper subject for a scientific meeting. Nevertheless, I think we really have to give some thought as to how to engage the minds of the future scientists who are going to be solving these difficult problems even just 12 years from now, let alone 25 years from now.

Johnson: I get very worried when people talk about storing data in a manner not completely faithful to the way it was taken. Of course, you have to distinguish between your archival storage for data, which may be hard to gather, from the data that one wants to use and maybe transmit to other people who don't have the same requirements you do or don't need the same level of accuracy. You should never throw away your original data and keep what you think is the useful thing out of the data. I've been involved with a very interesting program that utilized somebody else's data from more than 50 years ago. I can't emphasize enough, don't throw away original data.

Treinish: I think the idea of not throwing away any bits is important but on the other hand you do have the problem. If you keep all the bits, how are you going to manage those bits and allow a scientist to get at the arbitrary pieces of information that one needs to do the analysis that is necessary? Part of the problem, if you look at current and old technology for collecting the data, and in the case of NASA a great deal of that data has never really been truly analyzed. A lot of that is due to the fact that the computer technology and the software technology to get at that data even today may not exist.

Kleespies: I would like to make a comment about the display systems methodology. At the risk of sounding like a spoil sport the AMS pretty much has a conference devoted to the subject and SPIE has recently started up a series, and given the amount of material we have to cover in this workshop as it is, I question whether we should open up this other area.

Speaker: Just to sum everything up it seems like there is a finite amount of resources and whether it is best spent on saving vast vaults of data or best spent educating students and paying their salaries so that they can analyze some of that data.

Thompson: The general archiving question seems to be what must be saved for fear of not having it. Until we can agree on that, then we have the dilemma that everybody is going to want to save everything forever. As an example, losing data on a convective event means that we will have to wait until the next convective event to pursue our study. Losing data on the 100-year climatology means that we will have to wait 100 years to pursue that study.

Johnson: Once you start saying you are going to save the statistics rather than the original data itself, you have made the decision about what is important. You have made that decision based on what you know today and I guarantee that is not going to be true in 5 or 10 years. Somebody is going to want something additional.

Treinish: The reason archives such as ours exist is that almost all of the resources have been spent getting the platform into orbit and getting the data back down to earth. In comparison, essentially no resources are available for saving the data, management, providing data access. What is necessary is really a greater effort to think about how to manage and work with the data along the same lines as how to process data, how to acquire data, and how to retrieve data.

Deepak: I would like to end the discussion with a comment. When I used to teach History and Philosophy of Science and Technology, a question was often asked as to why the Egyptian science that flourished in around 4000 B.C. came to a relatively abrupt end. The answer given is that "the scientific knowledge was usually confined to a handful of people, mostly priests. The head priest was called by the Pharoah to predict the eclipse, and if the prediction was wrong, the priest usually lost his life. Thus, scientific knowledge became confined to an even lesser number of people." The corollary to the previous question is: Why has there been an explosion of scientific knowledge during the last 80 years? One of the reasons is that scientific knowledge has been made accessible to a very large population of common people. Whenever knowledge is confined to a few people, it usually dissipates. We should learn a lesson from this historical perspective and educate more people in retrieval methods.

OBSERVABILITY OF ALBEDO BY SHORTWAVE WIDE FIELD-OF-VIEW RADIOMETERS IN VARIOUS ORBITS

G. Louis Smith
Atmospheric Sciences Division
NASA Langley Research Center
Hampton, Virginia 23665, USA

David Rutan
PRC Kentron, Inc.
Hampton, Virginia 23666, USA

ABSTRACT

The interpretation of shortwave wide field-of-view radiometer data requires the solution of an integral equation for albedo. The limits on the information which can be obtained are investigated for the NOAA 9 and NOAA 10 operational meteorological satellite orbits.

1. INTRODUCTION

Wide field-of-view (WFOV) radiometers have been used for the Earth Radiation Budget (ERB) instruments which flew on the Nimbus 6 and Nimbus 7 spacecraft (W. L. Smith et al., 1977; Jacobowitz et al., 1979) and for the Earth Radiation Budget Experiment (ERBE) instruments which flew on the NOAA 9 and NOAA 10 operational meteorological spacecraft and on a dedicated ERBE spacecraft (Barkstrom and Smith, 1986). Thus, the nearly unbroken high-quality WFOV Earth radiation budget measurements have continued since 1975. G. L. Smith and Green (1981) developed an analysis procedure for WFOV measurements which applies to the case of Earth-emitted radiation, and this method has been applied to a 1-year set of Nimbus 6 ERB data by Bess et al. (1981) and more recently to a 10-year set of Nimbus 6 and 7 ERB data by Bess and Smith (1987a and b).

The problem of analysis of reflected solar radiation was formulated by G. L. Smith (1981). This case is more difficult than that of the Earth-emitted radiation, for which there is an analytic solution. The equations for the reflected solar radiation case can be discretized, resulting in a set of matrix equations. G. L. Smith (1987) recommended the solution of these equations by the method of singular decomposition because the problem is ill-posed. That method is demonstrated by application to data from the Nimbus 7 spacecraft for December 1978. They have shown that the albedo distribution has an unobservable component which dominates in the polar regions and near the terminator for the Nimbus 7 WFOV data. Fortunately for this case, the incident solar radiation is small at these latitudes, and the effects of these unobservable components on determination of absorbed solar energy are small.

The ERBE instruments are on the NOAA spacecraft as spacecraft of opportunity. As such, the orbits are determined by NOAA operational

RSRM '87: ADVANCES IN
REMOTE SENSING RETRIEVAL METHODS
A. Deepak, H.E. Fleming, and J.S. Theon (Eds.)   475

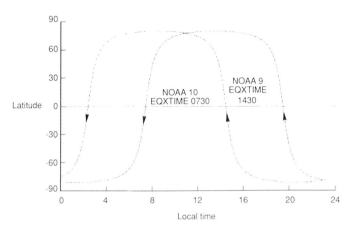

Figure 1.   Orbit tracks in terms of local time for NOAA 9 and 10.

needs.   The NOAA 10 spacecraft is in a Sun-synchronous orbit, with an inclination of 81° retrograde and Equator crossing times of 0730 and 1930 for the descending and ascending nodes, respectively, as shown in figure (1).   Thus, it moves near the terminator for much of its orbit, and never has views of the mid and high latitudes of the Southern Hemisphere with high solar incidence.   The question arises as to how much albedo information for the Southern Hemisphere is contained in the shortwave WFOV radiometer measurements from NOAA 10.   Similarly, the NOAA 9 is in a Sun-synchronous orbit with Equator crossing times of 1430 and 0230 for the ascending and descending nodes, respectively.   Thus, it gets good solar incidence within its FOV for all but the polar regions of the Northern Hemisphere, but lacks high latitude coverage in the Southern Hemisphere.   There is thus a question as to the albedo information for these regions for the NOAA 9.   In the present paper, we consider these questions by use of the analysis procedure of G. L. Smith (1987).

2.   THEORY

The measurement equation for reflected solar radiation, with the constraints imposed by the orbit, may be rewritten as (G. L. Smith, 1981)

$$m(\eta,\Phi_S) = \int_{FOV} K(\eta,\Phi_S,\theta_T,\Phi_T) \, A(\theta_T,\Phi_T) \, d\Omega \qquad (1)$$

where m is the measurement in $W\text{-}m^{-2}$ at the sensor, which is located by its longitude Φ_S and orbit angle η, the angular distance from the Equator to the spacecraft.   The kernel function K includes the angular response of the sensor, the viewing geometry and the bidirectional function describing the anisotropy of the reflected radiation.   The albedo A at a point at the top of the atmosphere is a function of colatitude θ_T and longitude Φ_T , and is assumed to be constant with local time.   Equation (1) is a two dimensional integral equation, and can be solved for the albedo by separation of variables under certain conditions:

$$A(\theta_T, \Phi_T) = \sum_{n=0}^{N} \exp(in\Phi_T) \, f_n(\theta_T) \tag{2}$$

affording a considerable theoretical and computational simplification. The resulting equations are discretized with respect to a grid system, resulting in

$$B_n f_n = g_n \tag{3}$$

where $g_n(\eta)$ is the complex Fourier transform in longitude of the measurement map. The behavior of the B_n matrices is discussed by G. L. Smith (1981).

The resolution with which the albedo A can be described by data from a WFOV instrument is limited, causing the B_n matrices to be singular. G. L. Smith (1987) presented the use of singular value decomposition for this system. This approach delineates the parts of the solution which are retrievable and those which are not. The solution of equation (3) may be written as (Twomey, 1965; 1977):

$$f_n = \sum_{j=1}^{M} \lambda_{nj}^{-1}(v_{nj}g_n) \, u_{nj} + \sum_{j=M+1}^{J} \sigma_{nj} \, u_{nj} \tag{4}$$

where M is the number of non-zero singular values and the σ_{nj} are the coefficients of the singular vectors which have zero singular values. The σ_{nj} cannot be determined from the solution of equation (3). These terms were shown by G. L. Smith (1987) to consist of two classes: those corresponding to short spatial wavelength features and those corresponding to albedo features where the solar illumination is small during observation. Because of errors, it is necessary to consider terms in the first summation of equation (4) for which λ_{nj}

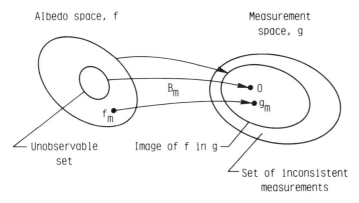

Figure 2. Schematic of function space relations for inversion problem.

all or zero as being unobservable. Results using only the observable components are unrealistic, as they go to zero in regions which are unobserved or have low lighting during observation, as shown by Smith and Rutan (1988). The problem is illustrated in figure (2), which shows that the albedo space is mapped into the measurement space such that a subspace of albedo maps into a zero measurement. This subspace of albedo is thus unobservable. It has been found that the unobservable components of the albedo are short spatial wavelength features and regions of low solar incidence during observation. In order to solve this problem, a priori information is used for the unobservable components. The compilation of Ellis and Vonder Haar (1976) is used in this study.

3.   APPLICATION

The theory discussed in the previous section can be used to delineate the unobservable regions of albedo by examining the profile of the second summation, i.e., the unobservable part of the albedo distribution. The question then is what is the appropriate value of M to use, that is, how many terms are observable?

This question is related to the singular values for the WFOV radiometers, which are shown in figure (3) for Nimbus 7, NOAA 9 and NOAA 10 for the zonal average (axisymmetric) terms, n = 0. The singular values decrease exponentially with the latitudinal wave-number, showing that the effect of the corresponding singular vector components on the measurements decreases. The singular values for the NOAA 10 are about 60% of those of the NOAA 9. The singular values for the NOAA 10 spacecraft are smaller than those for the NOAA 9 because the NOAA 10 crosses the Equator at 0730 (local time), and thus spends much of its time near the terminator, whereas the NOAA 9 crosses the Equator at 1430. As a consequence, the NOAA 9 has a larger region of observable albedo than does the NOAA 10, as will be

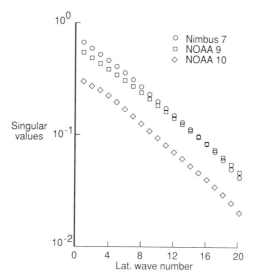

Figure 3.   Singular values for WFOV measurement of albedo.

seen. The reason for the values for the NOAA 9 and NOAA 10 having
the same slope as a function of latitudinal wavenumber is not clear
at present. However, we note that the slope increases with orbit
altitude, which is the same for both spacecraft. The effect of orbit
orientation is unknown.

Two criteria were used for M. The first was simply to use M =
18, which was the number of observable terms typically found for the
axisymmetric part of albedo for Nimbus 7 WFOV data. This is the
optimistic case. The second was to note that the limit of resolution
is caused by the decreasing effect of each term due to the decreasing
singular values. For Nimbus 7, the smallest measurable term was λ_{nj}
= .067; terms beyond this one were dominated by "noise," in which we
include the temporal variability of the albedo in a grid box as well
as instrument effects. For the NOAA 10 case, this gives M = 12.
This we regard as the pessimistic case. The resulting regions of
observability differed by 10° to 15° in latitude.

Figure (4) shows the unobservable component of albedo for NOAA 9
for the December, June, and September cases. The defining characteristic of these distributions is that when viewed by a shortwave WFOV
radiometer in a NOAA 9 orbit, the resulting signal is in the noise
level. Regions in which the unobservable component is large are
regions in which the albedo is unobservable. The region within which
the unobservable component is small is the region in which the albedo
is observable. Between the region of unobservability and the region
of observability is a transition region. In this paper, we take the
point of the minimum and the point of the local maximum of this curve
to define the transition region. The summer pole is always unobservable from a Sun-synchronous orbit because this latitudinal band reduces to a 10° circle which is only viewed by the radiometer from the
edge of the region. For September, the albedo is unobservable south-
ward of 60°S. This is because this region has a low solar incidence
at the local times of the NOAA 9 overflights, as seen in figure (1).
The same reasoning applies to the region south of 40°S for June, at
which time the terminator at local noon is at 67°S. A similar result
is seen for December northward of 40°N.

Figure (5) shows results for NOAA 10. Although the results are
similar to those for NOAA 9, it is noted that the regions of
unobservability are considerably larger. This is because the NOAA 10
orbit has its descending node at 0730 hours local time, so that it
stays near the terminator. It avoids good viewing conditions for a
shortwave WFOV in the winter hemisphere.

The variation with time of year of the boundaries of the regions
of observability, unobservability, and transition are shown in
figures (6) and (7) for NOAA 9 and 10, respectively. The NOAA 9 has
an unobservable region which is approximately 20° in latitude in each

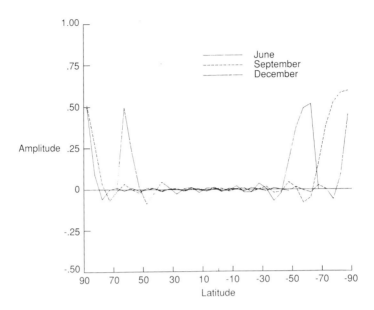

Figure 4.   Unobservable part of zonal albedo field for NOAA 9 for
December, June and September.

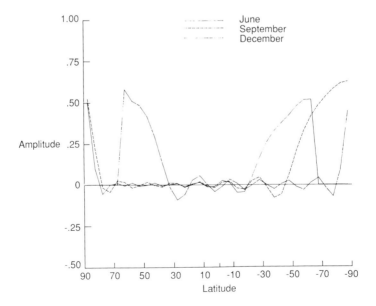

Figure 5.   Unobservable part of zonal albedo field for NOAA 10 for
December, June and September.

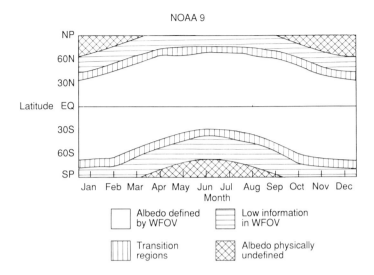

Figure 6. Observability regions for albedo for NOAA 9.

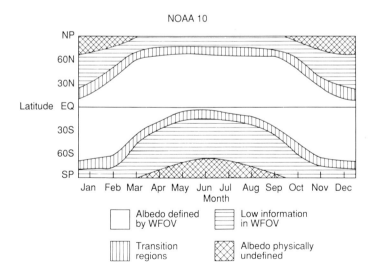

Figure 7. Observability regions for albedo for NOAA 10.

hemisphere, plus a transition region of approximately 10° width for most of the year. The unobservable region of the NOAA 10 is far more sensitive to time of year, with a width of 60° in the winter hemisphere.

Although the albedo is unobservable in the high latitudes, the more important question is how well do we know the absorbed solar energy? At the high latitudes, the solar incidence is small, so that albedo errors may be insignificant. The solar energy which would be reflected by the unobservable component of albedo is shown in figure (8) for the NOAA 9 and NOAA 10 WFOV radiometers for December. The incident solar radiation is also shown for comparison. Figure (9) shows the same quantities for June. Above 80° latitude at the summer pole, the albedo is unobservable and the solar incidence is high, so that the error in absorbed solar energy is high in that region.

4.   CONCLUSIONS

The theory of deconvolution of WFOV measurements of reflected solar radiation is used to establish the observability of albedo and absorbed solar radiation by the WFOV radiometers on the NOAA 9 and NOAA 10 spacecraft. The solution for albedo in terms of WFOV measurements of reflected solar radiation partitions the albedo field into an observable part and an unobservable part. This unobservability is due to limitations in the nature of the wide field-of-view measurement and is present in any analysis technique. The unobservable part includes high spatial frequency components and regions of low solar illumination. The error at high latitudes due to the unobservable part causes only small errors in absorbed solar radiation for the winter, but very large errors for the summer.

It is found that the NOAA 9 orbit is quite good for observing albedo when considered in terms of absorbed solar energy; however, the NOAA 10 orbit provides poor observations in the winter hemisphere. This is in agreement with the expectations formed by considering the orbit tracks of the spacecraft.

For a radiometer aboard the NOAA 9 spacecraft, the albedo has a large unobservable component in the Northern Hemisphere in December and a large unobservable component in the Southern Hemisphere in June and September. For a radiometer aboard the NOAA 10 spacecraft, the albedo has a large unobservable component in the Northern Hemisphere in December, and a large unobservable component in the Southern Hemisphere in June and December. For a radiometer aboard the NOAA 9 spacecraft, the unobservable component of reflected solar radiation is very small, except for the summer pole. For a radiometer aboard the NOAA 10 spacecraft, the unobservable component of reflected solar radiation is significant north of 25°N in December and south of 15° in June. Also, the error is large at the summer pole.

In this study, we have not considered the use of the ERBE medium field-of-view (MFOV) radiometers aboard the NOAA 9 and NOAA 10

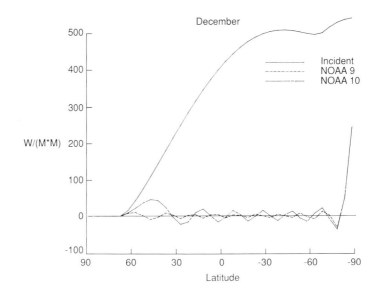

Figure 8.   Incident solar energy and absorbed solar energy due to
             unobservable part of albedo field for December for NOAA 9
             and NOAA 10.

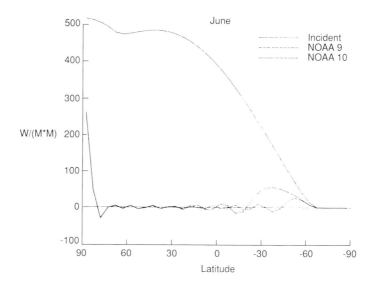

Figure 9.   Incident solar energy and absorbed solar energy due to
             unobservable part of albedo field for June for NOAA 9 and
             NOAA 10.

spacecraft. For MFOV radiometers, the observability question is not nearly so acute as for the WFOV radiometers.

5.   REFERENCES

Barkstrom, B. R. and G. L. Smith, 1986: The Earth Radiation Budget Experiment: Science and Implementation, Reviews of Geophysics, 24, 379-390.

Bess, T. D., R. N. Green, and G. L. Smith, 1981: Deconvolution of Wide Field-of-View Radiometer Measurements of Earth-Emitted Radiation. Part II: Analysis of First Year of Nimbus 6 ERB Data, J. Atmos. Sci., 38, 474-488.

Bess, T. D. and G. L. Smith, 1987a: Atlas of Wide Field-of-View Outgoing Longwave Radiation Derived from Nimbus 6 Earth Radiation Budget Data Set - July 1975 to June 1978, NASA Reference Publication 1185.

Bess, T. D. and G. L. Smith, 1987b: Atlas of Wide Field-of-View Outgoing Longwave Radiation Derived from Nimbus 7 Earth Radiation Budget Data Set - November 1978 to October 1985, NASA Reference Publication 1186.

Ellis, J. S. and T. H. Vonder Haar, 1976: Zonal Average Earth Radiation Budget Measurements from Satellites for Climate Studies. Atmospheric Science Paper 240, Colorado State U.

Jacobwitz, H., W. L. Smith, H. B. Howell, F. W., Nagle and J. R. Hickey, 1979: The First 18 Months of Planetary Radiation Budget Measurements from the Nimbus 6 ERB Experiment. J. Atmos. Sci., 36, No. 3, 501-507.

Smith, G. L., 1981: Deconvolution of Wide Field-of-View Satellite Radiometer Measurements of Reflected Solar Radiation. Proceedings, Fourth Conference on Atmospheric Radiation, American Met. Soc., June 16-18, 1981, Toronto, Canada.

Smith, G. L. and R. N. Green, 1981: Deconvolution of Wide Field-of-View Radiometer Measurements of Earth-Emitted Radiation. Part I. Theory. J. Atmos. Sci., 38, 461-473.

Smith, G. L., 1987: Deconvolution of Wide Field-of-View Measurements of Reflected Solar Radiation. Proceedings, Sixth Symposium Meteorological Observations and Instrumentation, American Met. Soc., Jan. 12-16. 1987, New Orleans, La.

Smith, G. L. and D. Rutan, 1988: Deconvolution Results for Wide Field-of-View Measurements of Reflected Solar Radiation. Proceedings, Third Conf. on Sat. Meteor. and Oceanography, American Met. Soc., Jan. 31-Feb. 5, 1988, Anaheim, Cal.

Smith, W. L., J. Hickey, H. B. Howell, H. Jacobowitz, D. T. Hilleary, and A. J. Drummond, 1977: Nimbus 6 Earth Radiation Budget Experiment. Appl. Opt., 16, No. 2, 306-318.

Twomey, S., 1965: The Application of Numerical Filtering to the Solution of Intergral Equations Encounted in Indirect Sensing Measurements. J. of Franklin Institute, 269, No. 2, 95-109.

Twomey, S., 1977: Introduction to the Mathematics of Inversion in Remote Sensing and Indirect Measurement. Elsevier Scientific Publishing Co.

EMPIRICAL ORTHOGONAL FUNCTION ANALYSIS OF A 10-YEAR DATA SET FOR OUTGOING LONGWAVE RADIATION

T. Dale Bess, G. Louis Smith, and Thomas P. Charlock
Atmospheric Sciences Division, NASA Langley Research Center
Hampton, Virginia 23665, USA

Fredrick Rose
PRC Kentron, Inc.
Hampton, Virginia 23666, USA

ABSTRACT

An empirical orthogonal function (EOF) analysis technique is applied to a 10-year data set of monthly averaged outgoing longwave radiation (OLWR). The data set is from the Nimbus 6 and Nimbus 7 fixed wide-field-of-view (WFOV) Earth Radiation Budget (ERB) instrument. The first eigenvector describes 66% of the variance in OLWR, and the first six eigenvectors account for 81% of the variance. The first two eigenvectors describe primarily the annual cycle. The third eigenvector is a semi-annual cycle describing monsoon activity, and the fourth eigenvector is an interannual cycle which describes much of the 1982-83 ENSO phenomenon.

1. INTRODUCTION

Outgoing longwave radiation (OLWR) is one of the primary quantities which govern our weather and climate, and thus has been the subject of a great deal of research. An understanding of the space and time variations of OLWR is necessary to the understanding of climate and its fluctuations at all scales. Because satellites are so well suited to the measurement of OLWR, they have been used for this since very early in the development of spacecraft (House et al. 1986). In order to obtain broadband measurements of the OLWR and solar radiation reflected by the Earth, an Earth Radiation Budget (ERB) instrument was flown on the Nimbus 6 spacecraft in June 1975 (W. L. Smith et al. 1977) and another on the Nimbus 7 spacecraft in October 1978 (Jacobowitz et al. 1984). The fixed wide-field-of-view (WFOV) radiometers on these spacecraft have provided a nearly continuous record of data from July 1975 to the present. Bess and Smith (1987a and b) have computed resolution enhanced monthly average OLWR using the deconvolution method of G.L. Smith and Green (1981) and compiled them in readily usable form for the 10-year period July 1975 through October 1985. This data set is expressed as spherical harmonic coefficients through degree 12 and, as such, is very compact. This data set is very well suited for studies of interseasonal, annual, and interannual variations.

The purpose of the present paper is to demonstrate the use of the EOF technique by applying it to the 10-year data set of Bess and Smith (1987a and b). This

RSRM '87: ADVANCES IN
REMOTE SENSING RETRIEVAL METHODS
A. Deepak, H.E. Fleming, and J.S. Theon (Eds.)   485

method computes from the data a set of patterns which can be used to describe most of the variation of the data. In doing so, many interesting features of time and space variability are brought out.

Heddinghaus and Krueger (1981) used the EOF approach to study the data set compiled by Winston et al. (1979). That set is based on the Scanning Radiometer (SR) aboard operational meteorological satellites. The SR has a window channel for monitoring sea surface temperature and to providing cloud imagery, so it is necessary to infer broadband OLWR from narrowband measurements. This procedure was used to compute a 45-month data set for the period June 1974 through March 1978. Heddinghaus and Krueger found several interesting results with this approach.

The present data record now extends to a full decade, whereas Heddinghaus and Krueger had less than 4 years of data with which to work. This additional period of data includes the 1982-83 El Niño/Southern Oscillation event, which is the strongest ever observed with good coverage by instrumentation. It is expected that the increased years of data will show additional interesting features.

2. ANALYSIS METHOD

In applying the method of EOFs to global fields of OLWR, it is intrinsically assumed that the monthly radiation fields are realizations of a random vector. This vector is the value of the OLWR field at the grid points in the map, so that it represents a field which is randomly distributed in space. The purpose of the EOF analysis is then to determine the statistical characteristics of the spatial structure.

The vector $\mathbf{y}$ is defined as the array of OLWR grid point values such that the OLWR at grid point p is the $p - th$ component of the vector. The dimensionality of the vector is the number of grid points, P. For the present case there are 120 months of data, so that there are I = 120 realizations. The mean $\langle \mathbf{y} \rangle$ of the data set is computed, and the deviation $\mathbf{x} = \mathbf{y} - \langle \mathbf{y} \rangle$ is computed for each month i. The covariance matrix is next computed as

$$C = \frac{1}{I} \sum_{i=1}^{I} \mathbf{x_i x_i^t} \tag{1}$$

Its eigenvalues λ_k and eigenvectors $\mathbf{u_k}$ of the C matrix are then computed:

$$C\mathbf{u_k} = \lambda_k \mathbf{u_k} \tag{2}$$

The covariance matrix is real and symmetric, thus the eigenvalues λ_k are real and its eigenvectors are real and orthogonal. The eigenvectors $\mathbf{u_k}$ thus form a basis set for the representation of the OLWR fields:

$$\mathbf{x_i} = \sum_{k=1}^{K} \alpha_{ik} \mathbf{u_k} \tag{3}$$

These coefficients are computed by the orthogonality property of the $\mathbf{u_k}$:

$$\alpha_{ik} = \mathbf{x_i^t u_k} \tag{4}$$

It can be shown that the EOF is the most economical basis for expressing the fields in the sense that for a given number of coefficients, more of the variance can be accounted for by this expansion than by any other. Thus, in order to study the time variation of a spatially varying field, the α_{ik} form the set which will be the smallest for a given level of accuracy in its description. Furthermore,

$$\frac{1}{I} \sum_{i=1}^{I} \left(\alpha_{ik} \right)^2 = \lambda_{\mathbf{k}} \tag{5}$$

that is, the variance explained by the $k - th$ coefficients in the series is the $k - th$ eigenvalue. The EOF expansion of Equation (3) may be thought of as analogous to a Fourier series or modal expansion, except that the functions are defined by the data set itself.

For the present case, the OLWR fields are described in terms of spherical harmonic coefficients. One approach to the EOF computation is to compute the grid point values and then compute the EOFs in terms of grid points. Another approach is to compute the EOFs in terms of the spherical harmonic coefficients and then to map them into grid points. This may be described as follows. The spherical harmonic coefficients for a given month i form a vector $\mathbf{h_i}$, and the deviation $\mathbf{z_i} = \mathbf{h_i} - \langle \mathbf{h} \rangle$ is computed. Next, the covariance S of the $\mathbf{z}$ is computed:

$$S = \frac{1}{I} \sum_{i=1}^{I} \mathbf{z_i z_i^t} \tag{6}$$

and finally the EOFs and eigenvectors using the spherical harmonic coefficients as a basis set can be computed:

$$S\mathbf{w_j} = \mu_j \mathbf{w_j} \tag{7}$$

The OLWR fields in the two systems are related by

$$y_{pi} = \sum_q Y_q\left(p\right) z_{qi} \tag{8}$$

where $Y\left(p\right)$ is the spherical harmonic function of order and degree denoted by the single index q, evaluated at point p, and z_{qi} is the $q - th$ coefficient for the $i - th$ month. For the $i - th$ month this relation may be written in matrix form as

$$\mathbf{y_i} = \mathbf{A z_i} \tag{9}$$

Because the $\mathbf{z_i}$ have a dimensionality of 169, their covariance matrix S will be 169x169. A 10 x10 grid will require 324 grid points for global coverage. It is quite advantageous to use the spherical harmonic coefficients as a basis for the computations and then to transform the results into the grid system. The question arises, Do we get the same results from both methods? The answer is that the results are the same if, and only if, we use a grid system of equal area boxes (to the level of accuracy of the mapping between the spherical harmonic coefficients and the grid system). For this case, the transformation of Equation (9) has the properties of a rotation. Equation (9) will also apply to the relation between the EOFs of the spherical harmonic coefficients and the EOFs of the grid system maps. However, the results are different if a latitude- longitude grid is used, as this system will weight the regions near the poles far more heavily than will an equal area grid system. Furthermore, in that case, Equation (9) will not relate the two sets of EOFs. Any EOF computations for global coverage should thus be based on an equal area grid in order to have physical significance.

3. RESULTS AND DISCUSSION

Empirical orthogonal functions were computed using the spherical harmonic coefficients and then transforming them into a latitude- longitude map. The percentages of variance associated with each eigenvector and the cumulative percentages of variance are listed in Table 1 for the first 10 EOFs. These variances are very near the results of Heddinghaus and Krueger (1981) for which they list the first five. (It is noted that they used a latitude-longitude grid from 60S to 60N for their investigations. In this manner, they avoided the regions further poleward, where the equal area consideration would become a major problem.) The first 10 terms account for 85% of the variance, and the tenth term is less than 1%. The convergence of the eigenvalues beyond these first 10 is quite slow. Inclusion of the first 40 terms gives 96% of the variance, an increase due to terms 11 through 40 of 11%, and the 40th term contains 0.17% of the variance.

Table 1

Percentages of Variance Explained by Each
Empirical Orthogonal Function

rank	Annual Cycle Present		Annual Cycle Removed	
	%variance	cumulative	%variance	cumulative
1	65.7	65.7	12.8	12.8
2	4.6	70.3	8.4	21.2
3	3.8	74.1	6.2	27.4
4	3.0	77.1	4.6	32.0
5	2.3	79.4	4.4	36.4
6	1.9	81.3	4.3	40.7
7	1.2	82.5	3.3	44.0
8	1.1	83.6	3.1	47.1
9	1.0	84.6	2.6	49.7
10	.9	85.5	2.4	52.1

Maps of the first 2 EOFs, their coefficient time histories, and temporal spectra are shown in Figures 1 and 2. EOF 1 is mainly an annual cycle, although there is a small nonannual part present. EOF 2 is largely an annual cycle, but with a semiannual part and an interannual part also. It is to be expected that two EOFs would be associated with the annual cycle in order to represent regions with differing phases. EOF 1 which represents 66% of the variance, follows the solar heating with a small phase lag. EOF 2 accounts for most of the annual cycle which is out of phase with EOF 1. These results are very similar to those found by G. L. Smith and Bess (1983) for the cosine and sine parts of the annual cycle. The largest variation in EOF 1 is over central Asia. The OLWR over central Asia, as at other points in the middle and high latitudes, responds to changes in surface temperature and boundary layer emission, which track solar heating. Over the Tropics and subtropics, the seasonal temperature changes are much smaller than at the higher latitudes. The low latitude OLWR responds to the movements of large cloud patterns which are in turn forced by seasonal changes in airflow and moisture. An increase in surface temperature produces an increase in OLWR, but an increase in cloudiness decreases OLWR; this accounts for the difference in sign between the low and high latitudes in EOF 1. The large difference between the emitting temperatures of the high clouds found in the Tropics and tropical surfaces explains the low latitude dominance in these maps.

EOF 2 appears as a seesaw between the central-eastern Pacific Ocean and the Indian Ocean-Indonesia regions. The negative component of the seesaw maps the fall position of the Intertropical Convergence Zone (ITCZ), which almost girdles the globe. The positive lobe in the Pacific Ocean signals a spring maximum in ITCZ cloudiness, corresponding to seasonal variations in sea surface temperature and low level convergence Horel (1982). EOF 2 has a noteworthy anomaly in months 93 and 94 of the record (March and April 1983), where it drops to -10 Wm^{-2} , rather than approximately -7 Wm^{-2} as for most March and Aprils.

Maps, coefficient time histories, and temporal spectra of EOFs 3 and 4 are shown in Figures 3 and 4. EOF 3 is dominated by a semiannual cycle. It has strong centers of monsoonal action located over India and the Timor Sea. Several areas are of moderate strength but out of phase, and much of the Earth is relatively unaffected. Figure 4 shows that EOF 4 has strong interannual and semiannual parts. It has strong centers of action over the central equatorial Pacific Ocean and New Guinea, which are out of phase. Other areas along the Equator also show significant changes, as does Antarctica. The peaks of its time history correspond to El Niño/Southern Oscillation (ENSO) episodes, so it seems reasonable to identify this EOF as the dominant EOF mode for describing an ENSO.

The time history of EOF 6(not shown) suggests a cycle of approximately a 1- decade period. Its time spectrum also shows a strong peak at the 1- decade period and another strong peak at the semiannual cycle. Time histories and spectra for EOF 7 through 10 are not too discernible from white noise. The maps, however, show major features in the Indian Ocean and Indonesia-New Guinea regions, indicating that the EOFs themselves are real patterns rather than spatial

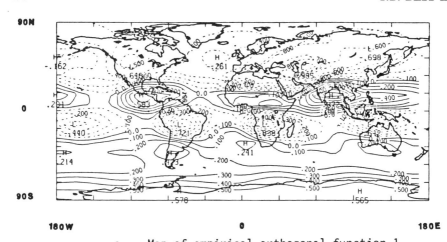

a.   Map of empirical orthogonal function 1.

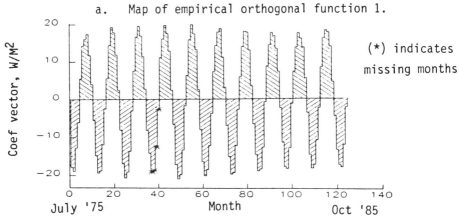

(*) indicates
missing months

b.   Time history of empirical orthogonal function 1.

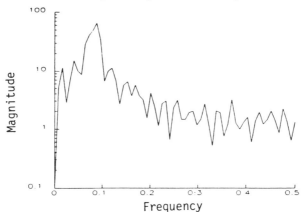

c.   Spectrum of empirical orthogonal function 1.

Figure 1.   Empirical orthogonal function 1. Map, time history,
             and temporal spectrum. In (b) July 1975 is month 1.

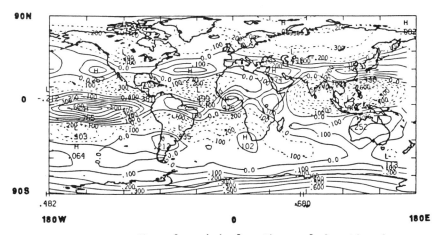

a.   Map of empirical orthogonal function 2.

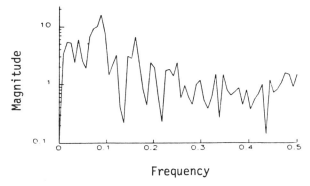

b.   Time history of empirical orthogonal function 2.

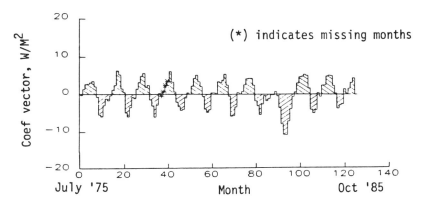

c.   Spectrum of empirical orthogonal function 2.

Figure 2.   Empirical orthogonal function 2.   Map, time history, and temporal spectrum.   In (b) July 1975 is month 1.

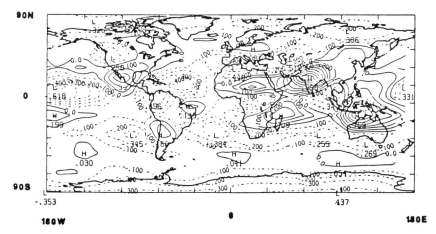

a.   Map of empirical orthogonal function 3.

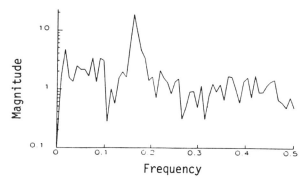

b.   Time history of empirical orthogonal function 3.

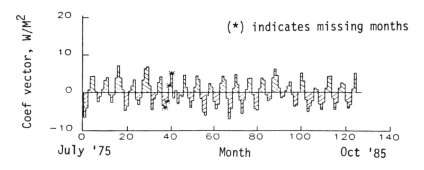

c.   Spectrum of empirical orthogonal function 3.

Figure 3.   Empirical orthogonal function 3. Map, time history,
            and temporal spectrum. In (b) July 1975 is month 1.

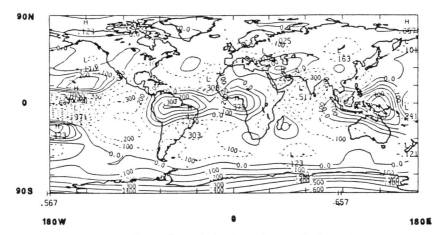

a.   Map of empirical orthogonal function 4.

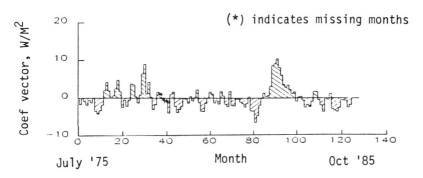

b.   Time history of empirical orthogonal function 4.

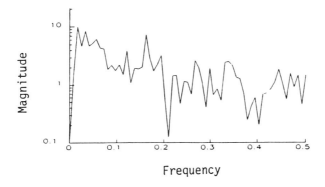

c.   Spectrum of empirical orthogonal function 4.

Figure 4.   Empirical orthogonal function 4.   Map, time history, and temporal spectrum.   In (b) July 1975 is month 1.

white noise, and that a large part of the variation in monthly averaged OLWR maps is in the Tropics.

In examining the maps of the first 10 EOFs, the major impression is that most of the EOFs of rank 3 and greater are describing variations primarily in the region near the Equator from India to the Pacific Ocean just east of New Guinea, and to a lesser extent their statistical relations to smaller changes elsewhere over the globe. There are no obvious peaks in the 26-month range. A part of the explanation of this may be that the 1982-83 ENSO episode was so strong that it stands out in Figure 4 as a unique feature. This feature is roughly a triangle in appearance, which would be described by a spectrum which is flat out to the frequency corresponding to the width of the feature, beyond which it would be red.

The eigenvalues are seen in Table 1 to become closely spaced with increasing rank. The significance of these numbers (which are statistics and thus random in their nature) must be examined. The noise level in the eigenvalues is given by $\lambda_k (2/N)^{-1/2}$, where N is the number of realizations(North et al. 1982). By this criterion, the first six eigenvalues and EOFs are significant, and the eigenvalues and EOFs from rank seven and beyond cannot be distinguished unambiguously.

In order to enhance the visibility of any interseasonal and interannual variations, the annual cycle was removed from the data sets in order to prewhiten the data so that the annual and semiannual cycles would not dominate the spectra. The eigenvalues and EOFs of these anomaly maps were then computed. The eigenvalues are listed in Table 1 also. It is found that the first four eigenvalues are significant, and rank five and beyond cannot be meaningfully distinguished. It is found that EOF 1 with the seasonal cycle removed resembles very closely EOF 4 with the seasonal cycle left in. Also, EOF 2 with seasonal cycle removed corresponds to the EOF 5 with seasonal cycle left in, and deseasonalized EOF 3 is very similar to EOF 6 with the annual cycle left in.

4.   CONCLUSIONS

The method of empirical orthogonal functions has been demonstrated by application to a 10-year data set of outgoing longwave radiation. It is seen that the EOFs provide a method for analyzing a time series of maps. Also, spherical harmonic functions provide a basis set which is suitable for producing equal area map results. The following findings are noted:

1. 66% of the variance is in EOF 1. After that, each EOF accounts for only a small percent of the variance, and the EOFs form a slowly converging series.
2. The first two EOFs describe mainly the annual cycle.
3. EOF 3 is primarily the semiannual cycle, although many other EOFs also contain significant semiannual parts.
4. A much stronger spring/fall mode was found in the central equatorial Pacific Ocean.
5. EOF 4 describes much of the 1982-83 ENSO phenomenon.

The first three results reaffirm those of Heddinghaus and Krueger (1981). There is typically a gap in the spectrum between a semiannual peak and the annual cycle for all but the first EOF. The occurrence of the 1982-83 ENSO event during this data period was of fundamental importance in interannual variation studies, as it skews all results.

5. REFERENCES

Bess, T. D., and Smith, G. L., 1987a: Atlas of Wide-field-of-view Outgoing Long-wave Radiation Derived from Nimbus 6 Earth Radiation Budget Data Set - July 1975 to June 1978, NASA Reference Publication 1185.

Bess, T. D., and Smith, G. L., 1987b: Atlas of Wide-field-of-view Outgoing Long-wave Radiation Derived from Nimbus 7 Earth Radiation Budget Data Set - November 1978 to October 1985, NASA Reference Publication 1186.

Heddinghaus, T. R., and Krueger, A. F., 1981: Annual and interannual variations in outgoing longwave radiation over the Tropics. *Mon. Wea. Rev.*, 109, 1208-1218.

Horel, J. D., 1982: On the annual cycle of the tropical Pacific atmosphere and ocean. *Mon. Wea. Rev.*, 110, 1863-1878.

House, Fredrick B., Gruber, Arnold, Hunt, Garry E., and Mecherikunnel, Ann T., 1986: History of Satellite Missions and Measurements of the Earth Radiation Budget (1957-1984). *Rev. Geophys.*, vol. 24, no. 2, May, pp. 357-377.

Jacobowitz, H., H. V. Soule, H. L. Kyle, F. B. House, and the NIMBUS 7 ERB Experiment Team, 1984: The Earth Radiation Budget (ERB) Experiment: An overview. *J. Geophys. Res.*, 89, No. D4, 5021-5038.

North, G. R., T. L. Bell, R. F. Cahalan, and F. J. Moeng, 1982: Sampling errors in the estimation of empirical orthogonal functions. *Mon. Weath. Rev.*, 110, 699-706.

Smith, G. L., and Bess, T. D., 1983: Annual cycle and spatial spectra of Earth-emitted radiation at large scales. *J. Atmos. Sci.*, 40, 998-1015.

Smith, G. L., and R. N. Green, 1981: Deconvolution of wide-field-of- view radiometer measurements of Earth-emitted radiation. Part I. Theory. *J. Atmos. Sci.*, 38, 461-473.

Smith, W. L., J. Hickey, H. B. Howell, H. Jacobowitz, D. T. Hilleary, and A. J. Drummond, 1977: Nimbus 6 Earth radiation budget experiment. *Appl. Opt.*, 16, 306-318.

Winston, J. S., A. Gruber, T. I. Gray, Jr., M. S. Varnadore, C. L. Earnest, and L. P. Manello, 1979: Earth-atmosphere radiation budget analysis derived from NOAA satellite data, June 1974- February 1978, vols. 1 and 2. NOAA, 34pp.

DISCUSSION

Neuendorffer: On the seasonalized version how much of the variance is taken care of with these first five orthogonal functions?

Bess: The first 6 are 81%.

Thompson: Could you define the deseasonalized eigenvector? You compute the eigenvector and then you take the eigenvector and remove the seasonal variation?

Bess: Remove seasonal variation and compute eigenvector.

Gautier: You said that there was a similarity between the eigenvector 4 non-deseasonalized and eigenvector 1 deseasonalized. When you showed them I saw some differences in terms of what is happening in the Atlantic. Which, in fact, my question is whether the difference comes from the way you outline your contours or if it is a real quantitative difference?

Bess: Quite possibly it might have been the way I outlined my contour because I was trying to just hit the highs and lows since it didn't show up on the contour map.

THE CONSTRAINED INVERSION OF NIMBUS-7 WIDE FIELD OF VIEW SHORTWAVE RADIOMETER MEASUREMENTS FOR THE EARTH ALBEDO

Richard Hucek[1], H. Lee Kyle, Philip Ardanuy[1], and Peter Cheng[1]
Space Data and Computing Division
NASA Goddard Space Flight Center
Greenbelt, Maryland 20771, USA

ABSTRACT

The diurnally-averaged Earth albedo is retrieved from the deconvolution of the wide field of view (WFOV) shortwave radiometer measurements collected on board the Nimbus-7 satellite. By using a truncated spherical harmonic series representation for the albedo field, the satellite measurement equations reduce to a set of linear equations in the unknown expansion amplitudes. Other linear equations in the expansion amplitudes are added to the system description. These include constraint equations which require the solution to exhibit features obtained from a priori knowledge about the albedo. Together, the measurement and constraint equations are solved in a weighted least-squares sense for the expansion coefficients.

The spherical harmonic representation used in this procedure is, in general, an expansion with nonorthogonal basis functions. This condition arises because the albedo field is undefined at the polar cap during the associated winter season. Spherical harmonic functions, however, although forming a complete basis set for any region of the globe, are only orthogonal when applied to the whole earth. The consequence is that the expansion amplitudes derived from the deconvolution procedure are subject to change by varying the length of series representation. This inconvenience makes it difficult to assign a physical interpretation to the separate amplitudes. Collectively, however, the series converges to a reproducible field.

A year of deconvolved Nimbus-7 albedos are presented as a sequence of 8-day period means. The data are displayed on a Hovmöller diagram of time vs longitude to consider the detection of the 40- to 60-day mode oscillations in the tropics.

1. INTRODUCTION

Satellite-borne wide-field-of-view (WFOV) radiometers directly sample the instantaneous field of upwelling flux at spacecraft altitude. This feature has made them an important component of the Earth radiation budget (ERB) experiments flown on board the Nimbus-6 and -7

---

[1]Research and Data Systems Corporation, 10300 Greenbelt Road, Lanham, MD 20706, under Contract No. NAS5-29373

spacecraft and most recently aboard the three satellites (viz. NOAA-9, NOAA-10, and ERBS) of the Earth Radiation Budget Experiment (ERBE). The measurement values, after application of space and time averaging methods, provide global estimates of emerging flux at satellite altitude. These we interpret as the radiative output of their source, the Earth.

In addition to their fundamental interpretation as instantaneous outgoing fluxes at satellite altitude, WFOV observations can be modeled by a two-dimensional integral across the visible Earth disc. Inversion of the integral equation leads to parameter field estimates at the top of the atmosphere (TOA). Smith and Green (1981) have shown how a set of WFOV measurement equations can be inverted by deconvolution to yield the TOA distribution of Earth-emitted (also termed longwave) radiation. There is, however, a fundamental limit to the spatial resolution achievable through deconvolution. By means of simulation studies, Hucek et al. (1987) and Ardanuy et al. (1987) have investigated the effect on resolution of spatial aliasing of temporal variability at the TOA during the course of a complete WFOV sampling cycle. For the Nimbus-7 satellite they have found that deconvolution leads to a spatial resolution corresponding to a spherical cap of about a 7° radius at the TOA. This is not as good as that obtainable by means of narrow-field-of-view (NFOV) radiometers which infer TOA fluxes from angular reflectance models and/or multiple radiance observations of the same region. Nevertheless, scanning NFOV radiometers have failed in the past on account of mechanical difficulties. The more reliable WFOV sensors then provide the only means for deriving long-term records of regional, TOA ERB parameters. A case in point is that of the ERB instrument package carried aboard Nimbus-7. After 18 months of operation, the NFOV scanning motor drive suddenly failed in June of 1980. During the succeeding years, only the WFOV radiometers have viewed such climatologically important episodes as the El Chichón eruption in April of 1982 and the El Niño/Southern Oscillation (ENSO) event of 1982-1983. To study climate changes related to these events, in-situ observations (i.e., at the TOA) of radiant exitances are most useful. This provides the major impetus for the deconvolution of Nimbus-7 WFOV observations.

In this paper we present a WFOV deconvolution strategy for the retrieval of TOA Earth-reflected (also termed shortwave) radiation. The shortwave problem differs in several aspects from the previously addressed longwave deconvolution. Among them are: (1) that the desired TOA parameter, the reflected shortwave field, varies within three orders of magnitude, whereas for longwave it is within one; (2) there is an Earth terminator present beyond which no solar illumination falls and associated parameter fields (viz. albedo) are undefined; and (3) strong diurnal and anisotropic reflection, especially for the latter near the terminator, preclude the use of isotropic reflectance models to describe the hemispherical distribution of outgoing radiances. Procedures for treating these distinctions are introduced together with a representative set of resultant TOA deconvolved fields. Finally, the utility of the deconvolved fields is tested by considering the detection of the 40- to 60-day mode oscillations in the tropics.

2.   MODEL OF A WFOV SHORTWAVE MEASUREMENT

2.1   IN TERMS OF INSTANTANEOUS TOA FLUX

A WFOV satellite observation is modeled as the convolution over the radiometer field of view (FOV) of the product of the incoming radiance with the instrument angular response function. This is expressed as

$$F_{sat} = \int_{FOV} R_{TOA} \cos(\eta) \, d\Omega \tag{1}$$

where F_{sat} is the measured flux at satellite altitude, R_{TOA} is the outgoing radiance field at the TOA, η is the incoming nadir angle at the satellite, and $\cos(\eta)$ is the flat plate cosine response characteristic of the Nimbus-7 WFOV radiometers. Since we desire the TOA distribution of reflected flux, the radiance from any scene is expressed as

$$R_{TOA} = \rho \, F_{TOA} \tag{2}$$

where F_{TOA} is reflected flux and ρ is a scene-dependent anisotropic reflectance model. It accounts for the nonuniform angular distribution of reflected flux over the outgoing hemisphere. Using Eq. (2) in Eq. (1), the instantaneous flux at satellite altitude is related to the instantaneous flux at the TOA by

$$F_{sat} = \int_{FOV} \rho \, F_{TOA} \cos(\eta) \, d\Omega \tag{3}$$

2.2   IN TERMS OF MEAN-DAILY ALBEDO

The measurement model of Eq. (3) is not the most convenient for deconvolution since the unknown, F_{TOA}, varies broadly in phase and amplitude with the large pole-to-pole gradient in solar insolation, I. Since the solar insolation is known from geometric considerations, it may be introduced into Eq. (3) by expressing F_{TOA} as

$$F_{TOA} = a \, I, \tag{4}$$

where a, the instantaneous albedo, becomes the unknown parameter field. However, the albedo field common to all measurements is not a, but rather $\bar{a}$, the diurnally averaged or mean daily albedo. These are related by

$$a = m \, \bar{a}, \tag{5}$$

where m is a field of diurnal models which depend both on the instantaneous scene type and its variation with time. Since scene type was not known continuously throughout the sampling period, the instantaneous m of Eq. (5) is expressed as the sum of a mean model, $\bar{m}$, and a deviation, ϵ. This we write as

$$m = \bar{m} + \epsilon \tag{6}$$

where the mean model is based upon a composite of five primary scene
types (viz. ocean, land, desert, cloud, and snow/ice) as defined by
Taylor and Stowe (1984, 1986); these are, combined in proportion to
their frequency of occurrence, to form an average scene type at each
field point. When the average scene is assumed to apply at each
observation, random errors with a nonzero mean will result. The
biases are due primarily to inadequacies of the 5-scene composite
models to describe all existing scene types and to correlations of
scene type with time of day such as the emergence of tropical cloud
fields during the early afternoon. In the latter instance, a series
of daily observations collected at the same local time for the same
subsatellite position will not view the variety of scene types that
occur throughout the diurnal interval and lead to an unbiased model.
Rather, that scene which is present at the local time of observation
is preferentially viewed. A systematic scene misidentification, with
respect to the mean model, occurs and a nonzero bias results.

Combining Eqs. (3), (4), (5), and (6) and using a mean model for
ρ as well as m, we obtain

$$F_{sat} = \int_{FOV} \bar{\rho} \ \bar{m} \ \bar{a} \ I \ \cos(\eta) \ d\Omega + \Delta\epsilon. \tag{7}$$

In Eq. (7) the model of a shortwave measurement has been transformed
to a representation in terms of $\bar{a}$. An advantage over a representation
in terms of $\bar{F}_{TOA}$ is presumed since the inversion will now recover only
information not already known by other means. The random scene-
related errors, $\Delta\epsilon$, that occur in the measurement model incorporate
the combined errors in $\bar{\rho} \ \bar{m}$ and are not further analyzed in our
formalism; they appear only as noise with a possible nonzero mean.

3. METHOD OF SOLUTION

3.1 EXPANSION IN SPHERICAL HARMONIC FUNCTIONS

The unknown albedo $\bar{a}$ of Eq. (7) is expressed as a truncated
spherical harmonic series over the illuminated portion of the Earth.
We write

$$\bar{a} = \sum_{k=1}^{N} A_k \ Y_k \tag{8}$$

where the A_k are unknown expansion amplitudes and the Y_k are spherical
harmonic functions. In the region of total darkness near the winter
pole, the albedo is undefined and the expansion of Eq. (8) does not
apply. As a result, the spherical harmonic functions, while forming a
complete set for any region of the globe, are a nonorthogonal basis
set when applied to only a portion of the Earth as in this study. A
consequence of this is that the expansion coefficients will depend on
the length, N, of the summation. Nevertheless, the series representation for $\bar{a}$ leads to a reproducible field for any choice of N. In this
paper, N includes all spherical harmonic terms in a rhomboidal expansion of 15 meridional nodes for each of 8 zonal waves and the mean.

When Eq. (8) is substituted for $\bar{a}$ in Eq. (7), a linear equation in the unknown expansion amplitudes results. We have

$$F_{sat,i} = \sum_{k=1}^{N} M_{ik} A_k \tag{9}$$

where M_{ik} is a matrix element for the ith equation and k^{th} unknown amplitude given by

$$M_{ik} = \int_{FOV_i} \bar{\rho} \; \bar{m} \; Y_k \; I \; \cos(\eta) \; d\Omega. \tag{10}$$

3.2 CONSTRAINT EQUATIONS

Due to the 6-day repeat cycle of the Nimbus-7 sampling pattern, a time-averaging error is committed by assuming that the average of the observed satellite fluxes is equal to the actual period mean flux as represented by the integral of Eq. (7). At low spectral resolution, the inversion of this noise is automatically suppressed and does not affect the solution. At high resolution, an auxillary method of noise suppression is required. This we accomplish by introducing con-straints in the form of other linear equations in the unknown expan-sion amplitudes. The constraints represent known properties of the solution and are included, along with the measurement equations, as part of the system description.

There are two objectives in the use of constraints in this problem. One is to control the inflation and instability of the solution vector **A** caused by the inversion of sampling noise. In this case we look for a solution vector with components of relatively small magnitude. This condition can be approached through ridge regression in which we demand for each unknown amplitude

$$A_k = 0 \qquad k = 1, N. \tag{11}$$

A second objective of the constraints is to ensure the proper meridional boundary behavior of the solution at the north and south extremes of the Nimbus-7 sampling pattern. Here we require that the meridional curvature of the zonal mean be reasonably small for the illuminated portion of the Earth. This characteristic is approached through the condition

$$d^2\bar{a}/d\theta^2 \;\big|_{\theta=\theta_i} = 0 \tag{12}$$

where θ is latitude and i represents one of 40 possible zonal bands. The zeros of the constraint equations are not meant to represent plausible values for these conditions, but only to provide a conver-gence limit in the direction of which a reasonable solution must lie.

3.3 WEIGHTED LEAST SQUARES SOLUTION

3.3.1 The Constraint Equation Weights

The measurement and constraint equations are solved simultaneously for the A_k by minimizing the weighted least-squares error given
by

$$E^2 = \sum_{i=1}^{M} w_i^2 \ (F_i - \sum_{k=1}^{N} M_{ik} A_k)^2 \tag{13}$$

In Eq. (13) the constraint equations, assumed to be written in the
matrix form of Eq. (9), are included in the outer summation along with
the measurement equations for a total of M equations, and the F_i are
generalized measurement values including both the satellite and
constraint values. Realistic solutions are chosen from a sequence of
results obtained by varying the weighting factors, w_i, of the constraint equations between zero (unconstrained) and infinity (the null
solution). Suitable values for these weights are determined by
comparing features of the retrieved solutions to a priori known
characteristics of the Earth albedo and to lower-resolution results
obtained by other methods (Jacobowitz et al., 1984). Figures 1, 2,
and 3 illustrate the empirical procedure that we have used to derive
the monthly mean albedo for December 1983 for the case of the ridge
constraints. Figure 1 shows the result of a lower-resolution solution
termed the WFOV archived albedo. It depicts patterns of high and low
albedo that appear within the tropics and mid-latitudes and that must
also exist in the deconvolved result. Figure 2 is a deconvolved
solution but without the application of ridge constraints. It brings
out the large-scale patterns of albedo ridges and troughs as were seen
in the archived albedo and, in addition, reveals several high resolution features not previously discernible. The emergence of an albedo
ridge across the equatorial Pacific, for example, is noted in the
deconvolved result. This new detail is acceptable as a valid feature
of the Earth albedo since it corresponds to the known cloud fields
along the Intertropical Convergence Zone (ITCZ). The unconstrained
result, however, exhibits instability near the terminator, as seen by
the large amplitude waves at 65°N. These have been subdued in Figure
3 with the application of the ridge constraints. This solution
preserves all the features of the previous unconstrained result within
the tropics and mid-latitudes, with the addition that it has a plausible behavior near the terminator where incident insolation drops to
the level of noise.

Table 1 contains a summary of the meridional and ridge equation
weight factors that we have used in obtaining 8-day mean fields of
albedo. They are given in the form

$$w_i = \lambda \ v_i \tag{14}$$

where λ is a common magnification factor and v_i is a relative factor
for the weights of a given constraint type. The variables m and n
that appear in the table are, respectively, the number of zonal waves
and meridional nodes contained in a given term of a spherical harmonic
series. θ is latitude.

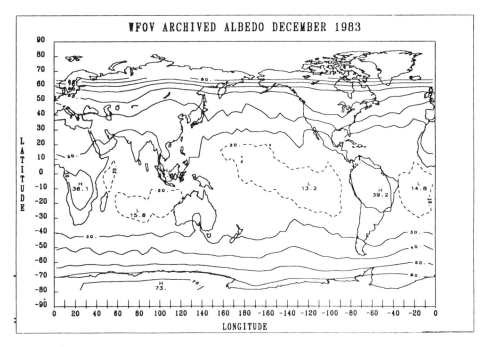

FIGURE 1.   WFOV archived, mean-daily albedo (%) for the month of
             December 1983.

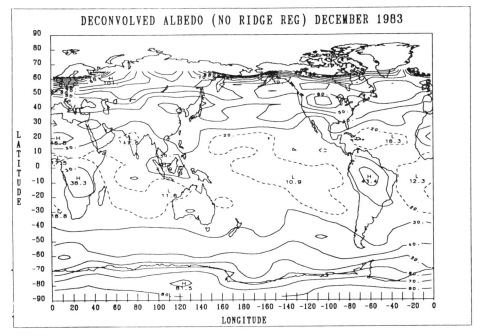

FIGURE 2.   Deconvolved, mean-daily albedo (%) without ridge regression
             constraints for the month of December 1983.

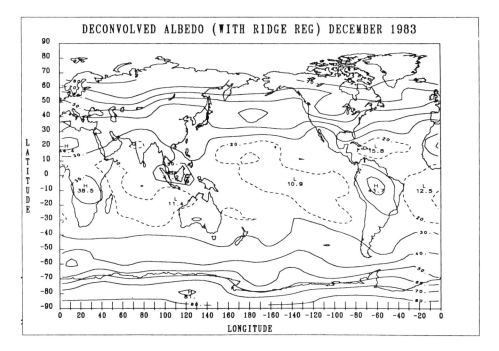

FIGURE 3.   Deconvolved, mean-daily albedo (%) with ridge regression
             constraints for the month of December 1983.

TABLE 1.   INVERSION WEIGHTS FOR THE CONSTRAINT EQUATIONS

Constraint Type	λ	v_i	w_i
Ridge	100	$m_i + n_i$	$100 \cdot (m_i + n_i)$
Meridional Smoothing	8	$\sin^2(\theta_i)$	$8 \cdot \sin^2(\theta_i)$

3.3.2   The Measurement Equation Weights

    For the satellite measurement equations, weight factors are
derived by noting that the mean-reflected flux is related to the mean
albedo and insolation, $\bar{I}$, by

$$\bar{F}_{TOA} = \bar{a}\ \bar{I}. \tag{15}$$

When Eq. (15) is used in Eq. (7), the satellite measurements may be
expressed in terms of $\bar{F}_{TOA}$. Furthermore, according to the mean value
theorem for integrals, a solution for $\bar{F}_{TOA}$ at an undetermined point,
x, within the FOV may be extracted from the integral and is given by

$$\overline{F}_{TOA}(\mathbf{x}) = F_{sat} \{\int_{FOV} \overline{\rho} \ \overline{m} \ (I/\overline{I}) \ \cos(\eta) \ d\Omega\}^{-1} \tag{16}$$

Equation (15) shows that a satellite measurement value may be converted to an equivalent mean flux measurement value at the TOA by scaling with the term in brackets. This is important because the factor $(\overline{I}/I)$ scales an observation in proportion to its mean-daily solar insolation. This allows the least-squares procedure to minimize the errors in the albedo solution as weighted by the mean-daily illumination rather than the instantaneous insolation at the time of the observations (which is satellite-dependent). Similarly, the model of a satellite measurement may be converted to the model of a TOA measurement by the same scaling factor. The TOA equivalent measurements and models are not useful individually since the point $\mathbf{x}$ remains unknown. However, the scaling factors can be applied to the measurements equations as least-squares weights. They transform the inversion problem from a minimization of flux error at satellite altitude to a minimization at the TOA.

4. REPRESENTATIVE DECONVOLUTION RESULTS

In Figures 4 and 5 we compare global TOA maps of the WFOV archived and deconvolved albedos, respectively, for the 8-day period from August 10, 1983 through August 17, 1983. The archived result exhibits the general pattern of peaks and valleys that can be discerned by the WFOV radiometers and which must appear in the deconvolved result. It, however, is plagued by an extremely tight gradient in albedo as we approach the region of total darkness in the southern hemisphere. There the albedo is seen to rise above the physically acceptable limit of 100 percent. Elsewhere the archived result appears noisy, especially in the southern hemisphere mid-latitudes between 30° and 60° south latitude.

The deconvolved albedo, on the other hand, has suppressed the noise apparent in archived solution and yields a physically admissible albedo even to the terminator boundary. This is, in part, due to the inclusion of anisotropic reflection in the measurement model which can account for relatively large WFOV measurements through the preferential forward scattering of radiation toward the satellite. This is particularly true when the satellite is near the terminator and able to view the large forward beam that occurs for reflection at high local-solar zenith angles of incidence. Throughout the tropics and mid-latitudes we notice the enhanced resolution of the deconvolved field by the magnification of the ridge-to-trough amplitudes that occur. Also evident is the clarification of finer structure such as the cut-off high over Greenland, the track of the ITCZ across the equatorial Pacific, and the shape of the South Pacific Convergence Zone from southeast Asia through Borneo and the New Guinea and east of Australia.

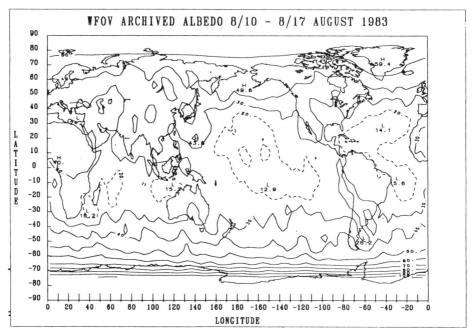

FIGURE 4.   WFOV archived, mean-daily albedo (%) for the 8-day period
             August 10, 1983 through August 17, 1983.

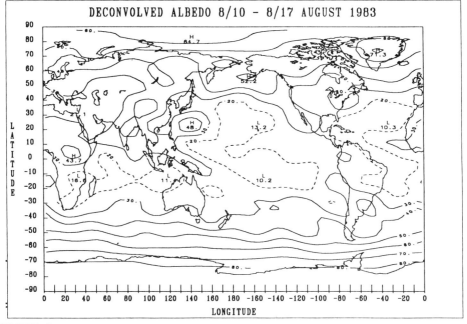

FIGURE 5.   Constrained, deconvolved, mean-daily albedo (%) for the
             8-day period August 10, 1983 through August 17, 1983.

5. TIME SERIES ANALYSIS

A time-longitude section along the equator of a one-year sequence of 8-day deconvolved albedo fields is shown in Figure 6 for the period July 1983 through June 1984. The time series consists of 46 8-day mean fields for a latitude band of 9° centered about the Equator. It represents a raw data field which, when time-filtered, will be analyzed for eastward propagating wave modes. From the figure we immediately recognize three zones of relatively high albedos extending from 10° to 30° east, 60° to 150° east, and 60° to 90° west longitude. These regions are characterized by deep convective activity and correspond, respectively, to the Congo River basin of Africa, the Indian Ocean south of India eastward through Indonesia (Asian mon-soon), and the Amazon River basin of South America. The large albedos in these cases arise from the copious high cloud amounts produced by convection. Especially notable are the periodic impulses of high albedo over the Amazon which may be linked to convective outbreaks occurring elsewhere (Lau and Chan, 1985). Between these areas are the relative quiescent regions of lower albedo corresponding to the south Arabian Sea, Pacific Ocean, and Atlantic Ocean.

ALBEDO (%)

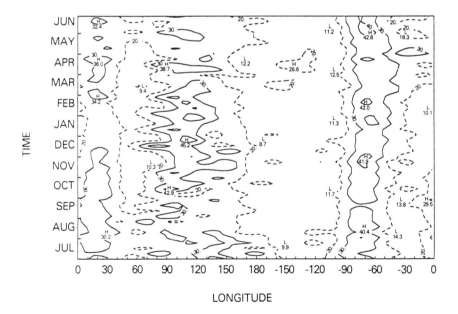

LONGITUDE

FIGURE 6.   Time-longitude section of 8-day mean fields of deconvolved albedo (%). The time series extends from July 1983 through June 1984 and represents the average albedo within a 9-degree latitude band centered about the Equator.

In Figure 7 are shown the results of applying a band pass filter
to the data of Figure 6. The time series at each of 80 equally spaced
longitudinal positions has been filtered to pass mainly 40- to 60-day
modes of oscillation while other frequencies are suppressed. The
precise frequency response of the first-order Butterworth filter is
included in Table 2. After filtering, the resultant time series are
reconstructed on a Hovmöller chart to consider the detection of
eastward propagating waves in the tropics. These correspond to
contours whose orientation lies in the direction from lower left to
upper right. In the convective regions of the Asian monsoon and
Amazon River basin, the contour patterns suggest a tendency toward
eastward motion over a limited longitudinal length. Especially near
the Amazon River the broken patterns display the presence of eastward-
moving orientations. We find that the filtered time series, which
retains power only in the 40- to 60-day range of periodicities,
explains almost 50% of the variance of the total signal near certain
longitudes. In particular, near the Congo and Amazon River valleys,
and over regions of the Pacific and Indian Oceans. The limited
spatial extent of the contours shown may be due to the movement of
such waves in the meridional direction and out to the 9° equatorial
band that we have analyzed.

FILTERED ALBEDO (%)

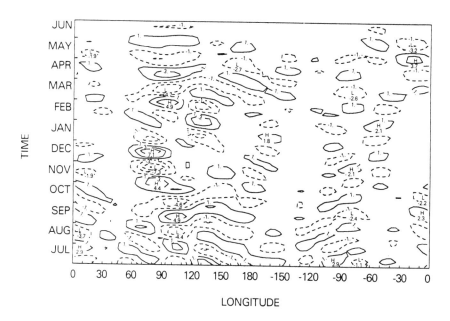

FIGURE 7.   Filtered time-longitude section of 8-day mean fields of
deconvolved albedo (%). The one-year time series at each
of 80 longitudinal positions has been filtered to pass
mainly 40- to 60-day frequency modes. The residual signals
are shown from July 1983 through June 1984 along a 9-degree
latitude band centered about the Equator.

TABLE 2.   BAND PASS FILTER RESPONSE

Period (days)	Response (% Amplitude Retained)
16	0
30	24
35	50
45	100
60	50
75	24
90	14
105	9
120	7

6.   CONCLUSIONS

The deconvolution of WFOV shortwave radiometer measurements has been demonstrated by using the ERB observation data set collected on board the Nimbus-7 satellite.   Three major aspects of the shortwave problem were treated which have no analogous significance for the more tractable longwave inversion.   First, the large-scale variations in magnitude of the unknown TOA-reflected shortwave field are strongly influenced by the known distribution of incident insolation.   This signal is removed from the reflected shortwave radiation by expressing the latter as the product of albedo and insolation.   The albedo becomes the unknown parameter field.   Second, the Earth's terminator is handled by the use of constraint equations which guide the solution to a physically reasonable result even though the signal-to-noise ratio approaches zero.   Finally, the strong directional and bidirectional reflectance properties of shortwave radiation are accounted for by means of scene identification and the application of corresponding angular models.   When the solution fields obtained by the deconvolution method are compared to lower-resolution results, we immediately notice the suppression of noise, a realistic boundary behavior at the terminator, magnification of ridge-to-trough amplitudes, and the modification of large-scale albedo patterns.

The deconvolution algorithm has been used to create a one-year sequence of 8-day mean albedos for the period from July 1983 through June 1984.   The evolving fields which emerge were analyzed for the detection of 40- to 60-day eastward-propagating modes in the tropics. It appears that these waves are discernible in the deconvolved data set in the regions of the Asian monsoon and the Amazon rain forest. They may also move meridionally north and south so that they appear within the 9° latitudinal zone that we have analyzed for only a limited time.   Similar analyses of adjacent latitudinal bands is required to demonstrate this supposition.

REFERENCES

Ardanuy, P. E., H. L. Kyle, R. R. Hucek, and B. S. Groveman, 1987: Nimbus 7 Earth Radiation Budget Wide Field of View Climate Data Set Improvement, 2, Deconvolution of the Earth Radiation Budget Products and Consideration of the 1982-1983 El Niño Event, J. Geophys. Res., 92, 4125-4143.

Hucek, R. R., H. L. Kyle, and P. E. Ardanuy, 1987: Nimbus 7 Earth Radiation Budget Wide Field of View Climate Data Set Improvement, 1, The Earth Albedo from Deconvolution of Shortwave Measurements, J. Geophys. Res., 92, 4107-4123.

Jacobowitz, J., H. V. Soule, H. L. Kyle, F. B. House, and the Nimbus 7 ERB Experiment Team, 1984: The Earth Radiation Budget Experiment: An Overview, J. Geophys. Res., 89, 5021-5038.

Lau, K. M. and P. H. Chan, 1985: Aspects of the 40-50 Day Oscillation During Northern Winter as Inferred from Outgoing Longwave Radiation, Mon. Wea. Rev., 113, 1889-1909.

Smith, G. L., and R. N. Green, 1981: Deconvolution of Wide Field-of-View Radiometer Measurements of Earth-Emitted Radiation, I, Theory, J. Atmos. Sci., 38, 461-473.

Taylor, V. R. and L. L. Stowe, 1984: Reflectance Characteristics of Uniform Earth and Cloud Surfaces Derived from Nimbus 7 ERB, J. Geophys. Res., 89, 4987-4996.

Taylor, V. R. and L. L. Stowe, 1986: Revised Reflectance and Emission Models from Nimbus-7 ERB Data, Extended Abstracts, Sixth Conference on Atmospheric Radiation, May 13-16, 1986, Williamsburg, VA (American Meteorological Society), 45 Beacon Street, Boston, MA 02108), J19-J22.

DISCUSSION

McCormick: You mentioned doing some analytical simulations with a matrix inversion and imagine it even up to 270 terms. Did you have any random error noise in that modeling?

Hucek: The random error, yes, there was random error in the modeling. What we did was we ran a simulation of a Nimbus-7 orbit over the course of 1 month. The orbital pattern allows you to sample any local region only 4 times a month. And the average values that go into the data are averages over the course of 30 days. So, we call it sampling noise and it was automatically included.

Kleespies: In the question and answers session could you address the last item on your conclusions, which looks to be an interesting conclusion?

Hucek: I sure would like to.

Rodgers: Your original basic resolution, I guess, is around 3000 kilometers. Have you a numeric estimate as to what the resolution is of the retrieved field?

Hucek: Yes. If you draw a spherical cap around the deconvolved field and then compare it with truth you will find that the minimum resolution of that spherical cap is about 10 degrees.

OPEN DISCUSSION

RECOMMENDATIONS FOR FUTURE MEETINGS

Deepak: I would like to have a list of recommendations for future meetings. For instance, what should be the frequency of this workshop? What major topics or themes should be considered?

King: I think it should be at least in the solicitation quite unstructured and following on your lines not restricted too closely, not constrained. And the only other suggestion is I like it bi-annually.

Fleming: Two categories that were emphasized this morning were classification techniques which should be stressed next time, and involvement of numerical forecasters in one way or another. Oh, and sequential estimation was another topic mentioned.

Mango: I would like to sort of pose a challenge to everyone here. Several of our colleagues have expressed the frustration of not being able to get data. And the way you put it, Adarsh, in terms of the "high priest"-we should consider how people usually get involved in areas such as data compression. Usually people like me are involved in a project like synthetic aperture radar with its enormous data rates and data volumes and we are forced into the business of looking into data compression techniques in order to be able to disseminate that information in a practical way. What I am suggesting is that in order to be able to really do justice to the dissemination of information, you really have to know the nature of your data stream. It is only the "high priests" that are here that can really specify that. So the challenge I pose to all of you is that by the time you meet next time, address how you would funnel down your own data and information so that it indeed could be sent in a widely disseminated form. It may sound like a very trivial thing, but it is not. The frustration of not being able to get data is one of not always being close enough to the characterization of it to be able to do a simple job.

Crosby: I'm going to switch gears entirely and make sort of a special pleading. One thing was sort of skipped over and that was the calibration problem. I would like to see something on this in the next remote sensing meeting. A much longer session on calibration and some questions that were not addressed like long-term monitoring. If you don't calibrate the instruments in succession they are useless for long-term monitoring.

Deepak: At the 1976 Workshop on Retrieval Methods, we invited Drs. Jacqueline Lenoble and Henk Van de Hulst to give papers on radiative transfer problem, the forward problem. Would you be interested in including papers on the forward problem?

Neuendorffer: I would just like to say that the biannual idea seems good and I enjoyed this conference very much. We had mentioned before about perhaps having some users such as modelers who actually use the data at the end. I think that would be helpful for the next time around.

Speaker: Just that I think you should increase the participation of active sensors, lidar and so on.

Fleming: Your idea of appointing chairpersons who can pull together enough papers I think is very important. Of course, emphasize retrieval techniques and the kinds of information that can be retrieved.

Deepak: For the next workshop we will announce the dates and issue a call of papers later in the coming year. In addition, we will select the chairmen for suggested topics in advance, who will have enough foresight to understand what topics will be important two years from now, and what problems need to be solved. These chairmen can then structure their sessions with 4 to 5 papers they invite or select.

Speaker: The other thing I suggest is that in connection with this computer-related facet I think that is a tool that everyone needs, everyone needs to know more about it. And one way to address that without having a specific session would be to get hold of computer vendors to display their wares in parallel sessions. Computer folks are always willing to do that and I think that is useful. And it's free.

Deepak: That is a good comment. We shall further explore how we can implement these suggestions. Any other comments? If there are none, then I would like to bring this workshop to a close. I would like to thank all the speakers, and all the participants for an excellent and stimulating workshop. I would like to thank the session chairmen, and members of the program committee for doing an excellent job. A successful conference is like a duck on water: calm on the surface but paddling like hell underneath. And there are a lot of people who have been doing the paddling behind the scenes to make this a smooth-running workshop. I would like to recognize them and ask you to join me in thanking them. First of all I would like to recognize the meetings staff headed by Ms. Carolyn Keen. They are Jan Wrobel, Linda Budd, and Judy Cole. And last, but not the least, I would like to thank Mr. Henry Fleming, who took the most active part in setting up the technical agenda for this workshop. With this happy note let us end this workshop, and start the plans for the next workshop.

LIST OF AUTHORS